T0137257

Springer Series in Materials Science

Volume 278

The Springer Series in Materials Science covers the complete spectrum of materials research and technology, including fundamental principles, physical properties, materials theory and design. Recognizing the increasing importance of materials science in future device technologies, the book titles in this series reflect the state-of-the-art in understanding and controlling the structure and properties of all important classes of materials.

More information about this series at http://www.springer.com/series/856

Suprakas Sinha Ray
Editor

Processing of Polymer-based Nanocomposites

Processing-structure-property-performance relationships

Editor
Suprakas Sinha Ray
DST-CSIR National Centre for
Nanostructured Materials
Council for Scientific and Industrial
Research
Pretoria, South Africa

and

Department of Applied Chemistry
University of Johannesburg
Johannesburg, South Africa

ISSN 0933-033X ISSN 2196-2812 (electronic)
Springer Series in Materials Science
ISBN 978-3-030-07404-3 ISBN 978-3-319-97792-8 (eBook)
https://doi.org/10.1007/978-3-319-97792-8

This Springer imprint is published by the registered company Springer Nature Switzerland AG
The registered company address is: Gewerbestrasse 11, 6330 Cham, Switzerland

Preface

Over the last few years, "nanomaterial" and "nanotechnology" have become well-known terms, not only among scientists, engineers, fashion designers, and architects, but also the general public. Owing to their extraordinary and unexpected behavior, nanomaterials have gained tremendous attention in fields such as automotive, electronics, aerospace, health care, and biomedical and have significant potential for many modern advanced technological applications. In this direction, a great deal of research and development effort has emerged around the hybrid organic–inorganic systems, and, in particular, attention has been given to those in which nanofillers (or nanoparticles, NPs) are dispersed in a polymer matrix. This class of materials is called polymer nanocomposites (PNCs) and shows unique value-added properties that are completely absent in neat matrices and conventional composites.

NPs can be made from a wide range of materials, the most common being layered silicates, carbon nanotubes, graphene or graphite oxide, and metal oxides. Over the last two decades, various types of NPs have been used for the preparation of PNCs with almost all types of polymers and polymer blends. However, layered silicates, carbon nanotubes, and graphene-containing PNCs have attracted in current advanced materials research, because these NPs can remarkably enhance the inherent properties of neat polymers after PNCs formation.

Control of the dispersion of NPs in polymer matrices and the interactions between NPs and macromolecular chains is necessary to explore PNCs full potential for a wide range of applications. This can be achieved by proper surface functionalization and modification of the NPs, which enhance their interaction with matrix. Functionalization of NPs consequently influences the colloidal stability, dispersion, and controlled assembly of NPs.

The performance of a PNC is dictated by three main factors: (i) the inherent properties of the components; (ii) interfacial interactions; and (iii) structure of the PNCs. The structure of a PNC depends on the dispersion and distribution of the NPs in the polymer matrix. However, improving the dispersion by mechanical means or via chemical bonding can influence the properties of the obtained PNCs.

Therefore, elucidating the dispersion and distribution characteristics and the associated mechanisms is important and can allow prediction of the final properties.

Over the past few years, great deals of advancements have been made in PNCs research and development, and the dispersion of the NPs in the polymer matrix remains a key challenge to their widespread application. Undoubtedly, to maximize the interfacial area in the PNCs, the dispersion should be on the scale of the individual NPs; otherwise, aggregation results in a lower specific surface area, and microcomposites are ultimately formed. A large portion of this book is dedicated to reviewing methods for manipulating the interface, as well as the kinetic aspects of the dispersion of nanoparticles in polymer matrices, for example, controlling extrusion parameters during melt processing.

Another pressing matter is the control of the structural morphology of the PNCs to achieve time-independent equilibrium. Often, further processing, for example extrusion and then injection molding, results in a different dynamic equilibrium, which makes it difficult to tune the emergent properties. A similar complex challenge involves scaling-up preparation of the PNCs to industrially viable quantities, especially for solvent-based systems. To secure economic and other societal impacts, sufficient volumes of NPs and PNCs need to be manufactured in order to be transformed into market-ready products. Most research and development laboratories around the world focus on processes with tightly controlled laboratory environments which are not necessarily appropriate for safe, reliable, effective, and affordable large-scale production. Funding to address these challenges is still required. In addition, as with other new technologies, societal aspects need to be considered in parallel with the development of the technology. Consumer perceptions regarding the risks of certain nanoparticles, whether proven or not, need to be addressed to ensure a smoother uptake of the technology.

In summary, the field of PNCs has shown disparate results, successes as well as challenges. Much work remains to be done in fundamental research to achieve better control of the desired properties, for example, processing PNCs to achieve the desired level of dispersion of the NPs. This book focusses the effect of process variables on the structural evolution of polymer composites and the production of either micro- or nanocomposites, as well as the effect of the dispersion state on the final properties at a fundamental level.

Processing these PNCs usually requires special attention as the resultant structures on the micro- and nanolevel are directly influenced by the polymer/NP chemistry and processing strategy, among others. The structure then affects the properties of the resultant composite materials. This book is structured into two volumes. The first volume introduces readers to nanomaterials and PNCs processing. After defining NPs and PNCs and discussing environmental aspects, Chap. 2 focuses on the synthesis and functionalization of nanomaterials with applications in PNC technology. A brief overview on nanoclay and nanoclay-containing PNC formation is provided in Chap. 3. Chapter 4 provides an overview of the PNCs structural elucidation techniques, such as X-ray diffraction and scattering, microscopy and spectroscopy, nuclear magnetic resonance, Fourier transform infrared spectroscopy and microscopy, and rheology. The last chapter provides an overview on how melt-processing strategy impacts structure and mechanical properties of polymer nanocomposites by taking layered silicate-containing polypropylene nanocomposite as a model system.

The second volume focuses heavily on the processing technologies and strategies and extensively addresses the processing–structure–property–performance relationships in a wide range of polymer nanocomposites, such as commodity polymers (Chap. 1), engineering polymers (Chap. 2), elastomers (Chap. 3), thermosets (Chap. 4), biopolymers (Chap. 5), polymer blends (Chap. 6), and electrospun polymer (Chap. 7). The important role played by NPs in polymer blends structures in particular is illustrated.

This two-volume book is useful to undergraduate and postgraduate students (polymer engineering, materials science and engineering, chemical and process engineering), as well as research and development personnel, engineers, and material scientists.

Finally, I express my sincerest appreciation to all authors for their valuable contributions as well as reviewers for their critical evaluation of the proposal and manuscripts. I also thank all authors and publishers for their permission to reproduce their published works. My special thanks go to Kirsten Theunissen, Aldo Rampioni, editor, and production manager at Springer Nature for their suggestions, cooperation, and advice during the various stages of manuscripts preparation, organization, and production of this book. The financial support from the Council for Scientific and Industrial Research, the Department of Science and Technology, and the University of Johannesburg is highly appreciated. Last but not least, I would like to thank my wife Prof. Jayita Bandyopadhyay and my son Master Shariqsrijon Sinha Ray, for their tireless support and encouragement.

Pretoria/Johannesburg, South Africa Suprakas Sinha Ray

Contents

Contributors

Orebotse Joseph Botlhoko DST-CSIR National Centre for Nanostructured Materials, Council for Scientific and Industrial Research, Pretoria, South Africa

Prasanna Kumar S. Mural DST-CSIR National Centre for Nanostructured Materials, Council for Scientific and Industrial Research, Pretoria, South Africa

Vincent Ojijo DST-CSIR National Centre for Nanostructured Materials, Council for Scientific and Industrial Research, Pretoria, South Africa

Suprakas Sinha Ray DST-CSIR National Centre for Nanostructured Materials, Council for Scientific and Industrial Research, Pretoria, South Africa; Department of Applied Chemistry, University of Johannesburg, Doornfontein, Johannesburg, South Africa

Reza Salehiyan DST-CSIR National Centre for Nanostructured Materials, Council for Scientific and Industrial Research, Pretoria, South Africa

Koena Selatile DST-CSIR National Centre for Nanostructured Materials, Council for Scientific and Industrial Research, Pretoria, South Africa

Chapter 1
Processing Nanocomposites Based on Commodity Polymers

Prasanna Kumar S. Mural and Suprakas Sinha Ray

Abstract Nanocomposites consisting of commodity polymers like polyethylene, polystyrene, polypropylene, and polyvinyl chloride have demonstrated good thermo-mechanical behavior and electrical properties. Common routes for producing polymer nanocomposites (PNCs) with commodity polymers involves either melt mixing, in situ polymerization, or solution mixing. However, the common processing techniques cannot adequately disperse nanoparticles (NPs) in the commodity polymer matrix. The chapter describes various strategies for dispersing NPs in commodity polymers, such as functionalization of the polymer, or preparing a nanocomposite. In addition, this chapter describes the structure–property relationships of commodity polymers after incorporation of NPs, along with their performance for specific applications. Finally, an outlook regarding the challenges, opportunities, and future trends in commodity PNCs is presented, along a summary of the chapter.

1.1 Introduction

Commodity polymers are those produced in high volume, which have been applied in large amounts for various applications. Commodity polymers generally include polyolefins, mainly polyethylene (PE), polypropylene (PP), polystyrene (PS), and poly(vinyl chloride) (PVC), which account for over two-thirds of polymer sales worldwide [1]. In addition, many commodity polymers have a softening temperature

P. K. S. Mural · S. Sinha Ray (✉)
DST-CSIR National Centre for Nanostructured Materials, Council for Scientific and Industrial Research, Pretoria 0001, South Africa
e-mail: rsuprakas@csir.co.za

P. K. S. Mural
e-mail: pmural@csir.co.za

S. Sinha Ray
Department of Applied Chemistry, University of Johannesburg, Doornfontein 2028, Johannesburg, South Africa
e-mail: ssinharay@uj.ac.za

S. Sinha Ray (ed.), *Processing of Polymer-based Nanocomposites*, Springer Series in Materials Science 278, https://doi.org/10.1007/978-3-319-97792-8_1

below 100 °C, restricting application temperature to several tens of degrees below this temperature, where the materials have poor mechanical properties. Commodity polymers have applications mostly in consumer goods (such as food storage) which generally require lower mechanical properties and a low cost. However, it is well reported that the addition of microscale particles to commodity polymers usually improves thermal stability and provides stiffness. Similarly, addition of the nanomaterials tends to improve physical and mechanical properties in commodity polymers. In addition, nanoparticles (NPs) are available in all three dimensions, i.e., 1D materials include carbon nanotube (CNTs) and nanowires, 2D materials include nanoclays and graphene, and 3D materials include cubical and spherical NPs [2]. Given the brevity of this chapter and usefulness of these particles in commodity applications, we have restricted our discussion to nanoclays, CNT, and graphene.

A key challenge in polymer nanocomposite (PNC) preparation is proper dispersion of NPs in the polymer matrix. Further, NPs have strong secondary hydrogen bonds which promote agglomeration and make dispersion in the polymer matrix difficult. These agglomerates reduce the surface area, resulting in poor distribution and dispersion of the NPs. In addition, the agglomerates act a sites of stress concentration from where failure is initiated, resulting in poorer overall mechanical properties [3]. The distribution refers to the homogeneity of NPs throughout the sample, where the quality of dispersion determines the level of agglomeration of NPs. Figure 1.1 shows a schematic diagram of the state of distribution and dispersion of NPs in a polymer matrix. Figure 1.1a shows an example of uniform distribution with poor dispersion of NPs, while Fig. 1.1b shows a poor state of distribution and dispersion of NPs in the polymer matrix. Similarly, Fig. 1.1c shows a poor state of distribution and good dispersion of NPs, while Fig. 1.1d shows a state of good dispersion and distribution of NPs in the polymer matrix. Improved understanding of the various processing parameters and techniques has led to advancement in easy bulk manufacturing and commercialization of PNCs. However, the significant challenges of purification, dispersion, and bulk processing of the polymer remains to be addressed in a broader perspective.

This chapter describes fundamental aspects of processing commodity polymer-based PNCs, including details of various processing strategies, such as functionalization of polymers and preparing nanocomposites of commodity polymers (e.g., PE, PS, PP, and PVC). We also focus on the structure–property relationships, performance, and applications of commodity PNCs. A brief discussion of the challenges, opportunities, and future trends in commodity PNC development is also presented.

1.2 Processing of Commodity PNCs

The processing of PNCs involves amalgamation of NPs and a polymer matrix. Commonly employed processing techniques include melt mixing, in situ polymerization, and solution mixing. All processing techniques have challenges related to the dispersion, alignment, and functionalization of NPs, which directly or indirectly affect the properties of the PNCs.

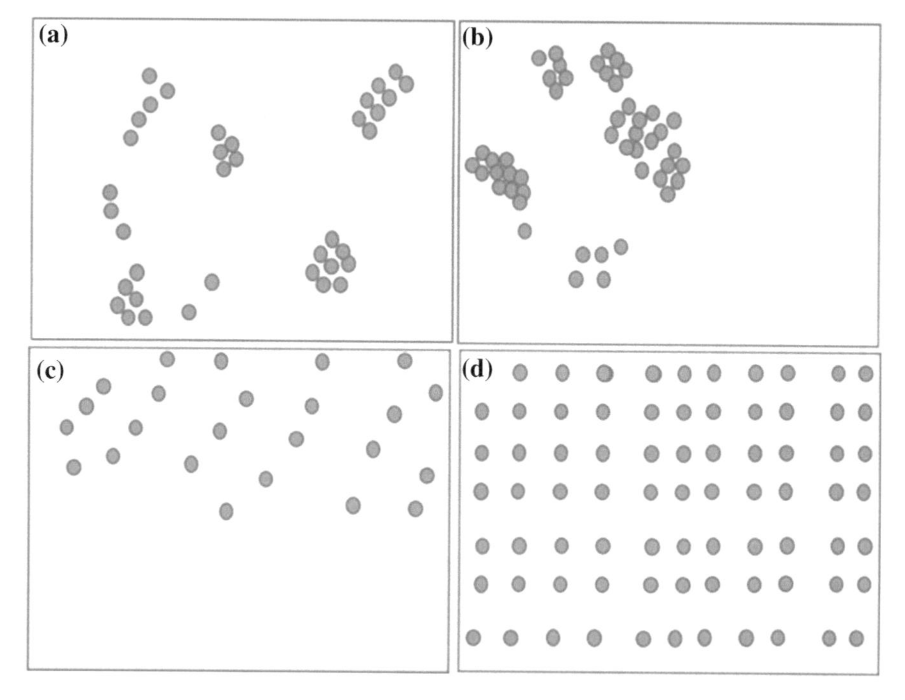

Fig. 1.1 Schematic of state of distribution and dispersion of nanoparticles in polymer

1.2.1 Melt Mixing

Melt mixing is an industrially viable technique which is useful for preparing thermoplastic PNCs using e.g., PE, PP, PS, PA-6, and PC. This method involves the direct mixing of the polymer and nanofiller in a top-down approach, which allows a high loading of NPs to be incorporated in the polymer matrix. Figure 1.2 shows a schematic of a typical melt mixing process, which involves melting of the polymer to form a viscous liquid, accompanied by shearing of the polymer. In general, amorphous and semicrystalline materials are heated to a processing temperature (softening temperature) which is above the glass transition temperature of the amorphous polymer and above the melting point of the semicrystalline polymer. A high-shear mixer or extruder imparts a shear stress on the viscous liquid, which transfers this stress to the nanofillers, resulting in nanofiller aggregates being disintegrated. This process achieves a homogeneous distribution and dispersion of NPs in the polymer matrix. The final desired shape of the PNC can be obtained by employing post processing operations such as injection molding, compression molding, or extrusion. Further, melt mixing is regarded as an environmentally friendly, cost effective, and industrially viable technique for manufacturing PNCs using commodity polymers.

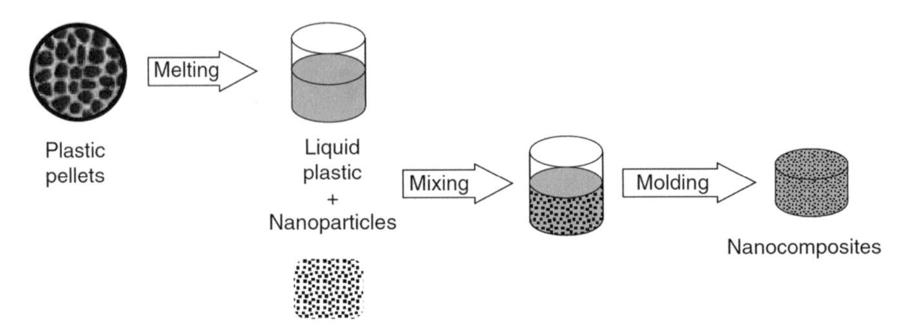

Fig. 1.2 Schematic of melt mixing process. Reproduced with permission from [4]. Copyright 2012, Woodhead Publishing

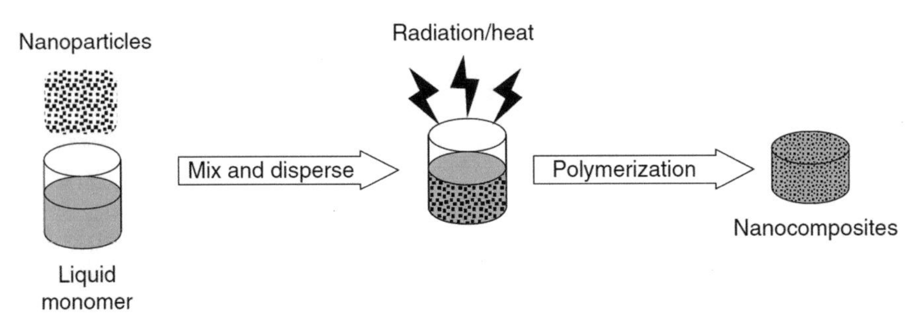

Fig. 1.3 Schematic of melt mixing process. Reproduced with permission from [4]. Copyright 2012, Woodhead Publishing

1.2.2 In Situ Polymerization

In situ polymerization involves dispersing NPs in liquid monomers, as shown in Fig. 1.3. Once the desired level of mixing is obtained, the polymerization of the monomers can be initiated either by a catalyst, external initiator, heat, or radiation [5]. After polymerization, the polymers are either wrapped around the NPs or covalently bonded to them, depending on the surface functionality of the NPs and type of polymer [6]. The desired good dispersion and distribution depends on appropriate dispersion of NPs in the monomer. The grafting of a polymer on the NP surfaces facilitates their incorporation at high loadings, along with their good miscibility in the matrix. In situ polymerization can be used to produce PNCs that cannot be processed using melt mixing (e.g., thermally unstable materials) or solution mixing. An advantage of this technique is that a molecular level of reinforcement can be obtained.

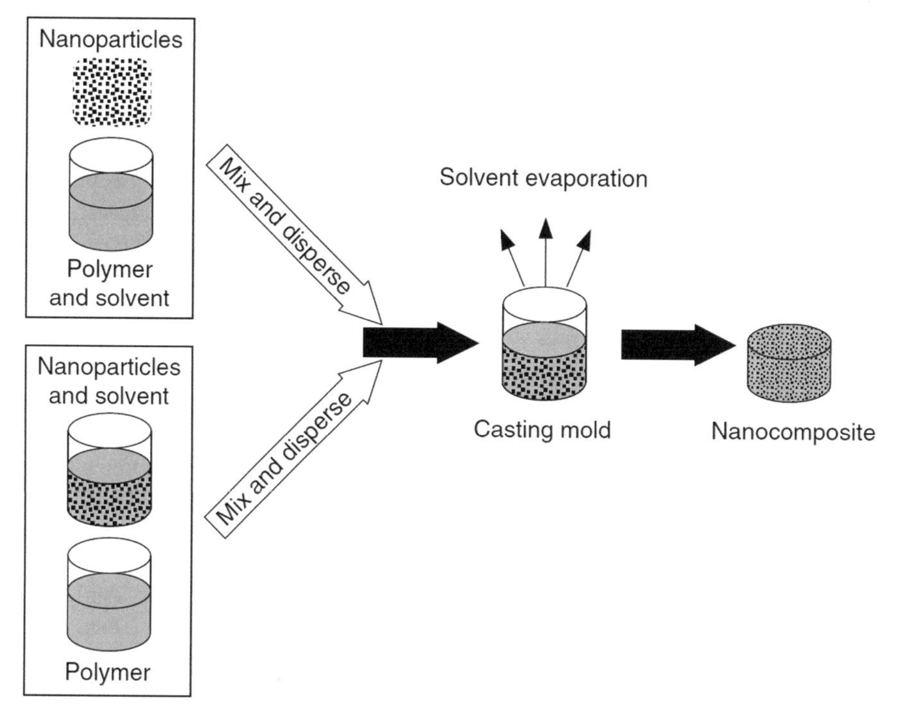

Fig. 1.4 Schematic of in situ polymerization process. Reproduced with permission from [4]. Copyright 2012, Woodhead Publishing

1.2.3 Solution Mixing

The solution mixing processing of PNCs has the advantage of using low viscosity fluids, which enhances the mixing and dispersion of NPs. Figure 1.4 illustrates the solution mixing process. NPs can either be dispersed in a solvent or in polymer solution. High-energy agitation, such as high-shear mixing, magnetic stirring, sonication, or refluxing is applied to ensure proper dispersion and distribution of the NPs. After homogeneous dispersion of the NPs in the solution, the solution in cast on a casting mold, then the solution is evaporated to obtain film or sheets in the mold. Subsequent post-processing operations (e.g., calendaring, or injection, blow, compression, and rotational molding) can be employed to obtain the desired shape.

1.3 Methods for Functionalizing Polymers for PNC Preparation

Most commodity polymers are nonpolar, except PVC. Since the incorporation of NPs into PVC has its own specific challenges, there is currently no major commercial nanocomposite preparation using this material [7]. Therefore, we restrict the majority of our discussion to nonpolar polymers, including PE, PP, and PS. However, a few studies of PVC-based nanocomposites have been included here to broaden the scope of the discussion of commodity PNCs.

Generally, NPs such as clays and layered double hydroxide (LDH) are polar, while NPs such as CNT and graphene are classified as carbonaceous. When polar or carbonaceous NPs are incorporated into nonpolar polymer matrices, the NPs tend to phase-separate in the composites due to thermodynamic effects. In addition, NPs have a large internal surface, which promotes agglomeration, rather than homogeneous dispersion, in a nonpolar polymer matrix. Further, the agglomeration of NPs and lack of interaction at the interface between polar NPs and nonpolar polymer chains inhibit the interaction between the particles and polymer matrix.

Further modification of NP surfaces with surfactant has proven to be highly successful for processing polymers. However, for the processing of polyolefins, it has shown only moderate success or failed completely [8]. As in most cases, the natural tendency for NP agglomeration is either difficult to overcome, or results in a mixture that is thermodynamically unstable. Therefore, a three-part system occurs in PNCs; generally, polar NP surfaces, nonpolar polymer chains, and the particle–matrix interface. To enhance the dispersibility of the NPs, the compatibility between the nonpolar polymer and polar NPs needs to be increased; this is achieved by either modifying the polar NP surface to increase the miscibility in the nonpolar polymer matrix or introducing functional groups onto the polymer chain to reduce the interfacial tension between the polymer and NPs.

Methods for functionalizing the polymer include radical polymerization, free-radical polymerization, and copolymerization, as shown in Fig. 1.5. Radical polymerization involves high-pressure copolymerization of ethylene with vinyl acetate, acrylic acid, and acrylates as comonomers. However, only copolymers of ethylene can be obtained using radical polymerization. Post-polymerization chemical modification involves melt grafting of functional groups (such as vinyl silanes, maleic anhydride, and methacrylates) with the help of free radical generating species (e.g., peroxides) via reactive extrusion. Finally, copolymerization involves the polymerization of monomers in addition to a functional monomer or reaction with intermediate monomers.

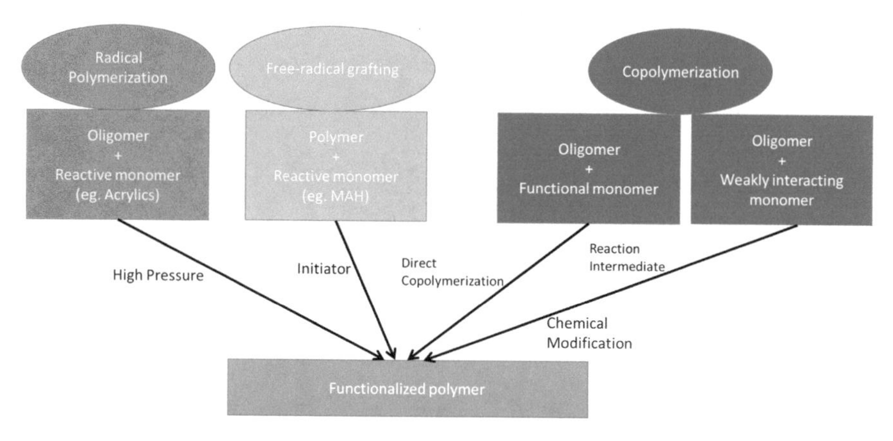

Fig. 1.5 Schematic of functionalization techniques of polymers

1.3.1 Modification of NPs

NPs form hard or soft agglomerates depending on the synthesis conditions and surface chemistry. Hard agglomeration occurs when small particles form a sintered-neck network that can only be broken by high-energy milling. Similarly, soft agglomeration occurs due to weak van der Waals or hydrogen bridge forces, which can be broken by mechanical shear forces. However, during mixing, NPs can re-aggregate in the PNC, rather than dispersing homogeneously in the nonpolar polymer matrix. Hence, modification of the NPs is required to suppress agglomeration and one method for addressing the agglomeration issue is surface modification of NPs.

1.3.1.1 Nanoclay Surface Modification

Surface modification of nanoclays is carried out by employing organic cations such as alkyl/aryl ammonium ions or anion surfactants including the sodium salt of dodecyl sulfate (SDS) or dodecylbenzene sulfonate (SDBS). Surface modification of nanoclay results in an increase of the interlayer gallery spacing and decreases the hydrophilicity of the clay, which enhances its dispersion in the polymer. Ding et al. [9] reported that the organic modification of clay resulted in substantial increase in the interlayer spacing compared to natural clay. In addition, incorporation of a compatibilizer (PP grafted with maleic anhydride) improved the compatibility between the organically modified clay and the nonpolar PP. Figure 1.6 illustrates various methods of dispersing organoclays in a polymer matrix, along with corresponding transmission electron microscopy (TEM) images, wide angle X-ray scattering (WAXS) data, and schematic diagrams showing the clay–polymer interactions. From the figure, it is clear that the immiscible clay existed as tactoids or aggregates of tactoids where there was no separation of platelets; this was further confirmed by WAXS, where no

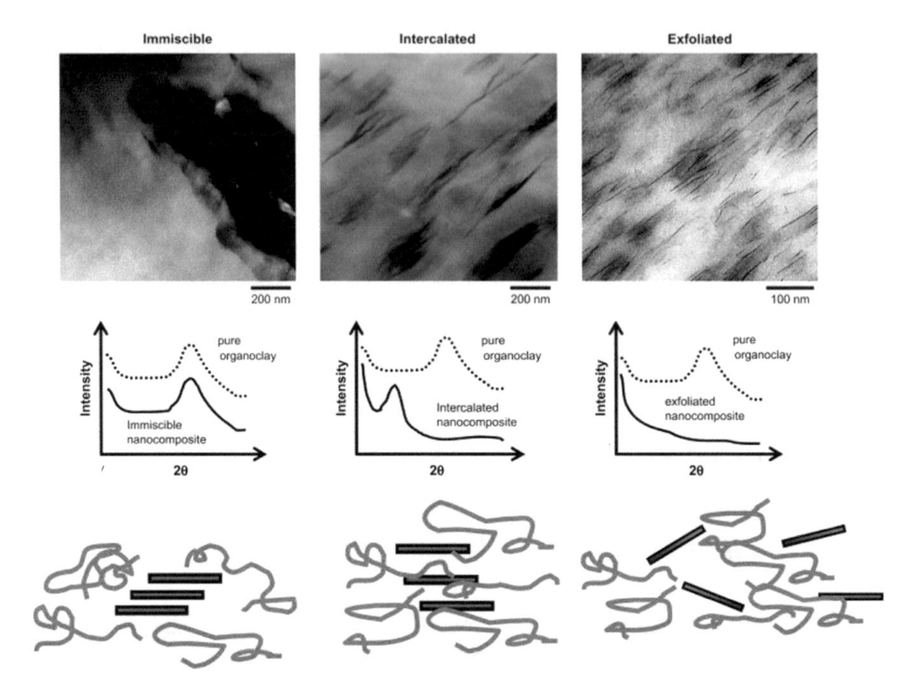

Fig. 1.6 Illustration of various situations of dispersion of organically modified clays in polymers with their analogous TEM, WAXS and cartoons representing clay and polymer interaction. Reproduced with permission from [10]. Copyright 2008, Elsevier Science Ltd.

shift in the peaks was observed after incorporation of the nanoclays. The possible platelet and chain interactions are illustrated in the schematic. In the case of intercalation, the polymer chains enter the clay structure, increasing the interlayer gallery spacing, as illustrated by the down-shift of the WAXS peak shown in Fig. 1.6; this was supported by the corresponding TEM results. Similarly, the exfoliation of clay is well illustrated in the schematic, where the attraction between the clay and polymer results in delamination of individual layers of the platelets stacks until all platelets are de-stacked and uniformly dispersed in the matrix provided sufficient time is given for this phenomenon to occur.

As illustrated in Fig. 1.7, during mixing, the polymer chains diffuse into the gallery spacings of the platelets due to the affinity between the nanoplatelets and polymer chains, which increases the interlamellar gallery spacing of the platelets. Further external shearing from the mixer provides sufficient energy for the platelets to overcome van der Waals forces of attraction or hydrogen bonding. Further shearing also provides energy for the polymers to be removed from the platelet layer. After peeling of a platelet layer, inner layers of the platelets are exposed for diffusion of the next polymer chain, which will result in peeling of the subsequent platelet given a sufficient mixing time. This peeling process continues until all the platelets are separated, leading to exfoliation of the nanoclay, resulting in a higher contact area

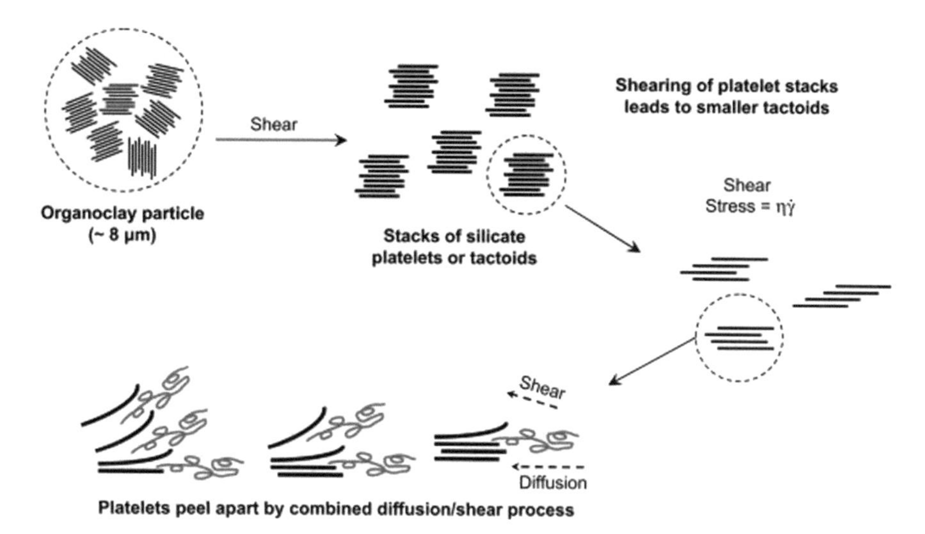

Fig. 1.7 The mechanism of organically modified clay dispersion and possible exfoliation of organically modified clay for the duration of melt processing. Reproduced with permission from [10]. Copyright 2008, Elsevier Science Ltd.

between the polymer and nanoclay and also increased load transfer/bearing capability of the nanocomposite. For processing of PVC-based nanocomposites, achieving a high level of exfoliation is difficult via melt mixing. However, solution mixing of PVC with clay resulted in a high degree of exfoliation, as evidenced by the TEM image shown in Fig. 1.8. In addition, in situ polymerization of the PVC and clay was reported to achieve good exfoliation, for instance in situ intercalative polymerization of Ɛ-caprolactone leads to good exfoliation in PVC based nanocomposites [11].

The surface modification of CNTs is carried out via covalent or non-covalent methods. The non-covalent modification involves the attachment of aromatic or small functional moieties to the CNT surfaces. Non-covalent modifications retain the CNT structure, preserving its inherent physicochemical properties; however, they show a weak force of attraction, which is not suitable for load transfer applications in composites. However, covalent modification occurs at defect sites that can attach to organic molecules. The functional groups bonded to CNTs are known to disperse well and offer load transfer potential, making the composites suitable for mechanical load transfer applications. Some materials used for the covalent modification of CNTs are diazonium nitrene, fluorination, and dichlorocarbene [4].

1.3.1.2 Surface Modification of Graphene

The most available form of graphene is graphene oxide (GO), which contains various functional groups such as carboxyl, hydroxyl, and epoxy groups; these groups can be tailored with different surface modifying agents, such as amino acids, organic

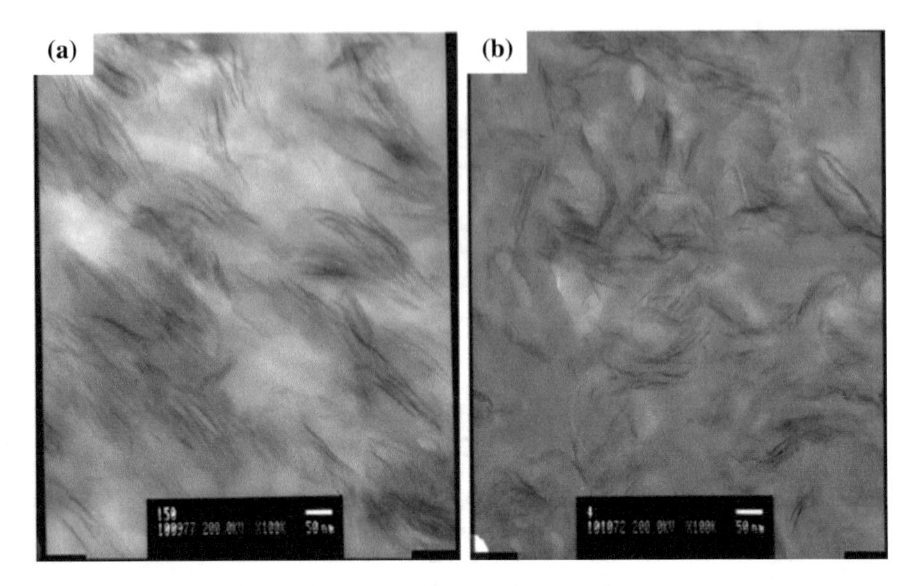

Fig. 1.8 The TEM micrographs of PVC/MMT nanocomposite prepared via melt mixing (**a**) and solution casting (**b**). Reproduced with permission from [7]. Copyright 2012, Elsevier Science Ltd.

amine, and amine terminated ionic liquids [12]. In addition, chemically reduced GO can be used as reinforcement. Recent studies showed that GO can be reduced to form reduced GO (rGO) in PNC during processing, which has attracted much attention as such materials provide improved electrical percolation at lower NP loading compared to GO [13].

1.3.2 Modification of Polymers

Commodity polymers that are generally encountered in daily life are nonpolar and do not contain functional groups that can interact with NPs to yield substantial improvement in reducing agglomeration of NPs. As most nanoclays are hydrophilic when they are naturally occurring, their dispersion in a nonpolar matrix is challenging. An alternative is to modify the NP surfaces, as discussed previously. Even after surface modification, NP dispersion is difficult due to poor compatibility between the modified nanoparticle and polymer chains. This can be addressed by adding a compatibilizer that consists of a nonpolar group (which is generally miscible in the polymer matrix) and a polar group to interact with the NPs during mixing. To compatibilize nonpolar polyolefins, a small amount of a polyolefin was grafted onto functional moieties (e.g., maleic anhydride, MAH) and polar monomers (e.g., methacrylic acid and vinyl acetate).

The various methods used to introduce functional moieties into nonpolar polymers are described in Fig. 1.5. Among the several functionalization methods and various functional moieties, only reactive extrusion and MAH, respectively, have shown potential environmental and economic benefits. Hence, we limit our scope to the introduction of MAH into PE and PP via reactive extrusion only. This process involves the addition of the reactive monomer (MAH) and an initiator (peroxide) to an extruder while processing polyolefins. The initiator decomposes during extrusion to yield free radicals that de-propagate the polyolefin chains, leading to chain scissoring or radical terminated polymer chains. The MAH reacts with the radical terminated polymer to form polymer-grafted MAH (present in the form of succinic anhydride) [14]. Thus, polyolefin-grafted MAH can be successfully extruded and is mainly used as a compatibilizer; it requires that the nanoclay is modified with surfactant to achieve some success as a PNC.

1.4 Structure–Property Relationships in PNC

The introduction of nanofillers into polymer matrices affects various physical properties (e.g., mechanical, electrical, and thermal properties), depending on the type of filler. Due to their small size (1−100 nm), the addition of NPs can improve the stiffness and thermal properties. However, the enormous surface area per unit mass of NPs results in strong van der Waals forces, leading to NP agglomeration, which reduces the aspect ratio of the NPs, resulting in slip when stress is applied due to poor interaction between the polymer and NPs. In addition, agglomerated NPs act as stress concentration points in the matrix, resulting in poor reinforcement behavior. The mechanical, electrical, thermal, and other properties depend on the shape and type of NPs. Moreover, the modification of the polymer and/or NPs can affect the PNC properties, as discussed in the following sections.

1.4.1 Mechanical Properties

NPs impart nanoscale reinforcement that depends on the specific surface area, functional groups on the NPs surface, interaction between the polymer and NPs, and the polymer chain stiffness [15]. With 0.01 vol.% filler loading with an aspect ratio of 1000 for platelets or nanotubes an enhancement of modulus of a factor of six was predicted using the Halpin-Tsai equation [16]. The addition of NPs to polymers results in a higher increment in the modulus than the strength. However, the strain at break decreased gradually with increasing NP content.

With the addition of nanoclay, an increase in the strength and modulus (compared to the pure polymer) of the PNC was accompanied by a decrease in the elongation at break. For nonpolar polymers, such as PP and PE, the interaction between the polar group of the clay and the nonpolar polymer is limited, resulting in a high interfacial tension between the polymer and nanoclay. Thus, increasing the nanoclay loading up to a critical wt% results in an increase in the modulus, above which, the improvement in modulus is minimal. Further increases in the nanoclay content resulted in minimal increments in the modulus [2]. This issue was addressed by introducing maleated PP (PP-grafted-MAH) to PNC. MAH contains a polar group which interacts with the clay, resulting in intercalation of the layered silicates; this helped increase the effective contact area between the polymer and silicate, resulting in an increase in both the tensile and impact strengths [17]. However, re-agglomeration of the NPs resulted in a subsequent decrease in both tensile and impact strength for a clay loading of 10 wt%.

In the case of nanotubes, a low filler volume fraction generally improves the tensile properties, while for higher volume fractions of nanotubes, the properties tend to degrade due to the dispersion and agglomeration issues. The nanotube diameter was reported to be scale inverse of reinforcement for multi-walled nanotubes, such as MWCNTs [18]. However, single-walled nanotubes, such as SWCNT exhibited poorer reinforcement due to poor dispersion or bundle formation. Nanotubes possess better tensile properties in the axial direction than in the transverse direction. Thus, a composite with aligned nanotubes (anisotropic nanocomposite) shows improved properties compared to an isotropic composite. However, attempts to improve the adhesion and dispersion can result in an isotropic composite with poorer mechanical properties than an anisotropic nanocomposite with aligned nanotubes. However, the anisotropic composite exhibits better tensile properties in the aligned direction than in the transverse direction. The development of most nanotube-based reinforcements focus on the alignment of MWNTs in a particular direction and measurement of the performance of the nanocomposite in the aligned direction [2]. Methods for aligning the nanotubes include polymer stretching [19], melt drawing [20], surface acoustic wave [21], and the application of alternating [22] or direct current electric, [23] or a magnetic field [24]. Table 1.1 summarizes the mechanical properties of selected PNCs.

Graphene-based PNCs show an increase in the modulus with increasing filler loading, as summarized in Table 1.2. Larger increases in the modulus are observed for elastomeric matrices as they have a lower intrinsic modulus. The tensile strength increases with increasing interaction between the graphene and polymer. However, the elongation at the break either decreased or remained unchanged with increasing graphene loading. However, most of the reported mechanical properties fell short of the theoretically predicted value. In addition, graphene-based PNCs showed better mechanical properties than those with carbon black or SWCNT fillers [32].

1.4.2 Effect on Glass Transition Temperature

It has been well established that the presence of NPs in a polymer matrix can polarize the configurational state of the polymer, which in turn changes the nature of molecular packing and subsequently the glass transition width [15]. Factors influ-

Table 1.1 Summary of mechanical properties of selected nanotubes based PNC

Matrix	Nanotubes type	Nanotubes loading wt%	Processing technique	% increment in tensile modulus	% increment in tensile strength	References
Low density Polyethylene (LDPE)	MWNTs	10.00	Melt mixing	89	56	[25]
Low density polyethylene (LDPE)	MWNTs	3.00	Ball milling	30	150	[26]
Polypropylene (PP)	MWNTs	0.25	Melt fiber spinning	130	10	[27]
Polypropylene (PP)	SWNT	1.00	Solution mixing—fiber spinning	200	150	[28]
Polystyrene (PS)	MWNTs	1.00	Solution casting	42	25	[29]
Polystyrene (PS)	MWNTs	40.00	Melt mixing	100	10	[30]
Polyvinyl chloride (PVC)	PBMA-grafted MWCNTs	0.20	Solution mixing	40	84	[31]

Table 1.2 Summary of mechanical properties of selected graphene based PNC

Matrix	Graphene type	Graphene loading wt%	Processing technique	% increment in tensile strength	% increment in tensile modulus	References
High density polyethylene (HDPE)	Expanded	3.00	Melt processing	4.00	100	[33]
Polypropylene (PP)	Graphene	5.00	Melt processing	22.72	20.59	[34]
Polystyrene (PS)	Solvent-exfoliated graphene	0.50	Solution mixing	6.00	77	[35]
Polyvinyl chloride (PVC)	Graphene	2	Solution mixing	160.00	150.00	[36]
Polyvinyl chloride (PVC)	Graphene oxide	3.00	Solution mixing	~28.50	~57.65	[37]

encing the glass transition temperature (T_g) include, the sample thickness, size of the NP, sample preparation and measurement technique, and chemical structure of the polymer. Importantly, the interaction between the polymer chain and particle surface determines the magnitude of change in T_g [38]. If the polymer chain and NPs strongly interact, T_g increases with respect to the bulk polymer. Similarly, NPs that do not interact with the polymer chain either decrease T_g or it remains unchanged. In addition, hydrogen bonding between the NPs and polymer chains can increase T_g. Thus, the presence of exfoliated or intercalated nanoclay can increase T_g of the polymer matrix. The interaction between the polymer and clay platelets contributes to increasing T_g. However, if the interlayer distance between platelets is shorter than the characteristic length of the polymer chain, T_g is either unaltered or reduced [39].

Adding nanotubes to a polymer can have either increase or decrease the T_g depending on the surface functionalization of the nanotubes and covalent bonding between the nanotubes and polymer chain. Similarly, for graphene NPs, the platelets have a high surface roughness and high aspect ratio which can increase T_g when the filler is well-dispersed in the polymer. An interesting feature reported by Liao et al. [40] is that covalent bonding between the polymer and graphene has a significant effect on the magnitude of the increase in T_g of the composite. In addition, they reported insignificant changes in T_g for composites prepared by solution mixing and melt blending. However, they reported significant improvement in T_g for a nanocomposite which was covalently bonded with a polymer matrix. In addition, they reported that the type of polymer can affect whether T_g is decreased, increased, or remains unchanged.

1.4.3 Thermal Stability

The addition of NPs into a polymer matrix imparts either physical or chemical crosslinking which results in increase of the initial degradation temperature [41]. In the case of nanoclays, the filler dispersion is critical for predicting the mechanism of polymer degradation. Studies indicate that the presence of nanoclay can prevent the free movement of free radicals and oxygen, resulting in stabilization of the nanocomposite. Blumstein proposed that the reduction in molecular movement and physical processes in the condensed phase can increase the thermal stability of nanoclay-based PNCs [42]. In addition, the nanoclay results in char formation or can catalyze decomposition of free radicals, which also results in increased thermal stability of the nanocomposite [43].

The mechanism of thermal stability enhancement depends on the dispersion state of the nanotubes. Physical adsorption of macromolecules on the nanotube surfaces leads to slower volatilization of polymer [44] or free radical absorption, which reduced mass loss during heating in an inert atmosphere [45]. Further, interfacial interactions between the polymer and nanotubes can increase the activation energy of degradation, thereby enhancing the thermal stability of the PNC [46]. In addition, chemical modification of the nanotubes and polymer chains leads to finer disper-

Table 1.3 Summary of the selected thermal conductivity results for commodity PNCs

Polymer	Nanoparticle	Nanoparticle loading (wt%)	Thermal conductivity ($Wm^{-1} K^{-1}$)	References
Polyethylene	Graphene	10.00	1.84	[49]
Polypropylene (PP)	Graphene	10.00	1.53	[49]
Polystyrene (PS)	MWNT	10.00	0.42	[50]

sion of the nanotubes, which substantially changes the degradation pathways and increases thermal stability.

Graphene-based PNCs exhibit substantially slower decomposition than the pure polymer. Graphene forms a platelet-like structure which reduces the chain mobility of the polymer near the graphene surface. In addition, they tend to form a jammed network of char layers which acts as a barrier for transportation of decomposition products, thus increasing the thermal stability [32].

1.4.4 *Thermal Conductivity*

Most polymers have a thermal conductivity in the range of 0.2–0.5 W/m K. Incorporation of a filler with a high intrinsic thermal conductivity, such as CNT and graphene, into the polymer has been shown to increase the thermal conductivity. The primary mode of thermal conduction in the polymer matrix is via phonon transportation [47]. However, the challenge lies in obtaining a thermal conductivity higher than the 4 W/m K at a low filler content. This can be addressed by functionalizing the CNT to improve the dispersion of the CNTs. However, the key issue of phonon scattering at the interface of the polymer and CNT surface remains unaddressed. A key improvement in the thermal conductivity of a PNC was achieved by in situ polymerization of chemically modified graphene filler. The thermal conductivity scales linearly with filler content [48] in PNCs. Table 1.3 summarizes selected thermal conductivity results for commodity PNCs.

1.4.5 *Electrical Properties*

The incorporation of materials with an intrinsically high electrical conductivity, such as CNTs and graphene, into polymer matrices has led to electrically conducting PNCs of dielectric polymers. However, the conductivity of the PNC increases with an increase in filler content up to the point where an electrically conducting network is formed. The critical filler concentration at which the PNC starts conducting is termed

Table 1.4 Summary of the selected electrical properties of commodity polymers-based PNCs

Polymer	Nanoparticle	Processing method	Nanoparticle loading (wt%)	Conductivity of nanocomposite (S m^{-1})	References
High density polyethylene (HDPE)	Expended graphite	Melt mixing	3.00	1×10^{-6}	[33]
High density polyethylene (HDPE)	Purified SWCNT	Melt mixing	5.00	1×10^{-1}	[31]
Polyethylene	MWCNT	Melt mixing	10.00	1×10^{-3}	[31]
Polypropylene (PP)	MWCNT	Melt mixing	5.00	1×10^{-7}	[31]
Polypropylene (PP)	Expended graphite	Solution mixing	0.67 (vol.%)	1×10^{-1}	[53]
Polystyrene (PS)	SWCNT	Solution mixing	2.00	1×10^{-3}	[31]
Polystyrene (PS)	SWCNT	Solution mixing	1.50	1	[31]
Polyvinyl chloride (PVC)	MWCNT	Solution mixing	1.4	1×10^{-4}	[54]

the percolation threshold. Beyond this point, further increases in the loading of the conducting filler in the nanocomposite results in an increase in the conductivity up to certain concentration, where the conductivity then plateaus. Generally, the factors that significantly affect the percolation threshold are the NP aspect ratio and shape, processing technique, state of NP dispersion in polymer matrix, and functionalization of the NPs. Theoretical predictions showed that rod-like structures percolate at a lower concentration than disk-shaped NPs [51]. In situ polymerization to produce the PNC was reported to produce a lower percolation threshold than melt mixing. Graphene was reported to have a low percolation threshold of ~0.07–0.1 vol.% due to the high level of exfoliation of thin graphene sheets with large aspect ratios. Similarly, the addition of 0.5 vol.% graphene to polystyrene increased the electrical conductivity from 10^{-16} to 0.01 S/m [52]. Table 1.4 summarizes the electrical properties of selected commodity PNCs.

1.4.6 Fire Retardancy

Polymers are highly flammable as they release volatile compounds above a certain temperature which readily react with oxygen, leading to combustion of the material. Well-dispersed NPs in a polymer matrix forms a continuous network-like structure which prevents diffusion of volatile compounds to the surface and their exposure to surface oxygen, impeding burning/flaming of the polymer. The aspect ratio and processing technique have a strong influence on the fire retardancy [55, 56]. If NPs are poorly dispersed and the filler loading is low, a discontinuous network may be formed, which leads to poor flame resistance. In the case of MWCNTs, their high aspect ratio ensures the formation of a network structure that acts as a protective layer in the condensed phase, physically shielding the heat [57], as well as localizing the heat, which increases the ignition time. The addition of graphene to a polymer matrix resulted in dense, compact, and uniform charring during combustion, resulting in inhibited flammability compared to pure polymer [58].

1.4.7 Membrane Separation and Barrier Properties

As discussed previously, NPs like nanotubes etc., added to a polymer matrix form a network owing to their high aspect ratio. This network creates a tortuous pathway for permeation in the composite. A decrease in gas permeability of a polymer was observed with incorporation of NPs, which was independent of the type of gas [59]. The permeability of a nanocomposite decreased with increasing aspect ratio of the NPs; the incorporation of MMT on LDPE-g-MA led to enhanced clay dispersion and improved barrier properties [60]. Defect-free graphene sheets were shown to impede permeation of gas molecules [61], while the use of graphene oxide sheets in PNCs significantly reduced the permeation of carbon dioxide and oxygen [62].

CNTs can act as membranes due to the nanometric control of pore dimensions [63]. Simulations performed by Skoulidas et al. [64] showed that vertically aligned CNTs exhibited good transportation and permeability of light gases, better than a crystalline zeolite membrane. The addition of aligned MWCNTs in a polystyrene-based nanocomposite resulted in higher gas permeability and improved selectivity [65].

1.5 Performance and Applications

It is estimated that the global consumption of nanocomposites was 118,768 mT in 2010, with a market value of 800 million USD [66]. PNCs accounted for a market share greater than 50% of the total consumption in 2010. Automotive parts, packaging, and coatings accounted for 41, 32 and 16%, respectively, of the market share

Table 1.5 Summary of commodity polymers-based PNCs and their suppliers

Matrix	Product name	Applications	Supplier
Thermoplastic polyolefins		Automotive	DuPont
Polyethylene (PE)	Nanolok	Packing	
High density polyethylene (HDPE)		Packing	LG Chemicals
Polyethylene (PE)	Maxxam LST	Packaging, flame retardance	
Polypropylene (PP)		Packing	Clariant
Polypropylene (PP)	Forte	Automotive parts, appliance, office furniture	Noble Polymer
Polypropylene (PP)	Nanoblend	Automotive	PolyOne
Poly vinyl chloride (PVC)	NanoVin	PVC paste application	Solvin

of PNCs in 2012. A major advantage of PNCs is their superior mechanical properties compared to conventional polymer composites. However, commercialization of this technology is limited as the cost reductions and performance enhancements are required before the technology is economically viable. Table 1.5 summarizes commercially available commodity PNCs.

1.5.1 Automotive Parts

In the 1990s, the Toyota Motor Corporation first introduced a PA6-clay nanocomposite for timing-belt covers. The nanoclay replaced the conventional filler at a 3:1 ratio, while the PA6-clay nanocomposite was a perfect lightweight substitute for conventional PA6 + 30% glass beads. The new material reduced the weight of the component by 25%, and showed good rigidity and thermal stability, and no warpage. Due to improved thermomechanical and barrier properties, this material is now widely used in the automobile industry, especially in oil reserve tanks, engine covers, and fuel hoses with lower filler loading. In 2002, General Motors began using a polyolefin-based thermoplastic material filled with 3 wt% nanoclay which showed a 25% weight reduction. There are many more applications of exfoliated clay NP-reinforced nanocomposites, but most have not been published or are patent protected.

Table 1.6 List of commodity polymers-based PNCs for packaging application with oxygen and water vapor percent increase in permeability data

Matrix	Reinforcement	Filler loading (wt%)	% decrease in permeability of oxygen	% decrease in permeability of water vapor	References
Polystyrene (PS)	Modified MMT	16.7	180		[68]
High density polyethylene (HDPE)	Modified MMT	5.0	180–190	80–140	[69]
Low density polyethylene (LDPE)	Modified MMT	4.8	120		[70]
Poly vinyl chloride (PVC)	Modified MMT	10	22		[71]

1.5.2 *Packaging Applications*

Packaging applications require materials that provide good physical protection and appropriate physicochemical conditions so that the shelf life of the product is maintained during storage and handling. Therefore, preventing mechanical damage and exposure to certain gases (e.g., O_2, CO_2) and water vapor is necessary to maintain physicochemical or biological activity of the product at time of storing. PNC-based packaging materials provide a barrier to moisture, water vapor, solutes, and gases, while minimizing microbial growth; this helps extend the shelf-life of the product and maintain product quality [67]. There are two main types of layered nanomaterials reported in the literature that effectively decrease permeability of gases in the polymer, nanoclay (layer of silicates) and graphene sheets (GO and graphene) [2].

Most literature on PNCs for packaging applications discusses nanoclay as a nanoscale filler/reinforcement. The nanoclays used for these applications are montmorillonite, hectrite, saponite, and kaolinite due to their high aspect ratio and high surface area. Nanoclay dispersed in a polymer matrix forms a network-like structure that forces the gas to follow a tortuous path, increasing the effective mean free length for diffusion of the permeate. Table 1.6 lists representative examples of commodity polymers explored for packaging applications, along with their moisture and oxygen permeability data. In addition, flame retardance is an added advantage for packaging made from nanoclay-based PNCs.

Recent studies have reported that graphene-based NPs such as graphene and GO have shown enhanced mechanical, optical, and electrical properties, along with enhanced gas barrier behavior. Graphene has a higher surface-to-volume ratio than other nanocarbon-based materials such as CNTS and fullerenes, which allows them to achieve longer gas diffusion pathways at low filler loadings. However, the large-scale synthesis of defect-free, large-surface-area, monocrystalline graphene at an acceptable cost is challenging. This could be overcome by employing either GO or rGO, which can be produced on a large scale with a high aspect ratio. However, dispersing graphene in the polymer matrix is a remaining challenge for the use of graphene for packaging applications.

1.5.3 Flammability

The incorporation of microparticles as flame-retardant and intumescent agents is known to reduce the flammability of polymers, but has the disadvantage of degrading the mechanical properties. However, with the incorporation of NPs, the flammability of the polymer is reduced, while maintaining or enhancing the mechanical properties.

1.5.4 Other Applications

Intrinsically conducting carbonaceous NPs are used to enhance the electrical conductivity of insulating polymers. The diclectric and conducting behavior of PNCs makes them suitable for e.g., charge-storage devices, self-controlled heaters, and current limiters.

PNCs are expected to find general applications, such as power-tool housings, vacuum cleaner impellers and blades, mower hoods, body covers for portable electronic devices, and aerospace systems.

1.6 Challenges, Opportunities, and Future Trends

1.6.1 Challenges and Opportunities

- Dispersing nanofillers in a PVC matrix is hampered due to the complex morphology of PVC. However, addressing this issue would offer benefits in the area of smoke suppression, impact modification, and thermal and electrical properties.
- Simultaneous functionalization of NPs to assist in their proper dispersion during melt processing needs to be addressed. In addition, tailoring the dispersion of NPs based on the required properties needs to be clarified. For example, to achieve improved electrical properties, a percolating network is sufficient, while for enhancing the mechanical properties, good dispersion is required. Thus, a study addressing this issue is required.
- Obtaining organically modified clays that are stable at higher temperatures (~180 °C) to allow processing at higher temperatures [72].

- Enhancement of mechanical reinforcement via melt processing needs to address the limitations of the reinforcement that can be incorporated into the polymer matrix. For instance, the addition of high contents of SWNCTs results in agglomeration, which degrades the mechanical properties. However, a lower loading results in under-performance of the nanocomposite, limiting the SWNCTs' loading in the composite.
- Systematic studies regarding the purity, aspect ratio, and degree and type of functionalization of the NPs, along with the interactions between various processing additives and modified NPs is required to allow further enhancement of nanocomposites properties. For example, graphene with an aspect ratio of 2000 exhibited strength and modulus values twice those of graphene with an aspect ratio of 1000.

1.6.2 Future Trends

- Most of the modification of CNTs and carbon nanofibers results in significant improvement in the modulus and strength when forming unidirectional nanocomposites. However, cross-ply composites could be used for advanced composite applications like future aircraft and wind energy turbines.
- The design of nanoscale PNCs that can mimic biological systems.
- To maximize the electrical properties, isotropic NP dispersion may not be an ideal morphology; instead, a hierarchical morphology with a percolation pathway would be a beneficial PNC structure [73].
- PVC nanocomposites are likely to attract attention from the foam industry for their flame and smoke retardant behavior, good thermal stability, and ability to be recycled.
- Most of the PNC literature focuses on a single kind of NP in the matrix, while reports of hybrid structures with the simultaneous incorporation of two or more NPs are less common. Focusing on hybrid structures to achieve synergistic properties is required in the near future. For instance, incorporation of a 1D filler in stacking of 2D platelets.
- Noncovalent modification of CNTs and graphene is reported to improve electrical and mechanical properties. Thus, a better understanding of the interface between the matrix and non-covalently functionalized CNT/graphene is required to enhance the mechanical and electrical properties of PNCs.
- The literature shows that NPs act as a hetero nucleating agent, which affects the crystallinity of the polymer [74]. However, studies investigating the effect of changes in the crystallinity on the mechanical properties are required. In addition, specific particles and processing conditions are known to induce phase changes in polymers; hence, studies to further explore this phenomenon are required. Further, the T_g of the polymer matrix affects the incorporation of surface functionalized NPs and should be investigated in detail [40].

1.7 Summary

This chapter discussed the preparation of commodity-polymer-based PNCs. To date, the use of NPs has been limited to a few special applications. It is predicted that the current special applications could be future commodity applications. The use of NPs in the near future is expected for commodity applications using commodity polymers. Thus, the present chapter discussed various aspects of processing commodity PNCs to meet the exponentially growing demand of PNCs.

The incorporation of NPs with high aspect ratio has resulted in lower filler loading requirements for improving the strength and modulus of the composite per specific unit mass. The extent of improvement in the properties depends on the type and functionalization of the nanofiller, its aspect ratio and loading, the interactions between the polymer matrix and nanofiller, and the processing technique used. Higher nanofiller contents are more difficult to disperse in the polymer matrix, resulting in agglomeration and degraded mechanical properties. Covalent functionalization of nanotubes improves the stress transfer between the nanofiller and matrix. In addition, covalent modification assists in dispersion of the nanofiller in commodity polymers, which addresses the issue of dispersion of the nanofiller for large-scale production. However, covalent functionalization has a trade-off in the case of CNTs, as it deteriorates the electrical properties. Graphene has a percolation threshold equivalent to CNTs. The conductivity and barrier properties are dependent on the number of defect sites. In addition, graphene sheets have a low bulk density, making them difficult to handle during processing.

Nanoclays are extensively employed in the automotive sector. The high cost of graphene and CNTs makes it challenging for them to find commodity applications. However, continued research into these materials may result in price reductions, which may expand their range of commodity applications and allow them to replace carbon black and carbon fibers.

Acknowledgements The authors would like to thank the Council for Scientific and Industrial Research and the Department of Science and Technology, South Africa, for financial support.

References

1. Rosato DV, Rosato DV (2004) Reinforced plastics handbook. Elsevier, New York
2. Bhattacharya M (2016) Polymer nanocomposites—a comparison between carbon nanotubes, graphene, and clay as nanofillers. Materials 9(4):262
3. Ajayan PM, Schadler LS, Braun PV (2006) Nanocomposite science and technology. Wiley, New Jersey
4. Gou J, Zhuge J, Liang F (2012) Processing of polymer nanocomposites. In: Manufacturing techniques for polymer matrix composites. Woodhead Publishing, Sawston, Cambridge, pp 95–115
5. Ray SS, Okamoto M (2003) Polymer/layered silicate nanocomposites: a review from preparation to processing. Prog Polym Sci 28(11):1539–1641

6. Mallick PK (2007) Fiber-reinforced composites: materials, manufacturing, and design. CRC press, Boca Raton
7. Gilbert M (2012) Poly (vinyl chloride)(PVC)-based nanocomposites. In: Advances in polymer nanocomposites. Elsevier, New York, pp 216–237
8. Fischer H (2003) Polymer nanocomposites: from fundamental research to specific applications. Mater Sci Eng C 23(6):763–772
9. Ding C, Jia D, He H, Guo B, Hong H (2005) How organo-montmorillonite truly affects the structure and properties of polypropylene. Polym Testing 24(1):94–100
10. Paul D, Robeson LM (2008) Polymer nanotechnology: nanocomposites. Polymer 49(15):3187–3204
11. Pagacz J, Pielichowski K (2009) Preparation and characterization of PVC/montmorillonite nanocomposites—a review. J Vinyl Add Tech 15(2):61–76
12. Liu N, Luo F, Wu H, Liu Y, Zhang C, Chen J (2008) One-step ionic-liquid-assisted electrochemical synthesis of ionic-liquid-functionalized graphene sheets directly from graphite. Adv Func Mater 18(10):1518–1525
13. Mural PKS, Sharma M, Madras G, Bose S (2015) A critical review on in situ reduction of graphene oxide during preparation of conducting polymeric nanocomposites. RSC Adv 5(41):32078–32087
14. Berzin F, Flat J-J, Vergnes B (2013) Grafting of maleic anhydride on polypropylene by reactive extrusion: effect of maleic anhydride and peroxide concentrations on reaction yield and products characteristics. J Polym Eng 33(8):673–682
15. Jancar J, Douglas J, Starr FW, Kumar S, Cassagnau P, Lesser A, Sternstein SS, Buehler M (2010) Current issues in research on structure–property relationships in polymer nanocomposites. Polymer 51(15):3321–3343
16. Affdl J, Kardos J (1976) The Halpin-Tsai equations: a review. Polym Eng Sci 16(5):344–352
17. Xu W, Liang G, Wang W, Tang S, He P, Pan WP (2003) PP–PP-g-MAH–Org-MMT nanocomposites. I. Intercalation behavior and microstructure. J Appl Polym Sci 88(14):3225–3231
18. Cadek M, Coleman J, Ryan K, Nicolosi V, Bister G, Fonseca A, Nagy J, Szostak K, Beguin F, Blau W (2004) Reinforcement of polymers with carbon nanotubes: the role of nanotube surface area. Nano Lett 4(2):353–356
19. Haggenmueller R, Gommans H, Rinzler A, Fischer JE, Winey K (2000) Aligned single-wall carbon nanotubes in composites by melt processing methods. Chem Phys Lett 330(3):219–225
20. Thostenson ET, Chou T-W (2002) Aligned multi-walled carbon nanotube-reinforced composites: processing and mechanical characterization. J Phys D Appl Phys 35(16):L77
21. Strobl C, Schäflein C, Beierlein U, Ebbecke J, Wixforth A (2004) Carbon nanotube alignment by surface acoustic waves. Appl Phys Lett 85(8):1427–1429
22. Chen X, Saito T, Yamada H, Matsushige K (2001) Aligning single-wall carbon nanotubes with an alternating-current electric field. Appl Phys Lett 78(23):3714–3716
23. Kumar MS, Kim T, Lee S, Song S, Yang J, Nahm K, Suh E-K (2004) Influence of electric field type on the assembly of single walled carbon nanotubes. Chem Phys Lett 383(3):235–239
24. Camponeschi E, Vance R, Al-Haik M, Garmestani H, Tannenbaum R (2007) Properties of carbon nanotube–polymer composites aligned in a magnetic field. Carbon 45(10):2037–2046
25. Xiao K, Zhang L, Zarudi I (2007) Mechanical and rheological properties of carbon nanotube-reinforced polyethylene composites. Compos Sci Technol 67(2):177–182
26. Gorrasi G, Sarno M, Di Bartolomeo A, Sannino D, Ciambelli P, Vittoria V (2007) Incorporation of carbon nanotubes into polyethylene by high energy ball milling: morphology and physical properties. J Polym Sci, Part B: Polym Phys 45(5):597–606
27. Dondero WE, Gorga RE (2006) Morphological and mechanical properties of carbon nanotube/polymer composites via melt compounding. J Polym Sci Part B Polym Phys 44(5):864–878
28. Moore EM, Ortiz DL, Marla VT, Shambaugh RL, Grady BP (2004) Enhancing the strength of polypropylene fibers with carbon nanotubes. J Appl Polym Sci 93(6):2926–2933
29. Qian D, Dickey EC, Andrews R, Rantell T (2000) Load transfer and deformation mechanisms in carbon nanotube-polystyrene composites. Appl Phys Lett 76(20):2868–2870

30. Andrews R, Jacques D, Minot M, Rantell T (2002) Fabrication of carbon multiwall nanotube/polymer composites by shear mixing. Macromol Mater Eng 287(6):395–403
31. Spitalsky Z, Tasis D, Papagelis K, Galiotis C (2010) Carbon nanotube–polymer composites: chemistry, processing, mechanical and electrical properties. Prog Polym Sci 35(3):357–401
32. Kim H, Abdala AA, Macosko CW (2010) Graphene/polymer nanocomposites. Macromolecules 43(16):6515–6530
33. Zheng W, Lu X, Wong SC (2004) Electrical and mechanical properties of expanded graphite-reinforced high-density polyethylene. J Appl Polym Sci 91(5):2781–2788
34. Song P, Cao Z, Cai Y, Zhao L, Fang Z, Fu S (2011) Fabrication of exfoliated graphene-based polypropylene nanocomposites with enhanced mechanical and thermal properties. Polymer 52(18):4001–4010
35. Zhao J, Liu Y, Cheng J, Wu S, Wang Z, Hu H, Zhou C (2017) Reinforced polystyrene via solvent-exfoliated graphene. Polym Int 66(12):1827–1833
36. Vadukumpully S, Paul J, Mahanta N, Valiyaveettil S (2011) Flexible conductive graphene/poly(vinyl chloride) composite thin films with high mechanical strength and thermal stability. Carbon 49(1):198–205
37. Deshmukh K, Joshi GM (2014) Thermo-mechanical properties of poly (vinyl chloride)/graphene oxide as high performance nanocomposites. Polym Testing 34:211–219
38. Tate RS, Fryer DS, Pasqualini S, Montague MF, de Pablo JJ, Nealey PF (2001) Extraordinary elevation of the glass transition temperature of thin polymer films grafted to silicon oxide substrates. J Chem Phys 115(21):9982–9990
39. Krishnamoorti R, Vaia RA, Giannelis EP (1996) Structure and dynamics of polymer-layered silicate nanocomposites. Chem Mater 8(8):1728–1734
40. Liao K-H, Aoyama S, Abdala AA, Macosko C (2014) Does Graphene Change T_g of Nanocomposites? Macromolecules 47(23):8311–8319
41. Majka TM, Leszczyńska A, Pielichowski K (2016) Thermal stability and degradation of polymer nanocomposites. In: Polymer nanocomposites. Springer, Heidelberg, pp 167–190
42. Blumstein A (1965) Polymerization of adsorbed monolayers. II. Thermal degradation of the inserted polymer. J Polym Sci Part A Polym Chem 3(7):2665–2672
43. Gilman JW (1999) Flammability and thermal stability studies of polymer layered-silicate (clay) nanocomposites. Appl Clay Sci 15(1):31–49
44. Yang J, Lin Y, Wang J, Lai M, Li J, Liu J, Tong X, Cheng H (2005) Morphology, thermal stability, and dynamic mechanical properties of atactic polypropylene/carbon nanotube composites. J Appl Polym Sci 98(3):1087–1091
45. Chatterjee A, Deopura B (2006) Thermal stability of polypropylene/carbon nanofiber composite. J Appl Polym Sci 100(5):3574–3578
46. Chipara M, Lozano K, Hernandez A, Chipara M (2008) TGA analysis of polypropylene–carbon nanofibers composites. Polym Degrad Stab 93(4):871–876
47. Han Z, Fina A (2011) Thermal conductivity of carbon nanotubes and their polymer nanocomposites: a review. Prog Polym Sci 36(7):914–944
48. Veca LM, Meziani MJ, Wang W, Wang X, Lu F, Zhang P, Lin Y, Fee R, Connell JW, Sun YP (2009) Carbon nanosheets for polymeric nanocomposites with high thermal conductivity. Adv Mater 21(20):2088–2092
49. Alam FE, Dai W, Yang M, Du S, Li X, Yu J, Jiang N, Lin C-T (2017) In situ formation of a cellular graphene framework in thermoplastic composites leading to superior thermal conductivity. J Mater Chem A 5(13):6164–6169
50. Yang Y, Gupta M, Zalameda J, Winfree W (2008) Dispersion behaviour, thermal and electrical conductivities of carbon nanotube-polystyrene nanocomposites. Micro Nano Letters 3(2):35–40
51. Garboczi E, Snyder K, Douglas J, Thorpe M (1995) Geometrical percolation threshold of overlapping ellipsoids. Phys Rev E 52(1):819
52. Stankovich S, Dikin DA, Dommett GH, Kohlhaas KM, Zimney EJ, Stach EA, Piner RD, Nguyen ST, Ruoff RS (2006) Graphene-based composite materials. Nature 442(7100):282

53. Chen X-M, Shen J-W, Huang W-Y (2002) Novel electrically conductive polypropylene/graphite nanocomposites. J Mater Sci Lett 21(3):213–214
54. Yazdani H, Smith BE, Hatami K (2016) Electrical conductivity and mechanical performance of multiwalled CNT-filled polyvinyl chloride composites subjected to tensile load. J Appl Polymer Sci 133(29)
55. Schütz MR, Kalo H, Lunkenbein T, Breu J, Wilkie CA (2011) Intumescent-like behavior of polystyrene synthetic clay nanocomposites. Polymer 52(15):3288–3294
56. Bartholmai M, Schartel B (2004) Layered silicate polymer nanocomposites: new approach or illusion for fire retardancy? Investigations of the potentials and the tasks using a model system. Polym Adv Technol 15(7):355–364
57. Cipiriano BH, Kashiwagi T, Raghavan SR, Yang Y, Grulke EA, Yamamoto K, Shields JR, Douglas JF (2007) Effects of aspect ratio of MWNT on the flammability properties of polymer nanocomposites. Polymer 48(20):6086–6096
58. Huang G, Gao J, Wang X, Liang H, Ge C (2012) How can graphene reduce the flammability of polymer nanocomposites? Mater Lett 66(1):187–189
59. Kojima Y, Usuki A, Kawasumi M, Okada A, Fukushima Y, Kurauchi T, Kamigaito O (1993) Mechanical properties of nylon 6-clay hybrid. J Mater Res 8(5):1185–1189
60. Jacquelot E, Espuche E, Gérard JF, Duchet J, Mazabraud P (2006) Morphology and gas barrier properties of polyethylene-based nanocomposites. J Polym Sci Part B: Polym Phys 44(2):431–440
61. Bunch JS, Verbridge SS, Alden JS, Van Der Zande AM, Parpia JM, Craighead HG, McEuen PL (2008) Impermeable atomic membranes from graphene sheets. Nano Lett 8(8):2458–2462
62. Yang YH, Bolling L, Priolo MA, Grunlan JC (2013) Super gas barrier and selectivity of graphene oxide-polymer multilayer thin films. Adv Mater 25(4):503–508
63. Peigney A, Laurent C, Flahaut E, Rousset A (2000) Carbon nanotubes in novel ceramic matrix nanocomposites. Ceram Int 26(6):677–683
64. Skoulidas AI, Ackerman DM, Johnson JK, Sholl DS (2002) Rapid transport of gases in carbon nanotubes. Phys Rev Lett 89(18):185901
65. Wu B, Li X, An D, Zhao S, Wang Y (2014) Electro-casting aligned MWCNTs/polystyrene composite membranes for enhanced gas separation performance. J Membr Sci 462:62–68
66. Ray SS (2013) Clay-containing polymer nanocomposites: from fundamentals to real applications. Newnes
67. Youssef AM (2013) Polymer nanocomposites as a new trend for packaging applications. Polymer-Plastics Technol Eng 52(7):635–660
68. Nazarenko S, Meneghetti P, Julmon P, Olson B, Qutubuddin S (2007) Gas barrier of polystyrene montmorillonite clay nanocomposites: effect of mineral layer aggregation. J Polym Sci Part B Polym Phys 45(13):1733–1753
69. Lotti C, Isaac CS, Branciforti MC, Alves RM, Liberman S, Bretas RE (2008) Rheological, mechanical and transport properties of blown films of high density polyethylene nanocomposites. Eur Polymer J 44(5):1346–1357
70. Dadbin S, Noferesti M, Frounchi M (2008) Oxygen barrier LDPE/LLDPE/organoclay nanocomposite films for food packaging. In: Macromolecular symposia. Wiley Online Library
71. Petersen H, Jakubowicz I, Enebro J, Yarahmadi N (2016) Development of nanocomposites based on organically modified montmorillonite and plasticized PVC with improved barrier properties. J Appl Polymer Sci 133(3)
72. Xie W, Gao Z, Pan W-P, Hunter D, Singh A, Vaia R (2001) Thermal degradation chemistry of alkyl quaternary ammonium montmorillonite. Chem Mater 13(9):2979–2990
73. Vaia RA, Maguire JF (2007) Polymer nanocomposites with prescribed morphology: going beyond nanoparticle-filled polymers. Chem Mater 19(11):2736–2751
74. Coleman JN, Cadek M, Blake R, Nicolosi V, Ryan KP, Belton C, Fonseca A, Nagy JB, Gun'ko YK, Blau WJ (2004) High performance nanotube-reinforced plastics: understanding the mechanism of strength increase. Adv Func Mater 14(8):791–798

Chapter 2
Processing Nanocomposites Based on Engineering Polymers: Polyamides and Polyimides

Vincent Ojijo and Suprakas Sinha Ray

Abstract Although polymer nanocomposites (PNCs) are now a relatively well-established technology, nanoparticles (NPs) such as carbon nanotubes (CNTs) and graphene are opening up new application areas in engineering PNCs. Therefore, research and development is increasingly being undertaken on the processing and performance of these nanocomposites in order to address keys challenges, including nanoscale dispersion in polymer matrices. This chapter discusses the processing techniques used in the fabrication of engineering PNCs. Emphasis is placed on two classes of engineering polymers: (i) polyamides (PAs) and (ii) polyimides (PIs). Similarly, we focus only on a limited number of NPs (clays, CNTs, and graphene). Apart from traditional methods, relatively new manufacturing processes, such as electrospinning and additive manufacturing, are highlighted and their applicability in the fabrication of PA- and PI-based nanocomposites is discussed.

2.1 Introduction

Engineering plastics are a category of high-performance polymers that have superior properties and are generally used for structural applications. In many instances, they are used to replace traditional materials, such as metals, as they possess both the good structural properties of metals, along with the ease of processing and chemical characteristics of polymers. When compared to other plastics, engineering plastics exhibit good performance when subjected to mechanical stress, chemical, impact, cold, high temperatures, and other forms of stress. Such plastics include

V. Ojijo (✉) · S. Sinha Ray (✉)
DST-CSIR National Centre for Nanostructured Materials, Council for Scientific and Industrial Research, Pretoria 0001, South Africa
e-mail: vojijo@csir.co.za; rsuprakas@csir.co.za

S. Sinha Ray
Department of Applied Chemistry, University of Johannesburg, Doornfontein 2028, Johannesburg, South Africa
e-mail: ssinharay@uj.ac.za

S. Sinha Ray (ed.), *Processing of Polymer-based Nanocomposites*, Springer Series in Materials Science 278, https://doi.org/10.1007/978-3-319-97792-8_2

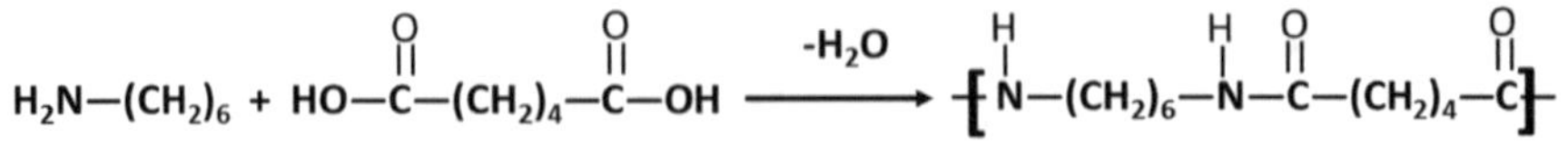

$$H_2N{-}(CH_2)_6 + HO{-}\overset{O}{\overset{\|}{C}}{-}(CH_2)_4{-}\overset{O}{\overset{\|}{C}}{-}OH \xrightarrow{-H_2O} \left[\overset{H}{\overset{|}{N}}{-}(CH_2)_6{-}\overset{H}{\overset{|}{N}}{-}\overset{O}{\overset{\|}{C}}{-}(CH_2)_4{-}\overset{O}{\overset{\|}{C}}\right]$$

Fig. 2.1 Condensation synthesis of PA6,6

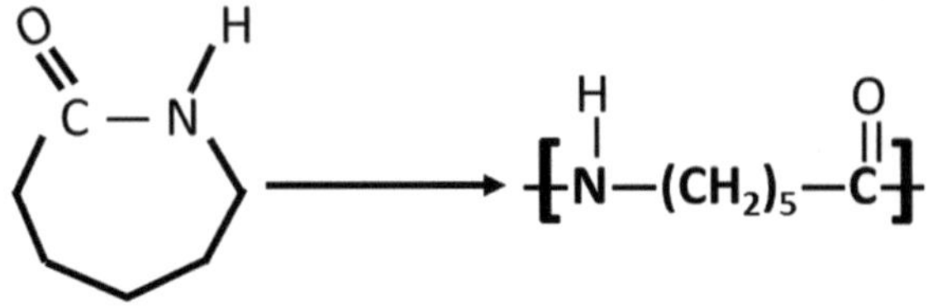

Fig. 2.2 Ring-opening polymerisation of ε-caprolactam

polyamides (PAs), polyacetals (polyoxymethylenes (POMs) or polyformaldehydes), thermoplastic polyesters, polycarbonates (PC), poly(phenylene ether) (PPE), polysulfones (PSU), polybiphenyldisulfones, liquid crystal polymers, poly(phenylene sulfide), polyetherimide (PEI), polyimides (PI), polyamide imides, polyacrylonitrile (PAN), aromatic polyketones, polyarylates, e.g. poly(methyl methacrylate) (PMMA), aliphatic polyketones, and polyphynylene. Comprehensive coverage of the these polymers can be found in the literature [1, 2]. In this chapter, we focus on PAs and PIs, where a brief overview of these two engineering polymers is presented next.

2.1.1 PA

PA, also known as nylon, was the first commercial engineering polymer. Development began in the beginning of the synthetic polymer age in the 1930s [3], with the introduction of nylon 6,6 (PA6,6). This material stemmed from early work by Wallace Carothers in 1928 on the condensation polymerization of six carbon diacid (adipic acid) and diamine (hexamethylenediamine) monomers, at DuPont. The condensation reaction is shown in the scheme in Fig. 2.1.

These PAs are denoted AABB, referring to the diamine and diacid components, and referred to as PA m, n, where m is number of carbon atoms in the diamine, while n is the number of carbons in the diacid. Apart from PA 6,6, other commercially relevant compositions include PA46, PA610, and PA 612. Another form of polyamides is derived from ring-opening polymerization. The commercial production of PA6 occurred at IG Farben, where Paul Schlack continued the work of Carothers and synthesized PA6 from a single monomer with both carboxyl and amine functions, as opposed to two monomers via catalytic polymerization of ε-caprolactam. The ring-opening polymerization of ε-caprolactam to prepare PA6 is depicted in the scheme in Fig. 2.2.

Subsequently, many different types of PAs have been developed, principally to fill niche markets. PAs are categorized into different groups, depending on the type of monomers and their combinations [3]. These include: *aliphatic PAs* (e.g. PA6, PA66); *aromatic PAs (aramids)*, in which at least 85% of the amide groups are bound directly to two aromatic rings (e.g. poly(*p*-phenylene terephthalamide), marketed as Kevlar®); *semi-aromatic PAs* and PAs from cycloaliphatic monomers, PAs with at least one monomer that has an aromatic ring separated by one or more methylene groups from the carboxyl and/or amine end groups. The different monomer components and configurations define the characteristics of the PAs and dictate the processing techniques suitable for converting them into final products, including melt processing and solution-based processing. A few examples of these PAs, their monomers, melting temperatures, and some key applications are shown in Table 2.1.

In the case of aliphatic PAs, a higher amide frequency in the structure (defined as 100 times the ratio of amide links to the total atoms along the backbone) results in a higher melting point. Generally, such materials can be melt-processed and spun from solution, and offer good strength, low abrasion and wear, and good chemical resistance. The two aliphatic PAs, PA6 and PA6,6 are still the most widely used, constituting 90% of total global PA usage [3]. In the subsequent sections, emphasis is placed on processing nanocomposites of these two PAs, owing to their widespread usage.

While aliphatic polyamides can be melt processed, fully aromatic PAs cannot, and need to be processes by solution spinning. The rigid aromatic amine linkages and the efficient directional inter-chain hydrogen bonds results in a material with extremely high cohesive energy, resulting in the outstanding mechanical, chemical, and thermal properties [3–5]. Therefore, aromatic PAs are considered high-performance engineering polymers and find applications for extreme temperatures (e.g., firefighting apparel), impacts (bullet proof vests), and structural applications.

2.1.2 PIs

PIs are a class of engineering polymers principally synthesized from a condensation reaction of a dicarboxylic acid anhydride with a diamine [1, 2]. Aromatic PIs are among the most important engineering polymers as they exhibit exceptional mechanical properties (high modulus, strength, and toughness) over a wide range of temperatures, ultra-high heat resistance, flame resistance, wear resistance, chemical resistance, resistance to hydrolysis, and good creep resistance. Due to these excellent properties, they have been used industrially in aerospace, electronics, and other industries over the past three decades.

Aromatic PIs are generally prepared via a two-step procedure: (i) synthesis of the precursor polyamic acid by ring-opening polyaddition of aromatic diamines (such as 4,4′-oxydianiline (ODA) and p-phenylene diamine (PDA)) to aromatic tetracarboxylic dianhydrides (e.g. biphenyltetracarboxylic dianhydride (BPDA)), usually in N-Methyl-2-pyrrolidone (NMP) or dimethylacetamide (DMAc); and (ii) thermal

Table 2.1 Some Polyamides, their monomers, melting points, commercial names and application

Type	Name	Monomers	Melting point (°C)	Pre-dominant processing technique	Some applica-tions
Aliphatic Polyamides	PA 6	ε-caprolactam	223	Melt and solution	Ropes, packaging, Clothing, automotive, electrical, electronic, etc.
	PA11	11-amino undecanoic acid	188	Melt and solution	
	PA12	Laurolactam	179	Melt and solution	
	PA 46	1,4 diaminobutane + adipic acid	295	Melt and solution	
	PA 66	Hexamethylenediamine +adipic acid	255	Melt and solution	
	PA 1212	1,12–dodecanediamine + 1,12 –dodecanedioic acid	185	Melt and solution	
	PA-MXD6	m-xylenediamine (MXDA) + adipic acid	237	Melt and solution	
Aromatic Polyamides	PPPT[a]	1,4-phenylene-diamine (para-phenylene diamine) + terephythaloyl chloride	–	Wet spinning	Bicycle tires, racing sails, body amour
	PMPI[b]	m-phenylene + isophthalamide	–	Wet spinning	Flame resistant apparel, military

[a]Poly(*p*-phenylene terephthalamide)-(e.g.Kevlar®);
[b]Poly(*m*-phenylene isophthalamide) (Nomex®)

imidization (cyclodehydration) to yield PI. Another class of aromatic PIs, including poly (pyromellitimide-1,4-diphenyl ether) (PDMA-ODA) can be synthesized from pyromellitic anhydride (PMDA) and ODA, as depicted by the scheme in Fig. 2.3.

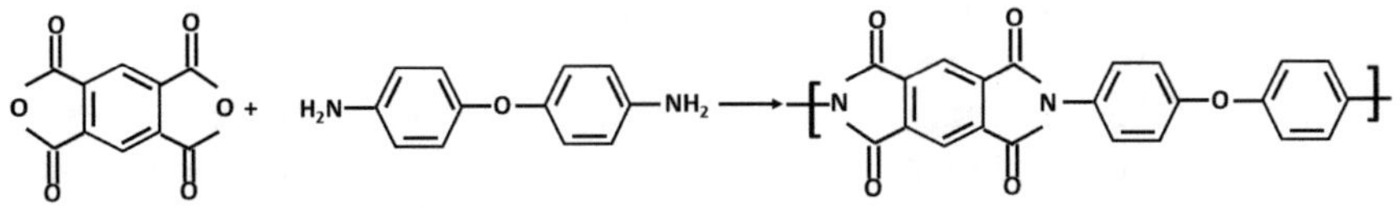

Fig. 2.3 Preparation of PMDA–ODA

To add additional functionalities or further improve the performance, engineering polymers can be modified with certain micro- or nanomaterials. We will now discuss some of the nanomaterials that are incorporated in the selected engineering polymers (PA, PI) and the available processing techniques to form nano- or micro-level dispersions. We also discuss the role of inclusions on material performance.

2.1.3 Overview of Engineering Polymer-Based Nanocomposites

Polymer nanocomposites (PNCs) are a relatively new composite technology compared to the traditional microscale composites. Due to the nanoscale dispersion of the NPs in the polymer matrices, this class of materials exhibits markedly improved properties compared to conventional composites. They gained prominence after groundbreaking work by the Toyota Central Research & Development Labs., Inc., that showed that an addition of only 4.7 wt% of molecular (exfoliated) montmorillonite (MMT) into N6 enhanced the thermomechanical properties, such as the heat deflection temperature [6].

In PNCs, the current key challenge is the ability to achieve consistent nanoscale dispersion of NPs in the polymer matrices [7]. It is well known that nanocomposites derive their emergent properties from polymer–particle interfaces, where larger interfacial areas (i.e., better dispersion) are anticipated to result in better performance. From theoretical considerations and numerous experimental results [8, 9], it has been shown that aggregation of particles results in a lower interfacial area and microcomposites with poorer performance. Research continues to investigate methods for improving nanoscale dispersion, including modifying the surface chemistry (enthalpic considerations) or optimizing the processing technique. Of the several types of engineering polymers, PA-based and PI-based nanocomposites have received the most attention, as indicated by the number of publications over time (Fig. 2.4). PET nanocomposites have also received substantial attention, but they are not discussed here.

2.1.3.1 Common Nanoparticles Used in Engineering PNCs

The nanoparticles (NPs) used in the formulation of engineering PNCs are divided into fiber-like particles (1D), platelets (2D), or particles (3D) depending on their geometry and dimensions. These are briefly discussed in the following sections.

(i) **Clays** have been widely used in formulation of engineering PNCs, both for scientific studies and attempts at commercialization, due to their availability and relatively low cost. Excellent discussion of the current literature can be found in various review articles [10–16]. The focus of such studies has been reinforcement at low filler concentrations (<5 vol.%), barrier property enhance-

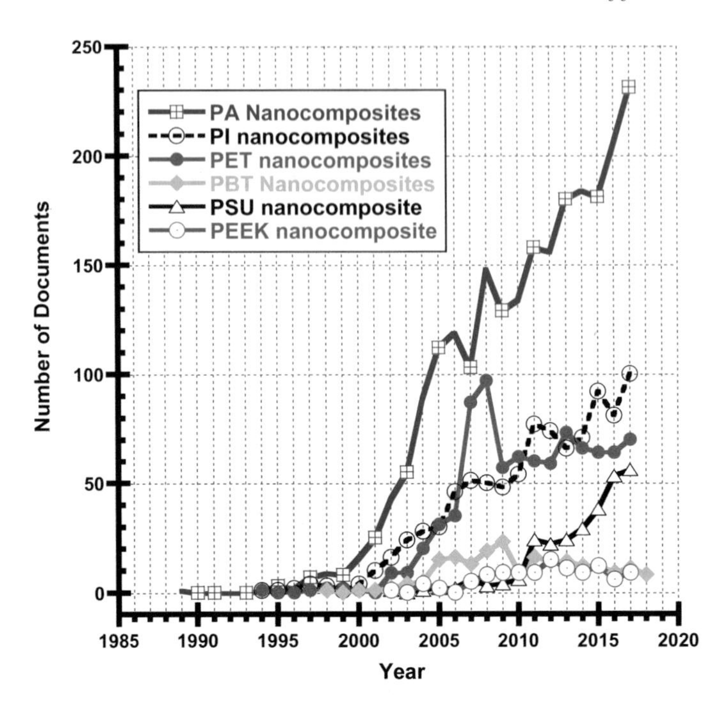

Fig. 2.4 Number of publications on nanocomposites. *Source* Scopus, using the keyword ("Polyamide Nanocomposite OR Nylon Nanocomposite OR PA Nanocomposites"); ("PET Nanocomposite"); ("PBT Nanocomposite"); ("Polysulfone nanocomposite") and ("PEEK nanocomposite")

ment, and crystal nucleation, among other topics. Hydrophilic clays are often modified with surfactants to make them more compatible with the hydrophobic polymers; reviews of such processes can be found in other texts [17–19]. The modified clays may disperse in the matrix to form a microcomposite, consisting of aggregated clay platelets (intercalated structures where the expanded gallery contains polymer chains), an intercalated/flocculated system [20], or an exfoliated system (where the clay platelets are dispersed uniformly within the matrix). For many applications, exfoliation is desirable. Processing strategies for engineering PNCs aim to achieve exfoliation of clays, where optimization of the screw design and speeds, and surface modification of the clays and polymer matrices are a priority. The optimization of screw configuration design entails the determination of proper sequence of elements, including the conveying and kneading elements. By varying the process conditions, i.e. temperature profile, screw speeds, sequence of processing, number of passes within the extruder, as well as surface chemistry, better dispersions of the clay can be achieved.

(ii) **CNTs** are used to enhance electrical conductivity, at a low percolation concentration (~0.1 wt%), and mechanical properties of engineering polymers [21–26]. CNTs are either single-walled or multi-walled [27]. Like other NPs, CNTs are incompatible with polymer matrices and surface modification to enable

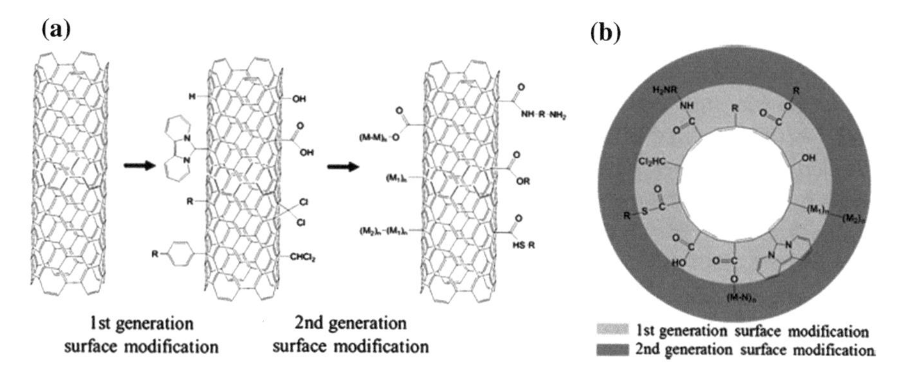

Fig. 2.5 Schematic representation of generation-based covalent CNT surface modifications. Reproduced with permission from [30]. Copyright 2012, Elsevier Science Ltd.

nanoscale dispersion is essential [21, 28–30]. Surface modification of CNTs is achieved via mechanical, physicochemical, or irradiation methods. Physiochemical methods involve covalent functionalization at the ends of the tubes or their sidewalls with e.g., COOH, –COH, and –OH groups; or non-covalent functionalization based on weak interactions, such as π–π and hydrophobic interactions, between CNTs and modifier molecules [30]. Kim et al. [30] proposed a new terminology to describe covalent modification of CNTs depending on the nature of the modified CNT surface. They refer to "*1st Generation (1G)*", molecules attached directly to the sidewall of CNTs. Subsequent modification of 1G materials is referred to as "*2nd Generation (2G)*", as depicted in Fig. 2.5.

CNTs have been incorporated in a number of engineering polymers in order to: modify electrical properties of PIs [31, 32] and polycarbonate [33]; enhance antifouling in polysulfone-based nano- and ultra-filtration membranes [34, 35]; enhance hydrophilicity and hence, flux in PA-based reverse osmosis membranes [36, 37]; and enhance reinforcement in poly(phenylene sulfide) [38, 39], and PET [40].

During processing of CNT-based engineering PNCs, efficient dispersion, and sometimes NP alignment, is critical, and can be partly ensured by controlling surface modifications and processing parameters. Care must always be taken to minimize particle breakage during processing. Typical processing procedures used to disperse CNTs in various engineering polymer matrices will be discussed later.

(iii) **Graphene** consists of a flat monolayer of carbon atoms tightly packed into a 2D honeycomb lattice. Graphitic materials of all other dimensionalities (0D, 1D, and 3 D) can be obtained from graphene [41]. Graphene has recently gained much attention [42–44] due to its unique intrinsic properties, including a very high surface area (calculated value of 2630 $m^2\ g^{-1}$), and excellent mechanical (Young's modulus of ~1100 GPa), electrical (mobility of charge carriers of 200,000 $cm^2\ V^{-1}\ s^{-1}$), thermal (conductivity of ~5000 $W\ m^{-1}\ K^{-1}$) and optical properties [45].

In many instances, engineering PNCs are fabricated from reduced graphene oxide (rGO) or graphene oxide (GO), or more generically, functionalized graphene (FG), due to the low cost and high yield production of GO and rGO [42, 46, 47]. FG loses some of the interesting properties of pristine graphene, such as high electrical conductivity. However, rGO retains acceptable physical properties, even with a partly damaged carbon structure. Moreover, the functionalized graphene surface enhances its dispersion in various polymer matrices due to interfacial interaction between the polymers and functionalized graphene [44, 48]. When formulated properly into various engineering polymer matrices, GO and rGO have been reported to enhance mechanical, electrical, thermal, and other properties of PIs [49–51], PAs [52–54], polycarbonates [55, 56], PET [57, 58], PBT [59–61], and polysulfones [62, 63], among others. Potential application areas for graphene-based PNCs include sensors [64], gas barriers, photovoltaics, and materials requiring high thermal and electrical conductivity [65].

Processing of graphene-based engineering polymer nanocomposites is achieved by various methods, including solution and melt mixing, layer-by-layer (LbL) assembly, vacuum-assisted routes, and in situ polymerization [47, 65]; these methods will be discussed later considering specific polymers. However, a current trade-off regarding graphene-based PNCs is achieving good dispersion (which requires surface functionalization) while maintaining the intrinsic properties of graphene, e.g. electrical conductivity, which is achievable only by removal of the functional groups. Other issues being investigated include controlling the reduction of graphite oxide in nanocomposites [65].

(iv) **Other NPs**, such as *nano-oxides* (TiO_2, ZnO, SiO_2) [66, 67] are used in research and commercial applications to tailor the rheological properties and enhance UV protection and self-cleaning properties [68]. Other functional fillers, such as polyhedral oligomeric silsesquioxane (POSS) [69], have shown significant modification of certain properties, mostly mechanical performance and flame retardance [70]. The incorporation of functionalized POSS in various engineering polymers, such as polycarbonate [71, 72], PAs [73–75], polyesters [76–78], polyether ketones [79], and PIs [80–84] has resulted in enhanced properties [69]. The processing technique and control of the polymer–particle interface significantly influence the dispersion of the particles, and subsequently the structure and performance of the nanocomposites. In the next section, we present an overview of some of the processing technologies applied for engineering PNCs.

2.2 Processing Techniques

Suitable processing techniques for engineering PNCs are determined by: the nature of the matrix (e.g., whether it can be heat-softened or not); the solubility of the polymer in certain solvents; the nature of the NPs; product, safety, and environmental considerations; and economic factors, among others. Two major routes are available for processing nanocomposites: (a) solvent-based methods and (b) non-solvent methods. A number of variants of these methods include: (i) in situ polymerization; (ii) solution casting; (iii) melt compounding; (iv) spinning; (iv) electrospinning; and (v) additive manufacturing. An overview of these processing techniques is presented here.

2.2.1 *In Situ Polymerization*

In this method, the NPs are firstly dispersed in a monomer or monomer solution. The dispersion can be enhanced by mixing techniques such as sonication and stirring. After dispersion, polymerization is initiated by heat, radiation, or a suitable initiator, resulting in the PNC material. This technique results in good dispersion of the NPs in the polymer matrix and minimizes damage/breakage of certain NPs, especially high-aspect-ratio CNTs and graphene derivatives [85–89]. A typical in situ polymerization process was illustrated by Zhu et al. [85] in the synthesis of PI/rGO nanocomposites, as shown in Fig. 2.6. This process involves dispersion of the GO in a solvent (NMP) aided by stirring. This solution was then mixed with PAA made from a mixture of ODA and PMDA. The PAA/GO composite was then cast into a film and dried, then thermally imidized under nitrogen to yield PI/rGO nanocomposites.

2.2.2 *Solution Casting*

Solution casting involves dissolving the polymer in a solvent system while simultaneously dispersing NPs in it [90]. Typically, the polymer is dissolved in the solvent, while the NP are separately dispersed in the same or a different solvent. The two solutions are then combined and further mixing is performed using sonication, stirring, or other methods. The mixture is finally cast and the solvent evaporated to produce the PNC.

2.2.3 *Melt Compounding*

Melt compounding involves mixing NPs with a molten polymer. This is usually done in a polymer processing vessel (an extruder or batch mixer). The shearing and or elongation stresses result in dispersion and distribution of the NPs in the polymer matrix. This method is preferred over solution methods as it is environmentally benign (no organic solvents are used). Moreover, it is compatible with the vast majority of industrial polymer processes (e.g., extrusion and injection molding). Hence, it is not surprising that the vast majority of the reported work involved melt-processing.

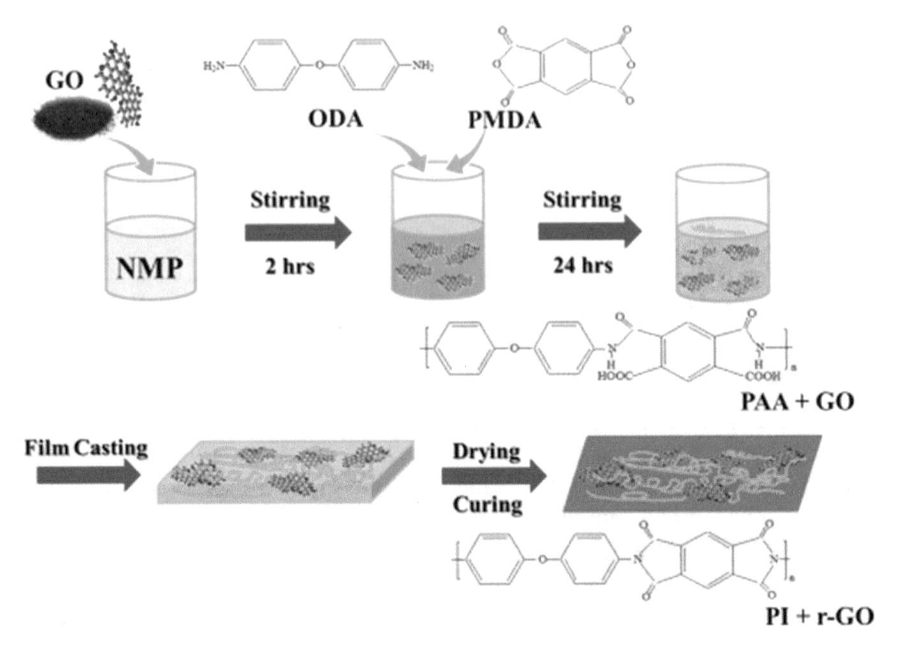

Fig. 2.6 Synthesis procedure for polyimide/graphene oxide composites via in situ polymerization. Reproduced with permission from [85]. Copyright 2014, Wiley Periodicals Inc.

In melt processing of engineering PNCs, key factors affecting the dispersion of the NPs are enthalpic (chemistry of polymer–NP interface) or kinetic (screw design, screw speed, processing conditions, sequence of processing). The drawback of this method when used with high-aspect-ratio NPs, such as CNTs, is particle breakage due to very high stresses present in the extruder. Attempts have been made, with varied outcomes, to enhance the performance of melt-processing during processing of engineering PNCs by: coupling the extruder with an *ultrasonication* system [91–96]; employing *water-assisted extrusion* [97], mostly for PAs [98–104], polyesters [105, 106], and polyethersulfone [107]; and the use of *supercritical fluid*.

Isayev et al. [91] attempted ultrasonic-assisted twin-screw extrusion of polyetherimide (PEI)/MWCNT nanocomposites (see Fig. 2.7a). No appreciable enhancements in the mechanical properties were realized, although the authors reported better dispersion of the CNTs in the PEI matrix. On the other hand, the same research group [95] reported that the electrical percolation threshold decreased substantially in a PC/CNT system with an increase in the ultrasonic amplitude, where the sonication unit was placed along the extruder barrel (see Fig. 2.7b). They observed better dispersion of the CNT as a result of ultrasonication, as shown by the TEM images of the PC containing 0.2 and 0.8 wt% CNTs (Fig. 2.8). Even though it is a promising technology, the ultrasound-assisted extrusion method and its mechanisms need further investigation in order to realize its full potential [94].

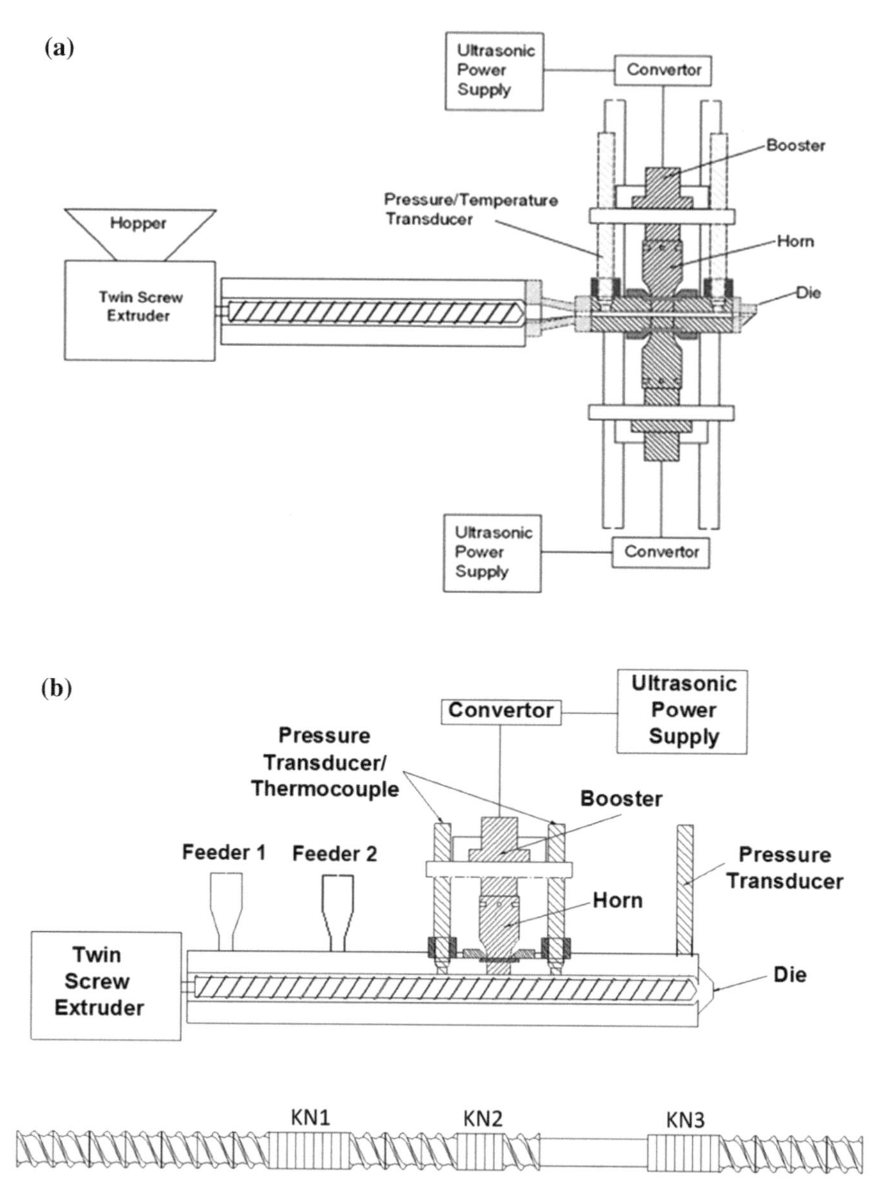

Fig. 2.7 The ultrasonic twin screw micro-extruder at different sections of the extruder. Reproduced with permission from [91, 95]. Copyright 2009/2016, Elsevier Science Ltd.

Water-assisted extrusion enables the processing of certain engineering PNCs without the need for surface modification, greatly reducing the cost while achieving similar levels of dispersion as in systems with surface-modified NPs. Mostly, clay NPs [98–102, 104, 108] have been incorporated into PA matrices without the need for

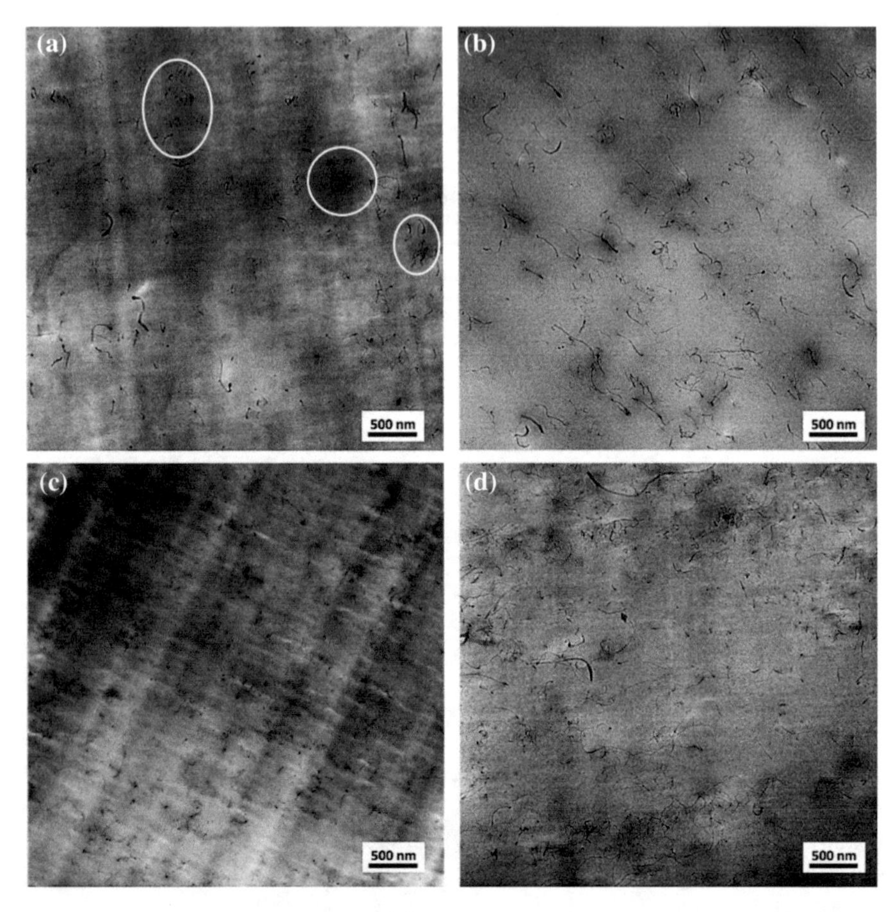

Fig. 2.8 TEM micrographs of PC/0.2 wt% CNT PC-0.2CNT **a** without sonication; and **b** with sonication at an amplitude of 13 μm; and PC/0.8 wt% CNT **c** without sonication and **d** with sonication at an amplitude of 13 μm (**d**). The circles highlight the agglomerates of CNTs. Reproduced with permission from [95]. Copyright 2016, Elsevier Science Ltd.

NP modification. In addition, bio-derived particles, such as cellulose nanocrystals [109], have been incorporated into PA via water-assisted extrusion.

Two configurations of water-assisted extrusion have been demonstrated: (i) feeding an NP slurry into a high-pressure zone, or any other suitable zone along the screw length; and (ii) injection of water into a polymer/NP stream along the length of the extruder. The schematic in Fig. 2.9 shows the water-assisted extrusion system, where the slurry (in this case, a clay slurry) is fed into a molten polymer (N6), where the vapor is extracted downstream [110]. In this process, Hasegawa et al. [110] proposed that the mechanism of clay exfoliation occurred in two steps. Firstly, the clays were swollen in the slurry, which were reduced into fine droplets containing

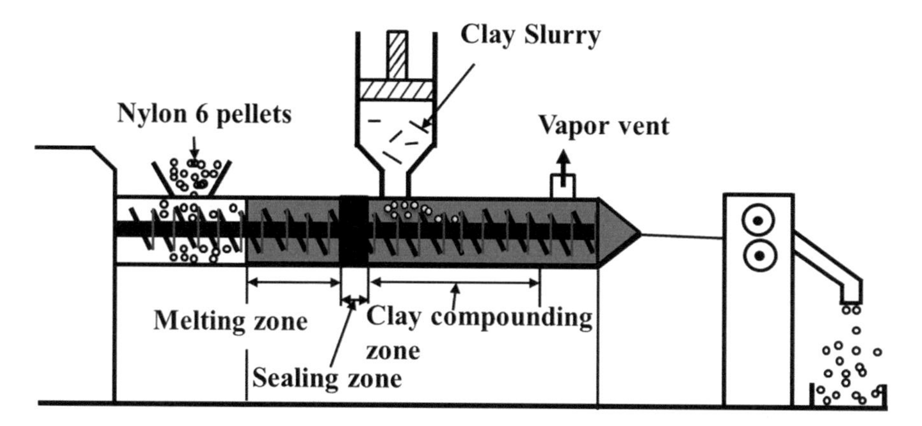

Fig. 2.9 Schematic depicting compounding process of PA 6/clay slurry. Reproduced with permission from [110]. Copyright 2003, Elsevier Science Ltd.

few clay platelets when the slurry was pumped into the molten PA. Then, the water evaporated, resulting in dispersed clay layers.

In the second configuration, the mechanism of exfoliation cannot be explained by pre-swelling of clays, as in the first configuration. Fedullo et al. [111] proposed the exfoliation model shown in the schematic of Fig. 2.10. An extruder with a special screw design to increase the pressure at the injection point to 125 bars was used to keep water in a liquid state. In this system, the water plays two functions. Firstly, it modifies the fluidity and polarity of the PA. At certain pressure and temperature conditions, water is miscible with PA, and hence, it plasticizes the PA. A decrease in the melting point due to the effect of the water (cryoscopic effect) was reported to be as high as 60 °C [111–113]. This resulted in an increase in the polarity and reduction in the viscosity of PA. Secondly, water diffuses between silicate layers and is adsorbed on the surface, thus increasing the gallery spacing. These two effects allowed intercalation of PA molecules in the clay gallery, with the aid of shearing by the screws. Water-assisted extrusion results in a similar dispersion level of pure clays in PA matrixes compared to functionalized clay, without the need for costly modification.

2.2.4 *Electrospinning*

Electrospinning is a facile process used to generate continuous ultrafine fibers (with dimensions down to the nanoscale) in the form of non-woven mats [114–118]. The first patent describing the operation of electrospinning appeared in the 1930s after Formhals [119, 120] disclosed an apparatus used in ‘electrical spinning’ of fibers.

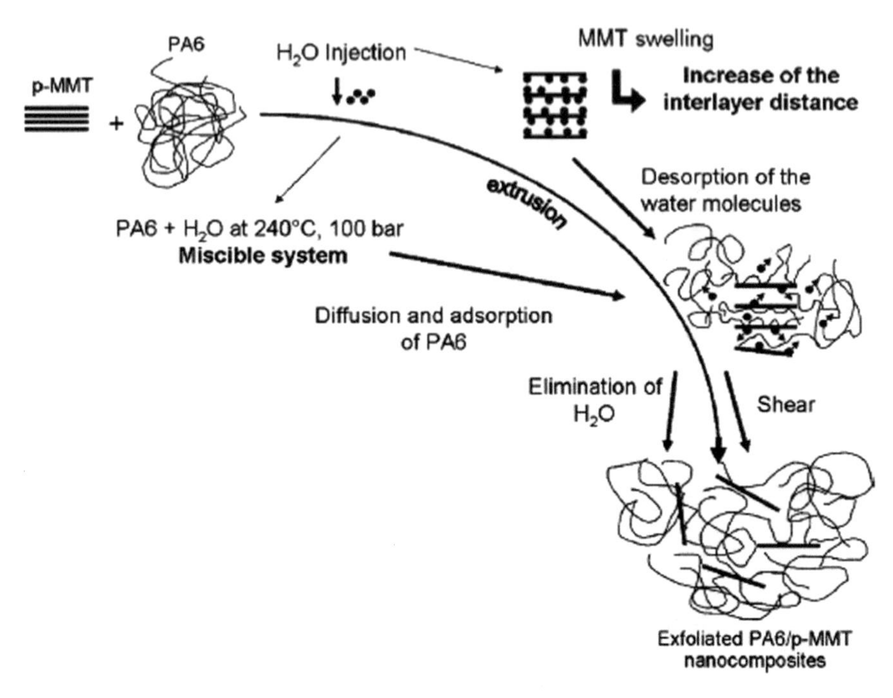

Fig. 2.10 Schematic description of the dispersion of untreated clay (MMT) in a PA6 matrix the water injection system. Reproduced with permission from [111]. Copyright 2006, John Wiley and Sons

More recently, a re-emergence of interest in electrospinning began in earnest after publication of the work performed in the 1990s by Reneker's group [121–123].

As depicted in Fig. 2.11, electrospinning is based on the uniaxial elongation of a viscoelastic jet derived from a polymer solution [124] or melt [125]. In a typical spinning process from a solution, a syringe pump forces the solution through a spinneret at a constant and controllable rate. An applied electric field between the spinneret tip and collector screen induces charges on the surface of the droplet, held together by its surface tension. Mutual charge repulsion causes a force that opposes the surface tension. Further increases in the electric field cause an elongation of the hemispherical droplet into a conical shape, known as the Taylor cone [126]. At a threshold where there is a sufficiently high electric field, the repulsive forces overcome the surface tension, causing an electrically charged jet to be ejected. As the electrified jet travels in the air towards the collector surface, the solvent evaporates, and the particle undergoes stretching and whipping, leading to the formation of a thin thread from hundreds of micrometers to tens of nanometers. This phenomenon, involving both Ohmic and convectional flow is depicted in Fig. 2.11b. From the described process, it is apparent that electrospinning is affected by solution parameters, such as viscosity, surface tension, and conductivity; process parameters, such as applied electric potential, feed rate, collector type, and distance between the tip and collector screen;

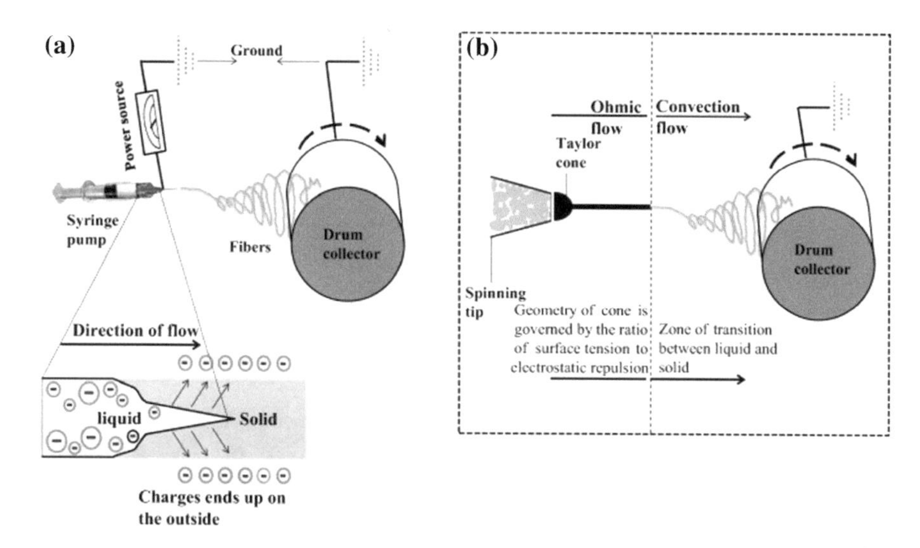

Fig. 2.11 Schematic depicting electrospinning setup and phenomenon of electrospinning. Reproduced with permission from [124]. Copyright 2015, Elsevier Ltd. (under creative commons license)

and environmental parameters, such as temperature, humidity, and air velocity [121, 124, 127].

Electrospun nanocomposites have current and future applications in a wide range of areas, including filtration media [117, 128], reinforcement of composites [118, 129], Li-ion battery separators, fuel cell proton exchange membranes, micro-electronics, and sensors [130]. In all these applications, the processing strategy employed to yield the desired fiber characteristics depends on the chemistry of both the NPs and polymer to be spun. Electrospinning of engineering PNCs can be achieved via two main strategies [128]: (i) electrospinning the composites from a pre-mixed dispersion of polymer/precursor and NPs; or (ii) post treatment and/or surface modification of the electrospun polymer fibers. Direct mixing before electrospinning has been used to produce various electrospun nanocomposite fibers, including: polysulfone/CNT [131, 132]; PI/CNT [133–135]; PAN/CNT [136–141]; PBT/CNT [142]; PMMA/CNT [143, 144]; PI/graphene; Fe-FeO NP-reinforced PI [145]; PI/$BaTiO_3$/CNT [146]; polyether amide-containing SiO_2, $BaTiO_3$, Si_3N_4, and boehmite NPs [147]; and polysulfone/TiO_2 [148].

The post treatment method involves decoration of NPs on the surface of the electro-spun fiber, either by physical or chemical means. This technique could be used to generate/grow NPs of Ag [149–151], Au [151], and ZnO [130] on the surface of electro-spun fibers. Recently, Bai et al. [151] demonstrated the use of plasma to directly decorate electrospun PAN fibers with Ag and Au. A metal salt solution was pre-mixed with a polymer solution to form a precursor solution from which electro-spun fibers were obtained. The fibers were treated with argon plasma to reduce the metal salt to metal NP of different sizes depending on the etching time.

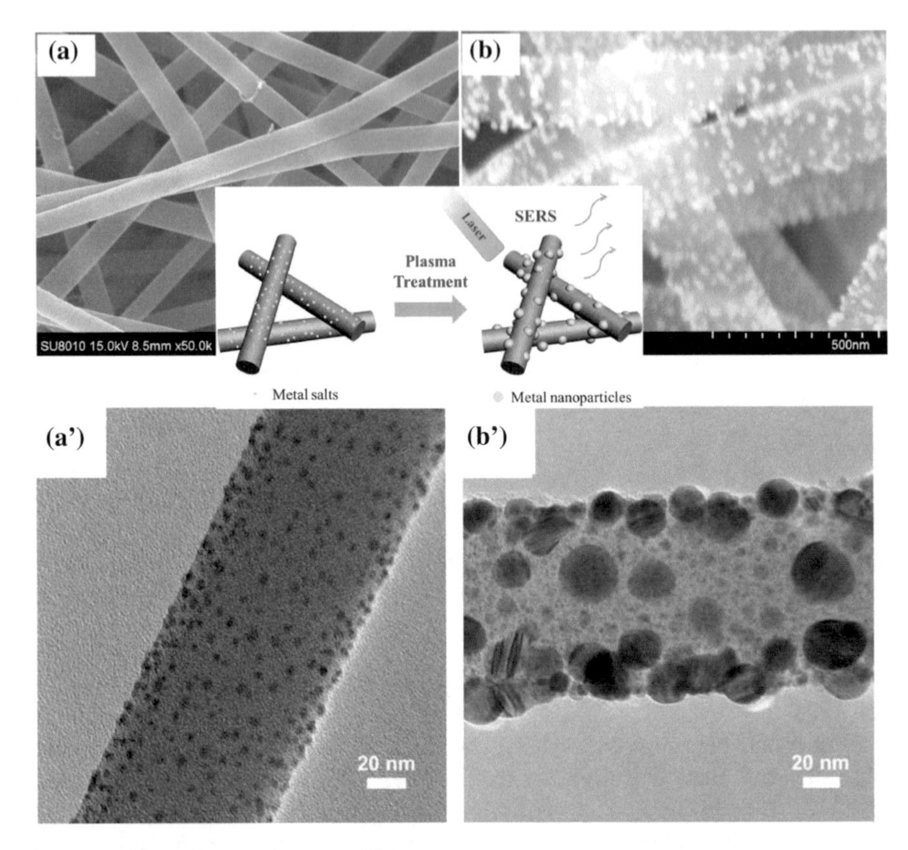

Fig. 2.12 Respectively, **a** SEM and **a′** TEM images of the PAN/$AgNO_3$ nanofibers before plasma treatment; and the corresponding SEM (**b**) and TEM (**b**) images after the plasma treatment. Reproduced with permission from [151]. Copyright 2017, Elsevier Science Ltd.

Figure 2.12 shows SEM and TEM images of PAN/$AgNO_3$ before and after plasma treatment, showing Ag NPs on the fiber surface [151]. In other variations of the post treatment technique, surfaces of the fibers can be chemically modified to enable localization of NPs at the surface. Carlberg et al. [150] opened the imide ring in PI fibers by hydrolysis in an alkaline KOH solution. The surface-confined cleavage at the dianhydride domain yielded a well-defined layer of potassium polyamate. By immersing the surface-modified PI fiber in aqueous $AgNO_3$ solution, Ag^+ ions were introduced via ion exchange with K^+ ions. The Ag^+ ions were then reduced into surface-confined Ag NPs via either thermal or chemical reduction.

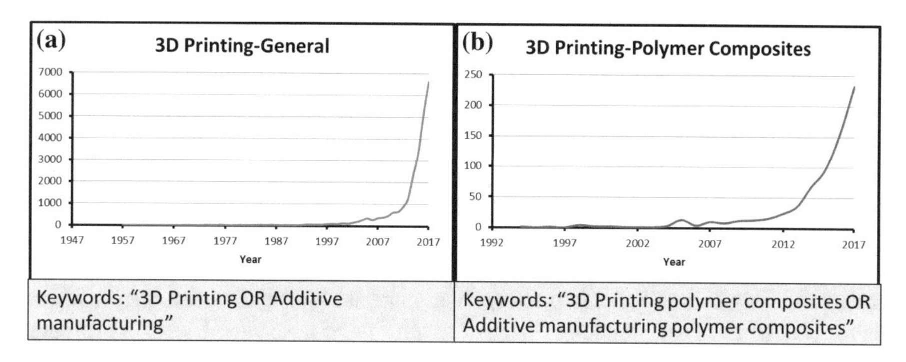

Fig. 2.13 Number of publications on **a** 3D printing in general, and **b** 3D printing of polymer composites *Source* Scopus

2.2.5 3D Printing

Additive manufacturing (AM), commonly known as 3D printing, is defined by the International Organization for Standardization (ISO)/American Society for Testing and Materials (ASTM) 52900:2015 standard as a "process of joining materials to make parts from 3D model data, usually layer upon layer, as opposed to subtractive manufacturing and formative manufacturing methodologies". One of the leading researchers in the AM field, Ian Gibson [152], proposes that AM and 3D need not be used interchangeably; whereas 3D printing may be used in conjunction with self-contained 3D printers that yield a product in its final usable configuration, "AM can perhaps be best viewed as a process, which in turn could be combined with other processes to create more complex combinations" [152]. The application areas of AM span aerospace, automotive, medical, and many other industry sectors and there are growing numbers of successful emerging products [152, 153]. As an enabling technology, 3D printing is quickly developing, with applications now well beyond the initial area of prototyping, to the production of practical parts with increasingly complex geometries [153–156]. There has been an exponential growth in research outputs and products developed using 3D printing over the last two decades. This is depicted by the number of publications in this field, as shown in Fig. 2.13a. Likewise, 3D printing applications in polymer composites have increased in a similar fashion (see Fig. 2.13b).

AM techniques are grouped into seven categories: (i) directed energy deposition (DED), (ii) binder jetting, (iii) material extrusion, (iv) material jetting; (v) sheet lamination, (vi) powder bed fusion and, (vii) vat photopolymerization. Detailed discussion of these techniques is beyond the scope of this book and can be found elsewhere [153, 157] and only a brief introduction to these techniques is shown here. Material extrusion (or fused deposition modelling; see Fig. 2.14a) is the most commonly used technique for fabrication of polymer composites. In this process, a thermoplastic material or composite is drawn from a spool through a heated nozzle, where it is heated and deposited layer by layer. It solidifies on the substrate and subsequent layers of material should fuse with previous ones. The binder jetting AM involves joining particles with a liquid bonding agent which is selectively dropped on a section, as illustrated in Fig. 2.14b. The platform is then moved downwards to add another layer of powder. This method is not suitable for the production of structural parts. Vat photopolymerization involves the use of UV laser to cure and harden a photopolymer resin in a bath. It is a general term that includes stereo-lithography, and shown in Fig. 2.14c. The material jetting process creates objects by jetting material onto a substrate, where it solidifies, followed by deposition of another layer of the material until the model is built up. The material layers are subsequently cured using UV. In the DED process, energy (usually from a powerful laser) is directed onto a small area to melt a material that is being continuously deposited, in many cases a metal power. The fused bed fusion (FBF) processes involves the use of laser (or other) energy to fully or partially fuse powder particles, before recoating with a fresh powder layer. The fusion can occur via sintering or melting. Examples of FBF processes include direct metal laser sintering, electron beam melting, selective heat sintering, selective laser melting, and selective laser sintering (SLS). The SLS process is illustrated in Fig. 2.14d. The 3D plotting AM process involves extrusion of a viscous material from a pressured syringe with a head that moves in three dimensions (see Fig. 2.14e), whilst the platform is kept stationary. Finally, sheet lamination processes include ultrasonic AM and laminated object manufacturing (LOM), where the material sheet is cut using a laser with or without ultrasound assistance. The bonding mechanism, including adhesive bonding, thermal bonding, and clamping further differentiates types of LOM.

After the initial wave of material development for polymer AM, some polymer products have served as conceptual prototypes rather than functional components, as they lacked functionality, and in some cases, suitable strength. These shortcomings are being solved partly by the incorporation of fillers, fibers, or nanomaterials [152, 155, 158]. A recent review of high-performance PNCs prepared using 3D techniques highlighted CNTs, graphene oxide, and nanoclays as common fillers used in such composites [159]. The common candidate engineering polymers include polysulfone, polyetherimide, poly(phenylene sulfide) polyether ether ketone (PEEK), PAs, and liquid crystalline polymers [159]. The targeted functional properties enhanced by these NPs include the mechanical strength, thermal stability, dimensional stability, electrical conductivity, and flame retardance. Application of AM techniques in processing PNCs based on various NPs and engineering polymer matrices will be discussed later. In the following sections, some important nanocomposites are discussed in detail, based on the selection of a few engineering polymer matrices and NPs, including PA-based, PI-based, and polyester-based nanocomposites.

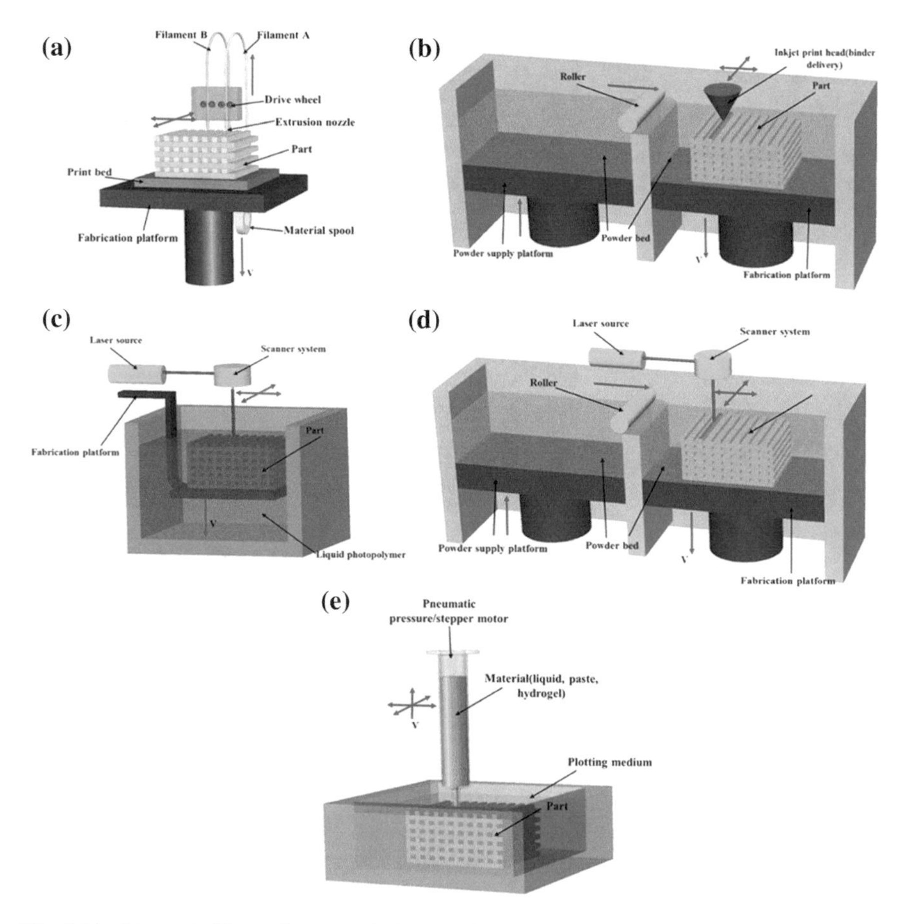

Fig. 2.14 Schematic illustration of typical setups for **a** material extrusion/FDM **b** binder jetting-3DP **c** SLA **d** SLS **e** 3D plotting setup. Reproduced with permission from [155]. Copyright 2017, Elsevier Science Ltd.

2.3 Polyamide Nanocomposites

2.3.1 Processing, Structure and Properties of PA/Clay Nanocomposites

From the onset of sustained research and development of PNCs, Toyota Central Research & Development Labs., Inc. reported a series of PA molecular composites (nanocomposites) based on MMT and saponite clays [6, 160–162], and later, hectorite and synthetic mica [163]. The predominant synthesis process used was *insitu* polymerization, where the clays were typically first intercalated with water and ε-caprolactam [161, 162] or 12-aminolauric acid [164] before polymerization of the

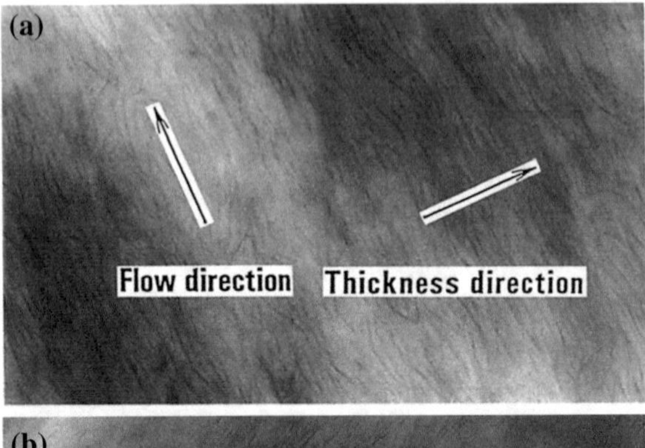

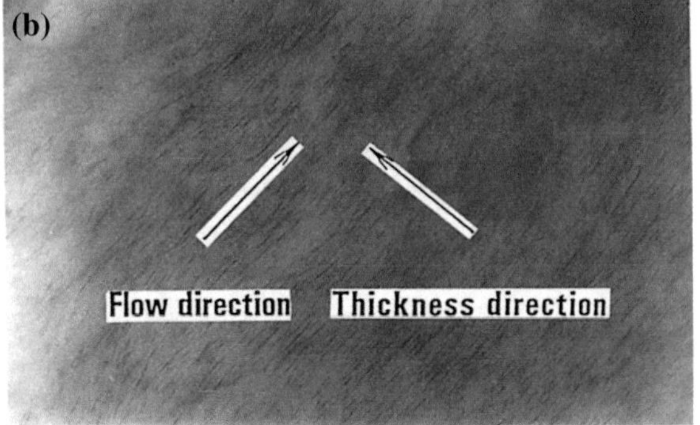

Fig. 2.15 TEM micrographs of PA6/MMT and PA6/Saponite nanocomposites. Reproduced with permission from [164]. Copyright 1993, John Wiley & Sons, Inc.

ε-caprolactam in one pot, at a reaction temperature of 260 °C for 6 h in a nitrogen environment, under atmospheric pressure. Molecular dispersion of the clays occurred, as demonstrated by a typical TEM micrograph shown in Fig. 2.15. Based on the high level of dispersion, it is not surprising that the properties of such nanocomposites were reportedly better than pure PA of a similar molecular weight; the molecular weights of the new nanocomposites were reportedly higher than the critical region in which the properties are sensitive to molecular weight [162]. These PA/clay nanocomposites exhibited remarkable clay-content-dependent improvements in mechanical properties [6] and other functional properties, such as a diffusion barrier to water molecules [164]. In an early study, Kojima et al. [6] reported that PA6/clay nanocomposites had superior strength and modulus, and comparable impact strength to pure PA6. The properties were significantly better for MMT-containing nanocomposites, where the heat distortion temperature improved by 87 °C, to 152 °C, for a 4.7 wt% MMT loading [6]. This was attributed to the contribution of the constrained regions, as calculated from the storage and loss moduli at the glass transition temperature (T_g).

Since then, numerous attempts have made to produce PA/clay nanocomposites using various techniques apart from in situ intercalation (e.g., melt compounding, casting, AM techniques, and electrospinning). The type of polymer has been broadened to include other PAs, such as PA12, PA11, PA66, and MXD6, while the clay types have expanded beyond the initial MMT, saponite, synthetic mica, and hec-

Table 2.2 Processing techniques for PA/clay nanocomposite

Clay-Type/Modification	Polymer type	Processing technique	Remarks	References
Neat clays	Aliphatic PA (PA6/PA12/PA66/PA11)	In situ polymerisation	On-set of polymer nanocomposites	[6, 161, 162, 165]
Surfactant modified clays		Melt Compounding	Based on organically modified clay	[166–175]
Neat clays		Water-assisted melt compounding	A more recent trend, and yields similar performances as those based on organo-clays.	[98–102, 104, 108]
Surfactant modified clays (<7.5 wt%)/neat MMT	PA6,	Electrospinning	Melt-compounded nanocomposite, followed by dissolution in formic acid	[176–178]
	PA6, PA66		In situ PA/clay polymerization followed by dissolution in formic acid	[179, 180]
	PA6		Solution blending of clay dispersion and polymer solution	[177, 181–184]
Organically modified clay	PA11, PA12	Additive manufacturing	SLS on dry-blended clay and PA powders	[185, 186]
			Melt compound PA and clay before pulverisation and SLS	[187, 188]
	PA6		PA/clay nanocomposite powder prepared dissolution-precipitation technique	[189]

torite to other categories, including vermiculite, sepiolite, laponite, bentonite, and attapulgite. Table 2.2 summarizes some of the techniques used to process PA/clay PNCs based on different clay and PA types.

Subsequent to the use of in situ intercalation techniques, Vaia et al. [190] developed a melt intercalation technique, which was deemed more versatile and industrially desirable. In their proposal, the use of an organically modified layered silicate achieved intercalation of polymers of varying degrees of polarity, based on a predominantly enthalpic mechanism [190]. In the melt intercalation model, unfavorable loss of entropy (associated with intercalation of polymer molecules within the clay galleries) is avoided by enhancing polymer–host interactions, predominantly by suitable modification of the clay surface prior to melt processing [190]. Later, a melt intercalation process for PA/clay nanocomposites was described in a patent by Maxfield et al. [166] and by other researchers including Liu et al. [167] and Cho and Paul [168], demonstrating the suitability of the technique; the properties of the nanocomposites were better than the pure PA. In melt processing, key factors determining good dispersion of the clays in the matrix include the screw type (twin screws are preferred compared to single screws [168]), shearing history (residence time), clay loading, and the type of organic modification of the clay [191]. Previous research studies on melt compounding have focused on improving the dispersion of the clays in order to realize improved properties of mostly, PA6 [192–194], PA66 [175, 195–197], PA11 [194, 198, 199], and PA12 [169, 172, 173, 200, 201]. A more recent trend in melt compounding of PA/clay PNCs is the use of water-assisted processes to avoid clay modification. With the aim of reducing costs and developing a more environmentally friendly method for clay dispersion, water-assisted extrusion has been developed and applied to PA/clay PNCs with encouraging results. The structure (and hence, performance) of such composites largely mirror those based on organically modified clays [98–102, 104, 108].

In addition to the enhancement of the *mechanical and barrier properties,* as already discussed, clays can retard the *flammability* of PA composites. PA/clay PNCs, especially those with delaminated or intercalated structures, exhibit substantial flame retardance compared to pure PA [192, 202–204]. These nanocomposites showed reduced peak heat release rates (HRR), which is the most important parameter for evaluating fire safety. The reader is referred to Gilman [203] for a more comprehensive coverage of the role of clays in flame retardation applications. As an example, the cone calorimetry flammability tests on PA6/clay nanocomposites showed a 63% reduction in peak HRR compared to pure PA6 [203], as shown in Fig. 2.16. These results are typical for other types of PA, such as PA12.

In addition to in situ polymerization and melt compounding, electrospinning has been employed to process PA/clay nanocomposites. Electrospinning is primarily used as an additional step to produce fibers. In PA/clay nanocomposites, three routes have been followed to obtain electro-spun mats: (i) dissolving a melt compounded PA/clay nanocomposite followed by spinning [176–178]; (ii) dissolving a nanocomposite prepared via in situ polymerization, followed by spinning [179, 184]; and finally, (iii) dissolving the matrix polymer and separately dispersing clays in a similar solvent, before mixing and electrospinning [177, 183, 205]. Agarwal et al. [205] recently claimed that the barrier properties of polypropylene improved when coated with an electrospun mat of PA/clay nanocomposite. Such claims ought to be investigated further as it is surprising that a porous mat could enhance barrier properties. Moreover, the technique used for dispersing the unmodified MMT into the PA matrix (overnight mixing of the PA solution and MMT dispersion in formic acid) is highly unlikely to yield a good dispersion of clay in the PA, which is a pre-requisite for enhanced barrier properties. Similarly, Saeed and Park [179] reported improved

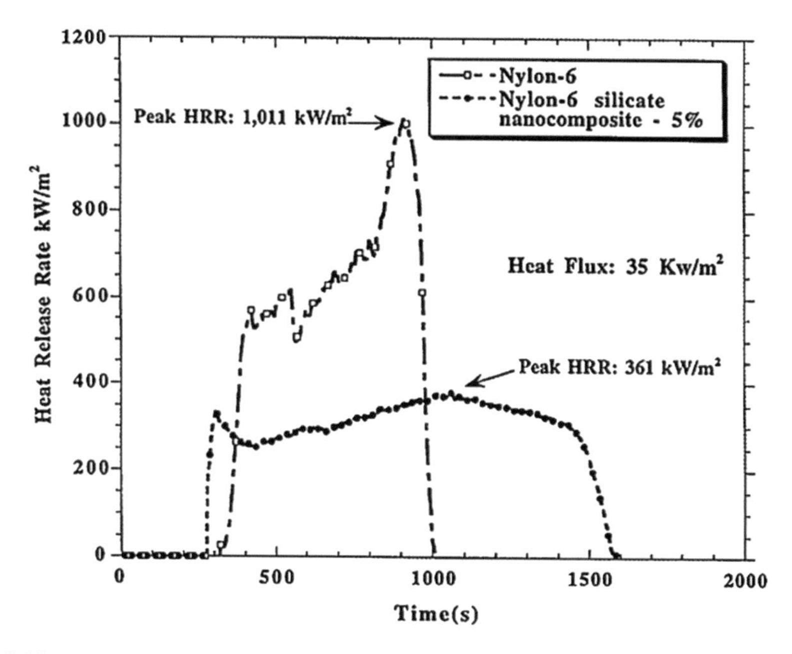

Fig. 2.16 Heat Release Rate (HRR) for neat PA6 and PA6/clay (5 wt%) nanocomposites at a heat flux of 35 kW/m^2. Reproduced with permission from [203]. Copyright 1995, Elsevier Science B.V.

thermal and mechanical properties for electrospun PA/clay PNCs. The nanocomposites were prepared via in situ polymerization before dissolution in formic acid for spinning. However, scrutiny of the standard deviations of the mechanical data showed that their conclusions were not supported by the data. We are of the opinion that electrospinning PA/clay nanocomposites may not be beneficial for enhancing the functionalities/properties of the fibers. However, 1D particles that can readily be aligned in the fiber direction (e.g., CNTs) may be a better candidate for property enhancement.

AM is a relatively new technique for developing products based on PA/clay PNCs. Of the various AM techniques, SLS has been most widely applied. Until 2011, two methods were available for the preparation of powders for SLS: (i) dry-blending of nylon and clays [185, 186], and (ii) melt compounding of PA/clay composites before cryogenic pulverization [188]. However, Yan [189] introduced a third method, namely the dissolution-precipitation process. In this process, suitable quantities of PA12 were dissolved and mixed with organically modified MMT (OMMT) in a reactor kept at 145 °C under nitrogen gas. After 2–3 h, the reactor was cooled to 105 °C at a rate of 10 °C/h, during which process, PA started to precipitate. The precipitated PA12/OMMT was vacuum dried before ball milling. This process reportedly yielded well-intercalated structures and more uniform powder particle sizes than the other two methods, solving the problems associated with the previous methods [189]. Wang et al. [185] dry-blended organically modified rectorite (OREC) with PA12, and then laser-sintered the material to fabricate specimens. Due to larger gallery spaces of the OREC compared to MMT, the authors claimed that it was possible for PA12 molecules to intercalate the OREC galleries during sintering. However, no TEM images were provided, even though XRD results suggested the possibility of intercalation. They noted that the laser power required for sintering the PA12/OREC composite was less than that of pure PA12. Similarly, PA12/OREC parts exhibited superior mechanical performances compared to pure PA12. On the contrary, Kim and Creasy [188] concluded that it is impractical to improve the strength of a part manufactured via SLS of PA6/clay nanocomposite powder. However, they neglected laser–polymer particle interaction and only considered thermal compact sintering.

2.3.2 *Processing, Structure, and Properties of PA/CNT PNCs*

PA/CNT nanocomposites are generally fabricated to realize enhanced electrical conductivity and mechanical properties, usually at very low percolation thresholds (< 1 wt%). To realize these benefits, suitable processing techniques are required that can achieve good dispersion of the CNTs. In order to enhance CNT dispersion in PA matrices, different surface functionalization techniques can be used, including encapsulation with reactive groups (e.g. styrene maleic anhydride copolymer [206], *n*-butyl acrylate [207], and methyl methacrylate [208]) or via acid functionalization [209, 210], plasma functionalization [211], or silanization [212].

Various techniques for processing PA/CNT nanocomposites have been demonstrated, including in situ polymerization [209, 210, 213], melt compounding [211, 212, 214–218], melt spinning [219–221], electrospinning [222, 223], and AM [224–226]. A summary of some of these techniques is given in Table 2.3. In the in situ polymerization technique, the polymerization of PA occurs in the presence of CNTs, mostly carboxylated CNTs [209, 210]. For instance, Xu et al. [210] prepared PA/CNT by polymerizing hexanolactam and 6-aminoaldehexose acid at 260 °C for 6 h in the presence of CNTs. However, melt compounding is a preferred technique, but care must be taken not to excessively shear the material and break the CNTs. In some cases, the use of hybrid fillers (CNT/graphene [227] or CNT/carbon black (CB)) was attempted in order to obtain synergistic benefits from each material. Socher et al. [228], [229] melt-compounded a hybrid system of CB and MWCNT with the intention of deriving synergistic effects from the interaction between the two electrically conducting NPs in a PA12 matrix. The authors did not observe any synergistic effects at loadings below the percolation threshold. However, at loadings above the percolation threshold, the hybrid system showed a higher volume conductivity than the system with single fillers, especially in PA12 with higher viscosity. They attributed this to a co-supporting network of CB upon MWNTs, as illustrated Fig. 2.17.

Ha et al. [215] elucidated the effect of hydrophobicity of the PA matrices on the performance of PA/CNT nanocomposites. They observed that higher hydrophobicity (longer methylene groups in the repeat unit) resulted in better MWCNT dispersion,

Table 2.3 Processing techniques for PA/clay nanocomposite

CNT/Graphene	Polymer Type	Processing Technique	Remarks	References
MWCNT	PA6, PA66	Melt-compounding	Various surface modifications of the CNTs	[211, 212, 214–218]
MWCNT	PA6, PA1010	In situ Polymerization	Mostly acid functionalisation of CNT before polymerisation	[209, 210, 213]
MWCNT	PA12	Melt-compounding Melt-compounding	CNTs are either functionalised or not	[215, 229, 232–234]
MWCNT	PA11			[215, 235, 236]
SWCNT	PA12	Melt-compounding	Encapsulated with Styrene maleic anhydride copolymer	[206, 237]
MWCNT	PA12, PA11	Melt-spinning	Role of spinning factors: speed, extrusion rate, and draw ratio	[219–221]
MWCNT (acid modified)	PA46, PA11	Solution mixing	The CNTS are modified or not	[238, 239]
MWCNT	PA11, PA12	Electrospinning		[222, 223]
MWCNT	PA12	AM	Coating CNTs on the surface of PA particles before SLS	[224–226]
	PA12		SLS-dry blending CNT and PA12	[230]
Carbon nanofibre	PA12		Pulverising PA/fibre followed by SLS	[231]

and hence, improved surface conductivity. Comparing PA6 and PA12, the greater number of hydrogen bonds in the former impedes interaction with CNTs, while van der Waals force between the hydrophobic methylene groups in PA12 and the CNT enhanced CNT dispersion.

PA/CNT composites can also be fabricated using AM, most commonly SLS. The SLS can be performed on PA/CNT powders obtained via three main routes: (i) coating CNTs on the surface of PA particles [224–226]; (ii) dry blending of CNTs and PA [230], and (iii) pulverizing PA/CNT composites before SLS [231]. In these examples, the CNT is used to enhance certain properties of the sintered part (e.g., mechanical or electrical), but also affects the melt flow properties of the polymer.

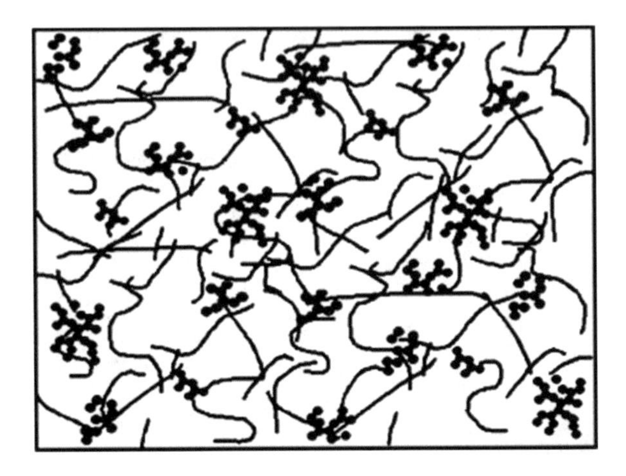

Fig. 2.17 Co-supporting network of CB and MWCNT Copyright © 2011, reproduced from [228] with permission from Elsevier Ltd.

Bai et al. [224] reported the use of AM to fabricate PA12/CNT nanocomposite parts and demonstrated enhanced mechanical performance with only 0.1 wt% MWCNT loading.

2.3.3 Processing, Structure, and Properties of PA/Graphene/Graphene Derivative PNCs

Like CNT-based nanocomposites, there has been a recent boom in the development of applications for graphene-based PA nanocomposites due to their exceptional electron transport and mechanical properties, and high surface area [42, 240]. Preparation methods include in situ polymerization [53, 241–244], melt compounding [245, 246], and solution techniques [247] (including electrospinning [248]). By far, the most common method for preparation of the graphene-based PA6 nanocomposites is in situ polymerization [240]. The process typically involves dispersion of graphene or its derivatives (e.g. GO [53, 241–243] and rGO [241, 244]) in a monomer or solutions of monomers. This is followed by the addition of a suitable initiator and then polymerization is achieved via control of the temperature and time. Xu and Gao [243] have described the synthesis of PA6/GO composites; in a typical case, suitable quantities of GO and ε-caprolactam were loaded into a reactor and the mixture was homogenized via sonication at 80 °C for 2 h to uniformly disperse the GO. Aminocaproic acid was added as an initiator, followed by mechanical stirring under a nitrogen atmosphere, firstly at 180 °C for 1 h, then at 250 °C for 9 h, to yield PA6/GO composites (Fig. 2.18).

Melt-compounding is the most economical and versatile method for preparing PA/graphene nanocomposites. It employs high shear forces and high temperature melting to mix the graphene and PA. A key aspect of the melt processing is the need for prior modification [249] of the graphene surface, which could involve in situ

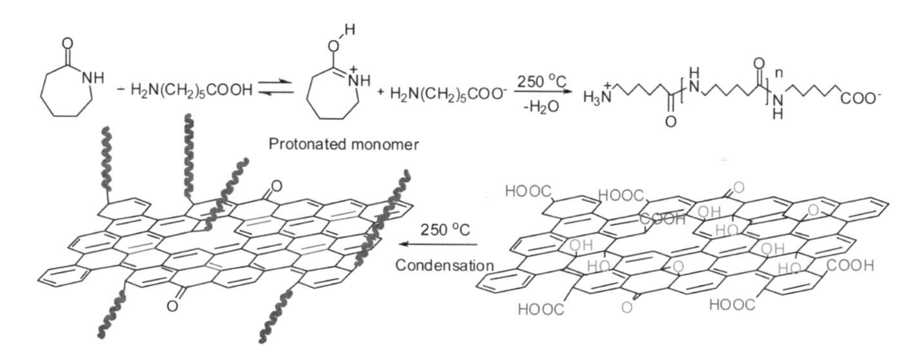

Fig. 2.18 In-situ ring-opening polymerization of ε-caprolactam in the presence of GO. Reproduced with permission from [243]. Copyright 2010, the American Chemical Society

polymerization on the graphene surface. Nguyễn et al. [245] made master batches (MB) of PA6/GO via ring-opening polymerization, before melt compounding the MB in a PA6 matrix to yield PA6/GO nanocomposites with enhanced mechanical properties. Similarly, Yan et al. [246] prepared PA12/graphene composites by melt-compounding PA and graphene in a twin-screw extruder at 220 °C with a rotational speed of 80 rpm for 15 min under a nitrogen purge. To enhance the electrical conductivity, a second polymer was added and localization of the graphene was tailored by tuning the process sequence.

Graphene and its derivatives can enhance thermal and electrical conductivity, dielectric properties, thermal stability, flame retardance, and mechanical properties of polymers [240]. One of the main benefits of incorporating graphene and its derivatives into PA is an increased electrical conductivity. The optimal conductivity would be achieved with a percolating graphene network within the polymer matrix. However, in most cases, handling pure graphene and dispersing it in polymer matrix is challenging. The alternative is to modify its surface or use GOs, which have lower conductivity values. Appreciable enhancement of the electrical conductivity has been reported by various researchers [53, 246]. Zheng et al. [53] synthesized PA6/GO nanocomposites via ring-opening polymerization and observed high electrical conductivity of ~0.028 S/m at a GO loading of ~1.64 vol.%. The nanocomposites exhibited a low percolation threshold of ~0.41 vol.%. In melt compounding, a novel strategy to lower the percolation threshold is to incorporate a second polymer phase and enhance localization of graphene in one of the phases. Yan et al. [246] used anhydride-grafted polyethylene-octene rubber (POE-g-MA) as a second polymer to improve the conductivity and toughness of PA12/graphene nanocomposites. They tailored the location of graphene in the ternary composite by varying the sequence of compounding the PA12, POE-g-MA, and graphene components. A low percolation threshold of 0.3 vol.% was achieved. These nanocomposites exhibited an increase in electrical conductivity from 2.8×10^{-14} S/m for PA12 to 6.7×10^{-2} S/m with the addition of ~1.38 vol.% graphene. This shows that the graphene sheets were homogeneously dispersed in the PA12 matrix.

In addition to the electrical conductivity, graphene and its derivatives can be incorporated in PA matrices to enhance thermal conductivity [244]. For instance, Ding et al. [244] reported enhancement of the thermal conductivity (λ) from 0.196 W m^{-1} K^{-1} of pure PA6 to 0.416 W m^{-1} K^{-1} of PA6/rGO nanocomposite with 10 wt% GO sheet loading. In a grafting strategy, the ε-caprolactam monomer was polymerized with 6-aminocaproic acid as the initiator in the presence of GO, where the GO was grafted to the latter, with simultaneous thermal reduction of the GO to rGO. They observed uniform dispersion and an interconnected structure of the rGO sheets in the matrix, which helped improve the thermal conductivity.

2.4 PI Nanocomposites

PI nanocomposites based on different NPs are prepared mostly via in situ polymerization and solution blending techniques (including electrospinning). Recently, most research has focused on CNT- and graphene-based PI nanocomposites due to their potential benefits for various applications, including the space industry. In situ polymerization is discussed here, followed by some post-polymerization shaping processes, such as electrospinning and AM.

2.4.1 In Situ Polymerization

In situ polymerization is by far the most commonly used method for PI nanocomposite processing of clays [250–256]; CNT [257–260]; graphene [261, 262]; ZnO [130]; $BaTiO_3$ [146, 263]; Ag [150]; Ba, Sr, Sn, and TiO_2 [264]. The initial step of mixing the precursor solution (PAA) with the NPs can be followed by various methods, including electrospinning [130], casting, freeze-drying (to form an aerogel) [265, 266], AM [267], and electrophoretic deposition [268] before thermal treatment to achieve imidization.

2.4.1.1 PI/Clay Nanocomposites Casting

PI/clay nanocomposites have achieved reduced dielectric constants [254, 269], enhanced mechanical properties [270, 271], thermal stability [270], corrosion inhibition [272], and other favorable performances compared to pure PI. To achieve good dispersion, most PI/clay nanocomposites are prepared via in situ intercalation using an intermediate polymer, poly (amic acid) (PAA). A typical process [253] starts with the synthesis of the PAA from diamine and dianhydride monomers in a nitrogen atmosphere. This is followed by dispersion of the NPs, either directly or in another solvent, and stepwise heating from 250 to 300 °C to imidize the material. In this procedure, OMMT was mixed with dimethylacetamide (DMAc) and stirred at

90 °C for 1 h. Separately, suitable quantities of 4,4′ diaminodiphenylether and DMAc were added and stirred at 30 °C for 30 min. Pyromellitic dianhydride is then added and stirred for 1 h at 30 °C to yield a DMAc solution of PAA, which was then mixed with the OMMT dispersion in DMAc for a period of 5 h at 30 °C. In this case, the solution was cast and the DMAc solvent was removed gradually over an extended period of 2 days. Finally, the imidization process was completed to yield PI/OMMT nanocomposites by heating to 100 °C and then 300 °C. This general approach has been followed by other researchers, with additional steps of prior intercalation of the clay galleries with reactive components. Tyan and colleagues [250, 273–276] reported the use of swelling agents with 1, 2, or 3 functional groups, for instance, only laurylamine, both di-functional 1,12 diaminododecane and 4,4′oxydianiline, and the trifunctional 4-(4-1,1-di[4-aminophenoxy) phenyl] ethylphenoxy) aniline. One end of the swelling agent ($-NH_2$) reacts and form an ionic bond with the negatively charged silicate surface. Other functional groups could possibly react with PAA containing dianhydride end groups. Another study demonstrated that trifunctional swelling agents result in improved dispersion of the clay in the resultant PI matrix due to the higher reactivity than single and bifunctional agents [276].

2.4.1.2 PI/CNT Nanocomposite Casting

Similar to clay-based PI nanocomposites, PI/CNT nanocomposites are prepared mostly by in situ polymerization, followed by casting a film before imidization via thermal treatment. CNTs can be used without modification, but the dispersion and subsequent performance (e.g. mechanical properties) is enhanced by surface functionalization [277]. Other properties enhanced by the CNT in PI include, the tribological properties [278], sensor sensitivity [257–260], and electromechanical actuation [279].

2.4.1.3 PI/Graphene Nanocomposite Casting

A typical process for preparing PI/graphene or PI/GO is illustrated in Fig. 2.19. The PAA/GO undergoes imidization via thermal treatment to give PI/GO nanocomposites. Graphene enhances various properties of PI, including the tribology [280, 281], flame retardance [282], mechanical performance [283], electrical properties [284], and moisture barrier performance [285, 286]. In one instance, Min et al. [281] prepared PI/GO nanocomposite films and studied their tribological properties; the composites showed improved tribological properties under seawater-lubricated conditions, and the wear resistance was enhanced compared to the pure polymer.

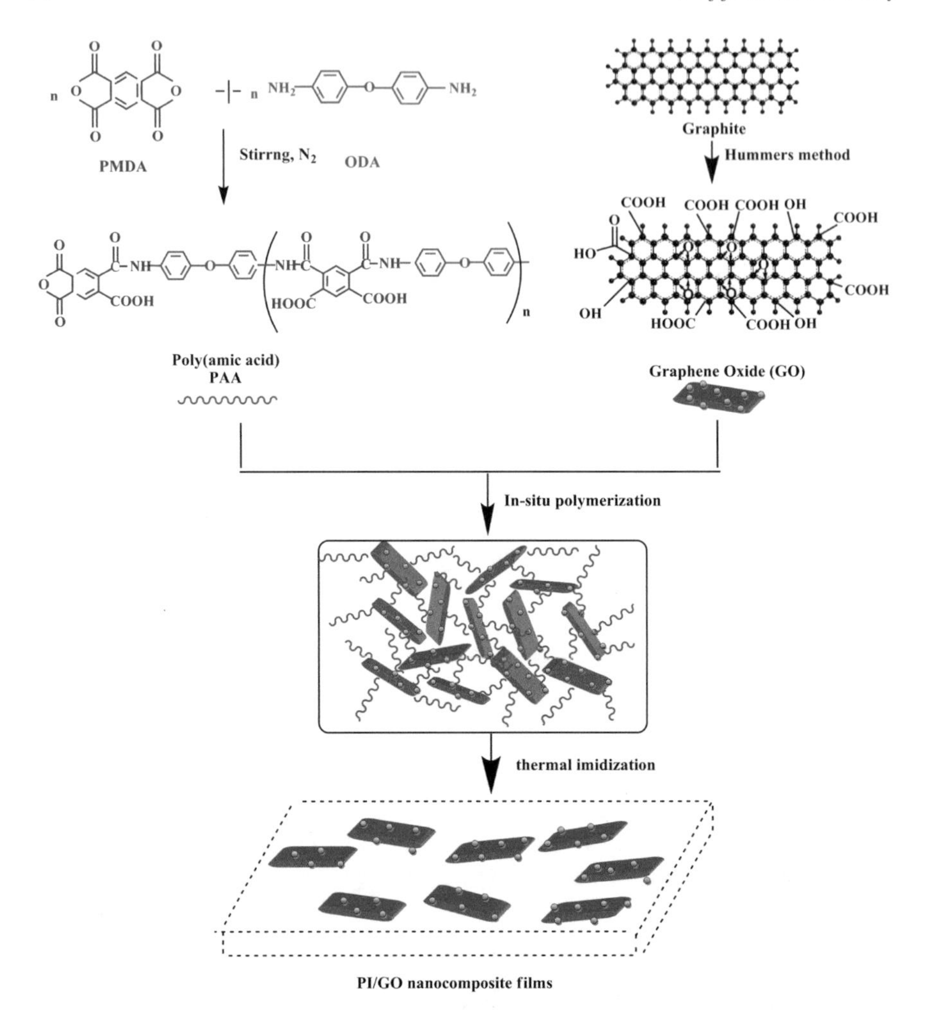

Fig. 2.19 The synthesis route for PI/GO nanocomposites case films. Reproduced with permission from [281]. Copyright 2014, Elsevier Science Ltd.

2.4.2 Additive Manufacturing

A new type of AM process known as "dual-material aerosol printing" (DMAP) was recently developed by Wang et al. [267] to fabricate PI/CNT with tunable intra-part CNT loading. The process (see Figs. 2.20 and 2.21) homogenously mixes aerosol from a CNT ink and a PAA solution, where the mixing ratio determines the CNT loading and subsequent conductivity of the film. The authors demonstrated the ability to control the CNT loading in different regions of the sample, as illustrated in Fig. 2.22. This method is scalable and could find industrial application soon.

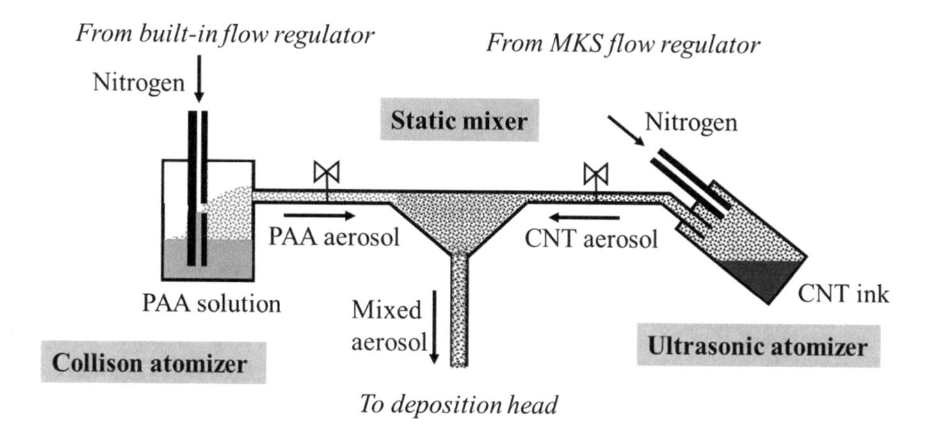

Fig. 2.20 Illustration of the dual-material aerosol jet printing (DMEJP) setup. Reproduced with permission from [267]. Copyright 2016, Elsevier Science Ltd.

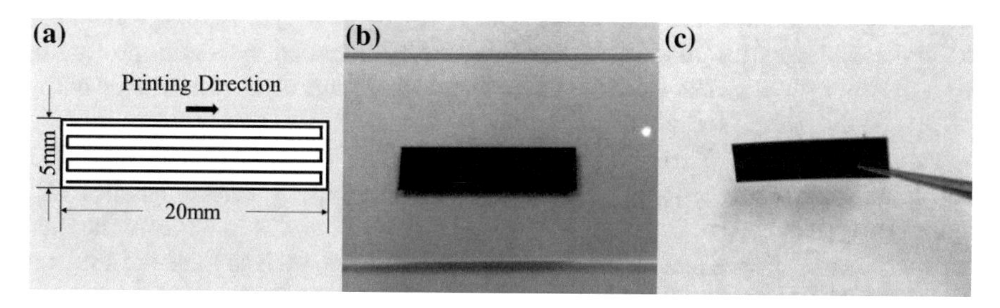

Fig. 2.21 DMAJP polyimide/CNT nanocomposite film **a** shows the sample dimensions and the direction of printing; **b** an printed PI/CNT film; **c** a freestanding nanocomposite sample. Reproduced with permission from [267]. Copyright 2016, Elsevier Science Ltd.

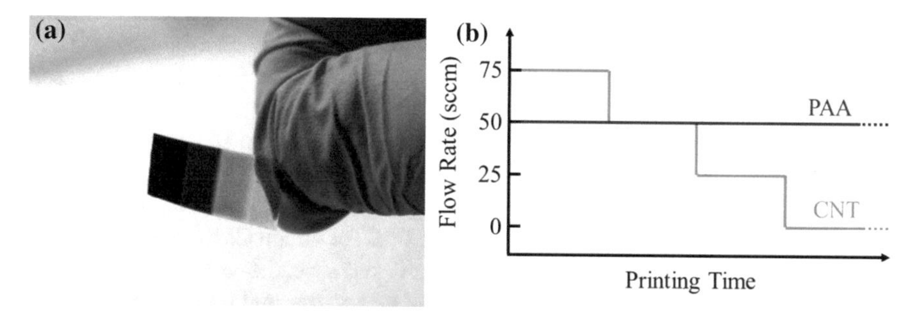

Fig. 2.22 **a** Printed PI/CNT nanocomposite film showing four regions with four different CNT loadings; **b** Corresponding flowrate from the two aerosols. Reproduced with permission from [267]. Copyright 2016, Elsevier Science Ltd.

2.4.3 Electrospinning

Electrospun fibers have a high specific surface area, which is advantageous for, e.g., sensing applications. PI nanocomposites based on Ag [150], MWCNT [135, 261, 287], GO [261, 287], ZnO [130], and Fe-FeO NP-reinforced PI [145] have been fabricated for various applications. Chen et al. [135] reported electrospinning of PI/MWCNTs for the first time in 1999. They investigated the effect of surface modification of MWCNT on their dispersion in a PI matrix. For MWCNT modification, carboxylic and hydroxyl groups were introduced by refluxing the pristine MWNTs in concentrated HNO_3. A typical process started with in situ polymerization of PAA containing 1–10 wt% MWCNTs in DMAc solution. The PAA/MWCNT solutions were then electrospun using a syringe with a spinneret of 0.5 mm diameter. The voltage difference was kept at 20-25 kV, with a spinneret–collector distance of 20 cm, while the feeding rate was 0.25 mL/h. After evaporation of most of the solvent, the fibers were then imidized in a nitrogen atmosphere with a controlled temperature ramp. The authors noted that PI fibers with acid-treated MWCNTs had a homogeneous dispersion, while pristine MWCNTs were agglomerated, with some protruding from the fiber surface. The thermal and mechanical properties were improved at low CNT concentrations (1–3.5 wt%).

Very recently, Guo et al. [261] used a sequential in situ polymerization and electrospinning-hot press technology to fabricate layered PI/GO nanocomposites with significantly enhanced thermal conductivity. The objective was to improve the thermal conductivity above that reported by other researchers, such as Li et al. [288] and Wu et al. [262]. Li et al. [288] reported an 21.9% increase in the in-plane thermal conductivity of PI/GO compared to pristine graphene film, while Wu et al. [262] modified graphene with a coupling agent ($KH\text{-}_{550}$) and observed a thermal conductivity three times higher in the $KH\text{-}_{550}$-modified graphene/PI nanocomposites compared to PI. In the approach presented by Guo et al. [261], GO was modified with hydrazine monohydrate and aminopropyl isobutyl polyhedral oligomeric silsesquioxane (NH_2-POSS), following the process illustrated in the schematic in Fig. 2.23, to fabricate CMG/PI nanocomposites.

Briefly, simultaneous chemical modification and reduction of the GO was performed to yield chemically modified graphene (CMG). Thereafter, suitable quantities of CMG, 1,3-bis(4-aminophenyloxy)benzene (APB), and bis-(3-phthalyl anhydride) ether (ODPA) were mixed and stirred for 3 h at 0 °C for obtain CMG/PAA. This solution was then electrospun at a voltage of 20 kV and a needle–collector distance of 30 cm. The obtained fibers were vacuum dried before thermal imidization via temperature ramps of 120 °C/1 h + 200 °C/1 h + 250 °C/1 h. The fibers were then layered up and hot-pressed at 320 °C and 10 MPa for 40 min to produce thermally conductive CMG/PI nanocomposites. The coefficient of thermal conductivity (λ), T_g, and heat resistance index for the nanocomposite fibers improved with increasing CMG loading. At 5 wt% CMG loading, λ was 1.05 W m^{-1} K^{-1} compared to 0.28 W m^{-1} K^{-1}.

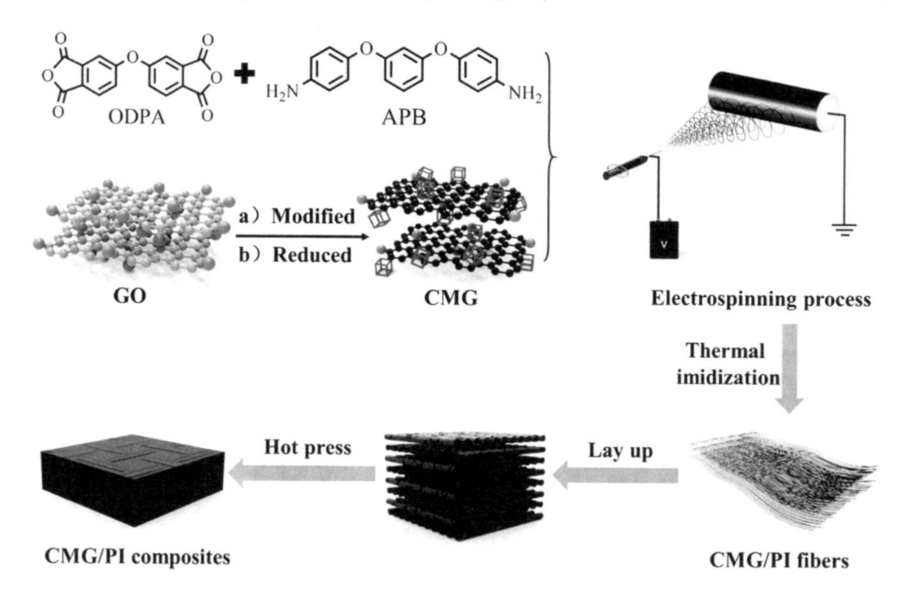

Fig. 2.23 Schematic of the process for fabrication of chemically modified GO/PI composites. Reproduced with permission from [261]. Copyright 2018, The Royal Society of Chemistry

Chang [130] claimed to have developed a simple novel method for fabricating firecracker-shaped ZnO/PI hybrid nanofibers, by combining electrospinning and a hydrothermal process to yield a material with high photocatalytic activity. As depicted in Fig. 2.24, the process involved three steps: (i) electrospinning PAA; (ii) immobilization of ZnO seeds on the fibers by dip-coating and imidization of PAA to PI; and (iii) growth of the ZnO nanorods from the seeds on the PI surface by immersion in an aqueous solution of equimolar mixture of zinc nitrate hexahydrate ($Zn(NO_3)26H_2O$; 0.025 M) and hexamethylenetetramine ($C_6H_{12}N_4$; 0.025 M). The hydrothermal process was performed for 6 h at 90 °C. SEM images of the obtained firecracker-shaped ZnO/PI hybrid nanofibers are shown in Fig. 2.24b, c, d, and d′.

2.5 Conclusions and Future Outlook

Even though many advancements have been made on PA and PI nanocomposites based on clays, particles such as CNTs and graphene are at the frontier of science and are yet to realize their full potential in polymer matrices. An issue still requiring intensive research is the need for nanoscale dispersion of the NPs in the polymer matrices, without comprising cost effectiveness, and human and environmental safety. Other important areas of further research include methods for aligning CNTs, and CNT–polymer interactions and their influence on intrinsic properties like the conductivity. The issue of alignment of the CNTs (for application such as sensors)

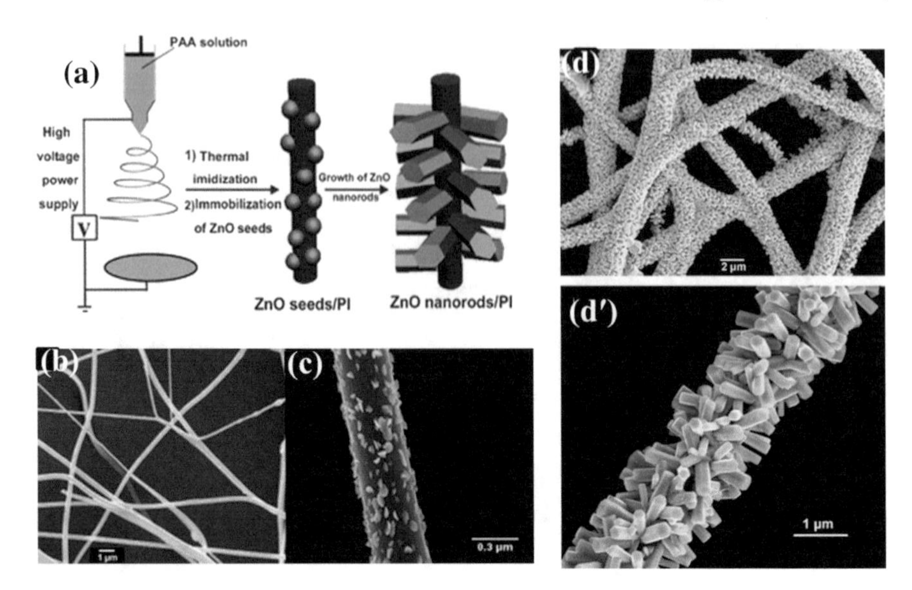

Fig. 2.24 **a** Schematic for growing ZnO nanorods on electrospun PI surface via hydrothermal method; **b** SEM image of pristine PI nanofibers; **c** SEM image of ZnO seeds on PI nanofiber; **d** SEM image of ZnO/PI hybrid nanofiber, and **d** high magnification of ZnO/PI hybrid nanofiber. Reproduced with permission from [130]. Copyright 2011, The Royal Society of Chemistry

can be partly addressed via electrospinning processes. Finally, development of engineering PNC materials suitable for AM and elucidation of the effect of such particles on the AM processes needs further exploration. The ability to customize properties such as the electrical conductivity on demand for a particular part is a goal for the future. The industry 4.0 concept of customizability and on-demand production of parts is offered by AM platforms; hence, further research should be focused on both the process itself and also optimizing the materials via the addition of NPs.

Acknowledgements The authors would like to thank the South African Department of Science and Technology (DST) and the Council for Scientific and Industrial Research (CSIR) for financial support.

References

1. Melton GH, Peters EN, Arisman RK (2011) Engineering thermoplastics. Applied plastics engineering handbook. Elsevier, Amsterdam, pp 7–21
2. Margolis J (2005) Engineering plastics handbook. McGraw Hill Professional
3. Marchildon K (2011) Polyamides—still strong after seventy years. Macromol Reaction Eng 5:22–54
4. García JM, García FlC, Serna F, de la Peña JL (2010) High-performance aromatic polyamides. Prog Polymer Sci 35:623–686

5. Reglero Ruiz JA, Trigo-López M, García FC, García JM (2017) Functional aromatic polyamides. Polymers 9:414
6. Kojima Y, Usuki A, Kawasumi M, Okada A, Fukushima Y, Kurauchi T et al (1993) Mechanical properties of nylon 6-clay hybrid. J Mater Res 8:1185–1189
7. Müller K, Bugnicourt E, Latorre M, Jorda M, Echegoyen Sanz Y, Lagaron JM et al (2017) Review on the processing and properties of polymer nanocomposites and nanocoatings and their applications in the packaging, automotive and solar energy fields. Nanomaterials (Basel, Switzerland) 7
8. Mittal V (2013) Modeling and prediction of polymer nanocomposite properties. Wiley, Weinheim
9. Mikitaev AK, Kozlov GV (2016) The role of interface surfaces in the formation of the properties of polymer nanocomposites. J Surf Investig X-ray Synchrotron Neutron Tech 10:250–253
10. Giannelis EP (1996) Polymer layered silicate nanocomposites. Adv Mater 8:29–35
11. Paul DR, Robeson LM (2008) Polymer nanotechnology: nanocomposites. Polymer 49:3187–3204
12. Sinha Ray S, Okamoto M (2003) Polymer/layered silicate nanocomposites: a review from preparation to processing. Prog Polymer Sci 28:1539–1641
13. Krishnamoorti R, Vaia RA, Giannelis EP (1996) Structure and dynamics of polymer-layered silicate nanocomposites. Chem Mater 8:1728–1734
14. LeBaron PC, Wang Z, Pinnavaia TJ (1999) Polymer-layered silicate nanocomposites: an overview. Appl Clay Sci 15:11–29
15. Okamoto M (2006) Recent advances in polymer/layered silicate nanocomposites: an overview from science to technology. Mater Sci Technol 22:756–779
16. Pavlidou S, Papaspyrides CD (2008) A review on polymer–layered silicate nanocomposites. Prog Polym Sci 33:1119–1198
17. Liu P (2007) Polymer modified clay minerals: a review. Appl Clay Sci 38:64–76
18. Chiu C-W, Huang T-K, Wang Y-C, Alamani BG, Lin J-J (2014) Intercalation strategies in clay/polymer hybrids. Prog Polym Sci 39:443–485
19. Kotal M, Bhowmick AK (2015) Polymer nanocomposites from modified clays: recent advances and challenges. Prog Polym Sci 51:127–187
20. Sinha Ray S, Okamoto K, Okamoto M (2003) Structure–property relationship in biodegradable poly(butylene succinate)/layered silicate nanocomposites. Macromolecules 36:2355–2367
21. Spitalsky Z, Tasis D, Papagelis K, Galiotis C (2010) Carbon nanotube–polymer composites: chemistry, processing, mechanical and electrical properties. Prog Polym Sci 35:357–401
22. Schartel B, Pötschke P, Knoll U, Abdel-Goad M (2005) Fire behaviour of polyamide 6/multiwall carbon nanotube nanocomposites. Eur Polymer J 41:1061–1070
23. Lee H-J, Oh S-J, Choi J-Y, Kim JW, Han J, Tan L-S et al (2005) In situ synthesis of poly (ethylene terephthalate)(PET) in ethylene glycol containing terephthalic acid and functionalized multiwalled carbon nanotubes (MWNTs) as an approach to MWNT/PET nanocomposites. Chem Mater 17:5057–5064
24. Nogales A, Broza G, Roslaniec Z, Schulte K, Šics I, Hsiao B et al (2004) Low percolation threshold in nanocomposites based on oxidized single wall carbon nanotubes and poly (butylene terephthalate). Macromolecules 37:7669–7672
25. Broza G, Kwiatkowska M, Rosłaniec Z, Schulte K (2005) Processing and assessment of poly (butylene terephthalate) nanocomposites reinforced with oxidized single wall carbon nanotubes. Polymer 46:5860–5867
26. Bauhofer W, Kovacs JZ (2009) A review and analysis of electrical percolation in carbon nanotube polymer composites. Compos Sci Technol 69:1486–1498
27. Iijima S, Ichihashi T (1993) Single-shell carbon nanotubes of 1-nm diameter. Nature 363:603
28. Sahoo NG, Rana S, Cho JW, Li L, Chan SH (2010) Polymer nanocomposites based on functionalized carbon nanotubes. Prog Polym Sci 35:837–867
29. Xie X-L, Mai Y-W, Zhou X-P (2005) Dispersion and alignment of carbon nanotubes in polymer matrix: a review. Mater Sci Eng R Reports 49:89–112

30. Kim SW, Kim T, Kim YS, Choi HS, Lim HJ, Yang SJ et al (2012) Surface modifications for the effective dispersion of carbon nanotubes in solvents and polymers. Carbon 50:3–33
31. Cai H, Yan F, Xue Q (2004) Investigation of tribological properties of polyimide/carbon nanotube nanocomposites. Mater Sci Eng A 364:94–100
32. Ounaies Z, Park C, Wise K, Siochi E, Harrison J (2003) Electrical properties of single wall carbon nanotube reinforced polyimide composites. Compos Sci Technol 63:1637–1646
33. Castillo FY, Socher R, Krause B, Headrick R, Grady BP, Prada-Silvy R et al (2011) Electrical, mechanical, and glass transition behavior of polycarbonate-based nanocomposites with different multi-walled carbon nanotubes. Polymer 52:3835–3845
34. Liu T, Tong Y, Zhang W-D (2007) Preparation and characterization of carbon nanotube/polyetherimide nanocomposite films. Compos Sci Technol 67:406–412
35. Sianipar M, Kim SH, Min C, Tijing LD, Shon HK (2016) Potential and performance of a polydopamine-coated multiwalled carbon nanotube/polysulfone nanocomposite membrane for ultrafiltration application. J Ind Eng Chem 34:364–373
36. Inukai S, Cruz-Silva R, Ortiz-Medina J, Morelos-Gomez A, Takeuchi K, Hayashi T et al (2015) High-performance multi-functional reverse osmosis membranes obtained by carbon nanotube polyamide nanocomposite. Sci Rep 5:13562
37. Kim HJ, Choi K, Baek Y, Kim D-G, Shim J, Yoon J et al (2014) High-performance reverse osmosis CNT/polyamide nanocomposite membrane by controlled interfacial interactions. ACS Appl Mater Interfaces 6:2819–2829
38. Yu S, Wong WM, Hu X, Juay YK (2009) The characteristics of carbon nanotube-reinforced poly(phenylene sulfide) nanocomposites. J Appl Polym Sci 113:3477–3483
39. Jiang Z, Hornsby P, McCool R, Murphy A (2012) Mechanical and thermal properties of polyphenylene sulfide/multiwalled carbon nanotube composites. J Appl Polym Sci 123:2676–2683
40. Anand K, Agarwal U, Joseph R (2007) Carbon nanotubes-reinforced PET nanocomposite by melt-compounding. J Appl Polym Sci 104:3090–3095
41. Geim AK, Novoselov KS (2007) The rise of graphene. Nat Mater 6:183
42. Kim H, Abdala AA, Macosko CW (2010) Graphene/polymer nanocomposites. Macromolecules 43:6515–6530
43. Potts JR, Dreyer DR, Bielawski CW, Ruoff RS (2011) Graphene-based polymer nanocomposites. Polymer 52:5–25
44. Ramanathan T, Abdala A, Stankovich S, Dikin D, Herrera-Alonso M, Piner R et al (2008) Functionalized graphene sheets for polymer nanocomposites. Nat Nanotechnol 3:327
45. Sungjin P, Ruoff RS (2009) Chemical methods for the production of graphenes. Nat Nanotechnol 4:217–224
46. Du J, Cheng HM (2012) The fabrication, properties, and uses of graphene/polymer composites. Macromol Chem Phys 213:1060–1077
47. Kuilla T, Bhadra S, Yao D, Kim NH, Bose S, Lee JH (2010) Recent advances in graphene based polymer composites. Prog Polym Sci 35:1350–1375
48. Cai D, Song M (2010) Recent advance in functionalized graphene/polymer nanocomposites. J Mater Chem 20:7906–7915
49. Tseng I, Chang JC, Huang SL, Tsai MH (2013) Enhanced thermal conductivity and dimensional stability of flexible polyimide nanocomposite film by addition of functionalized graphene oxide. Polym Int 62:827–835
50. Yoonessi M, Shi Y, Scheiman DA, Lebron-Colon M, Tigelaar DM, Weiss R et al (2012) Graphene polyimide nanocomposites; thermal, mechanical, and high-temperature shape memory effects. ACS Nano 6:7644–7655
51. Ha HW, Choudhury A, Kamal T, Kim D-H, Park S-Y (2012) Effect of chemical modification of graphene on mechanical, electrical, and thermal properties of polyimide/graphene nanocomposites. ACS Appl Mater Interfaces 4:4623–4630
52. Yin J, Zhu G, Deng B (2016) Graphene oxide (GO) enhanced polyamide (PA) thin-film nanocomposite (TFN) membrane for water purification. Desalination 379:93–101

53. Zheng D, Tang G, Zhang H-B, Yu Z-Z, Yavari F, Koratkar N et al (2012) In situ thermal reduction of graphene oxide for high electrical conductivity and low percolation threshold in polyamide 6 nanocomposites. Compos Sci Technol 72:284–289
54. Liu H, Hou L, Peng W, Zhang Q, Zhang X (2012) Fabrication and characterization of polyamide 6-functionalized graphene nanocomposite fiber. J Mater Sci 47:8052–8060
55. Yoonessi M, Gaier JR (2010) Highly conductive multifunctional graphene polycarbonate nanocomposites. ACS Nano 4:7211–7220
56. Gedler G, Antunes M, Realinho V, Velasco J (2012) Thermal stability of polycarbonate-graphene nanocomposite foams. Polym Degrad Stab 97:1297–1304
57. Bandla S, Hanan J (2012) Microstructure and elastic tensile behavior of polyethylene terephthalate-exfoliated graphene nanocomposites. J Mater Sci 47:876–882
58. Zhang H-B, Zheng W-G, Yan Q, Yang Y, Wang J-W, Lu Z-H et al (2010) Electrically conductive polyethylene terephthalate/graphene nanocomposites prepared by melt compounding. Polymer 51:1191–1196
59. Bian J, Lin HL, He FX, Wang L, Wei XW, Chang IT et al (2013) Processing and assessment of high-performance poly(butylene terephthalate) nanocomposites reinforced with microwave exfoliated graphite oxide nanosheets. Eur Polymer J 49:1406–1423
60. Fabbri P, Bassoli E, Bon SB, Valentini L (2012) Preparation and characterization of poly (butylene terephthalate)/graphene composites by in situ polymerization of cyclic butylene terephthalate. Polymer 53:897–902
61. Chen H, Huang C, Yu W, Zhou C (2013) Effect of thermally reduced graphite oxide (TrGO) on the polymerization kinetics of poly(butylene terephthalate) (pCBT)/TrGO nanocomposites prepared by in situ ring-opening polymerization of cyclic butylene terephthalate. Polymer 54:1603–1611
62. Ionita M, Pandele AM, Crica L, Pilan L (2014) Improving the thermal and mechanical properties of polysulfone by incorporation of graphene oxide. Compos B Eng 59:133–139
63. Rezaee R, Nasseri S, Mahvi AH, Nabizadeh R, Mousavi SA, Rashidi A et al (2015) Fabrication and characterization of a polysulfone-graphene oxide nanocomposite membrane for arsenate rejection from water. J Environ Health Sci Eng 13:61
64. Qin Y, Peng Q, Ding Y, Lin Z, Wang C, Li Y et al (2015) Lightweight, superelastic, and mechanically flexible graphene/polyimide nanocomposite foam for strain sensor application. ACS Nano 9:8933–8941
65. Hu K, Kulkarni DD, Choi I, Tsukruk VV (2014) Graphene-polymer nanocomposites for structural and functional applications. Prog Polym Sci 39:1934–1972
66. Rahman IA, Padavettan V (2012) Synthesis of silica nanoparticles by sol-gel: size-dependent properties, surface modification, and applications in silica-polymer nanocomposites—a review. J Nanomater 2012:8
67. Zou H, Wu S, Shen J (2008) Polymer/silica nanocomposites: preparation, characterization, properties, and applications. Chem Rev 108:3893–3957
68. Müller K, Bugnicourt E, Latorre M, Jorda M, Echegoyen Sanz Y, Lagaron JM et al (2017) Review on the processing and properties of polymer nanocomposites and nanocoatings and their applications in the packaging, automotive and solar energy fields. Nanomaterials 7:74
69. Kuo S-W, Chang F-C (2011) POSS related polymer nanocomposites. Prog Polym Sci 36:1649–1696
70. Zhang W, Camino G, Yang R (2017) Polymer/polyhedral oligomeric silsesquioxane (POSS) nanocomposites: An overview of fire retardance. Prog Polym Sci 67:77–125
71. Sánchez-Soto M, Schiraldi DA, Illescas S (2009) Study of the morphology and properties of melt-mixed polycarbonate–POSS nanocomposites. Eur Polymer J 45:341–352
72. Zhao Y, Schiraldi DA (2005) Thermal and mechanical properties of polyhedral oligomeric silsesquioxane (POSS)/polycarbonate composites. Polymer 46:11640–11647
73. Baldi F, Bignotti F, Ricco L, Monticelli O, Riccò T (2006) Mechanical and structural characterization of POSS-modified polyamide 6. J Appl Polym Sci 100:3409–3414
74. Wan C, Zhao F, Bao X, Kandasubramanian B, Duggan M (2009) Effect of POSS on crystalline transitions and physical properties of polyamide 12. J Polym Sci Part B Polym Phys 47:121–129

75. Yu H, Ren W, Zhang Y (2009) Nonisothermal decomposition kinetics of nylon 1010/POSS composites. J Appl Polym Sci 113:17–23
76. Yoon KH, Polk MB, Park JH, Min BG, Schiraldi DA (2005) Properties of poly(ethylene terephthalate) containing epoxy-functionalized polyhedral oligomeric silsesquioxane. Polym Int 54:47–53
77. Kim KJ, Ramasundaram S, Lee JS (2008) Synthesis and characterization of poly(trimethylene terephthalate)/polyhedral oligomeric silsesquixanes nanocomposites. Polym Compos 29:894–901
78. Zhou Z, Yin N, Zhang Y, Zhang Y (2008) Properties of poly(butylene terephthalate) chain-extended by epoxycyclohexyl polyhedral oligomeric silsesquioxane. J Appl Polym Sci 107:825–830
79. Geng Z, Huo M, Mu J, Zhang S, Lu Y, Luan J et al (2014) Ultra low dielectric constant soluble polyhedral oligomeric silsesquioxane (POSS)–poly (aryl ether ketone) nanocomposites with excellent thermal and mechanical properties. J Mater Chem C 2:1094–1103
80. Lee Y-J, Huang J-M, Kuo S-W, Lu J-S, Chang F-C (2005) Polyimide and polyhedral oligomeric silsesquioxane nanocomposites for low-dielectric applications. Polymer 46:173–181
81. Wahab MA, Mya KY, He C (2008) Synthesis, morphology, and properties of hydroxyl terminated-POSS/polyimide low-k nanocomposite films. J Polym Sci Part A Polym Chem 46:5887–5896
82. Iyer P, Coleman MR (2008) Thermal and mechanical properties of blended polyimide and amine-functionalized poly(orthosiloxane) composites. J Appl Polym Sci 108:2691–2699
83. Tamaki R, Choi J, Laine RM (2003) A polyimide nanocomposite from octa(aminophenyl)silsesquioxane. Chem Mater 15:793–797
84. Fan H, Yang R (2013) Flame-retardant polyimide cross-linked with polyhedral oligomeric octa(aminophenyl)silsesquioxane. Ind Eng Chem Res 52:2493–2500
85. Zhu J, Lim J, Lee C-H, Joh H-I, Kim HC, Park B et al (2014) Multifunctional polyimide/graphene oxide composites via in situ polymerization. J Appl Polym Sci 131
86. Qian Y, Wu H, Yuan D, Li X, Yu W, Wang C (2015) In situ polymerization of polyimide-based nanocomposites via covalent incorporation of functionalized graphene nanosheets for enhancing mechanical, thermal, and electrical properties. J Appl Polym Sci 132
87. Liu P, Yao Z, Li L, Zhou J (2016) In situ synthesis and mechanical, thermal properties of polyimide nanocomposite film by addition of functionalized graphene oxide. Polym Compos 37:907–914
88. He L, Zhang P, Chen H, Sun J, Wang J, Qin C et al (2016) Preparation of polyimide/siloxane-functionalized graphene oxide composite films with high mechanical properties and thermal stability via in situ polymerization. Polym Int 65:84–92
89. Wang Y, Wu X, Feng C, Zeng Q (2016) Improved dielectric properties of surface modified BaTiO3/polyimide composite films. Microelectron Eng 154:17–21
90. Ojijo V, Sinha Ray S (2013) Processing strategies in bionanocomposites. Prog Polym Sci 38:1543–1589
91. Isayev AI, Kumar R, Lewis TM (2009) Ultrasound assisted twin screw extrusion of polymer–nanocomposites containing carbon nanotubes. Polymer 50:250–260
92. Gao X, Isayev AI, Zhang X, Zhong J (2017) Influence of processing parameters during ultrasound assisted extrusion on the properties of polycarbonate/carbon nanotubes composites. Compos Sci Technol 144:125–138
93. Zhong J, Isayev AI (2015) Properties of polyetherimide/graphite composites prepared using ultrasonic twin-screw extrusion. J Appl Polym Sci 132
94. Ávila-Orta C, Espinoza-González C, Martínez-Colunga G, Bueno-Baqués D, Maffezzoli A, Lionetto F (2013) An overview of progress and current challenges in ultrasonic treatment of polymer melts. Adv Polym Technol 32:E582–E602
95. Gao X, Isayev AI, Yi C (2016) Ultrasonic treatment of polycarbonate/carbon nanotubes composites. Polymer 84:209–222

96. Isayev AI, Jung C, Gunes K, Kumar R (2008) Ultrasound assisted single screw extrusion process for dispersion of carbon nanofibers in polymers. Int Polym Proc 23:395–405
97. Karger-Kocsis J, Kmetty Á, Lendvai L, Drakopoulos S, Bárány T (2015) Water-assisted production of thermoplastic nanocomposites: a review. Materials 8:72
98. Touchaleaume F, Soulestin J, Sclavons M, Devaux J, Lacrampe M, Krawczak P (2011) One-step water-assisted melt-compounding of polyamide 6/pristine clay nanocomposites: an efficient way to prevent matrix degradation. Polym Degrad Stab 96:1890–1900
99. Stoclet G, Sclavons M, Devaux J (2013) Relations between structure and property of polyamide 11 nanocomposites based on raw clays elaborated by water-assisted extrusion. J Appl Polym Sci 127:4809–4824
100. Yu ZZ, Hu GH, Varlet J, Dasari A, Mai YW (2005) Water-assisted melt compounding of nylon-6/pristine montmorillonite nanocomposites. J Polym Sci Part B Polym Phys 43:1100–1112
101. Molajavadi V, Garmabi H (2011) Water assisted exfoliation of PA6/clay nanocomposites using a twin screw extruder: Effect of water contact time. J Appl Polym Sci 119:736–743
102. Stoeffler K, Utracki LA, Simard Y, Labonté S (2013) Polyamide 12 (PA12)/clay nanocomposites fabricated by conventional extrusion and water-assisted extrusion processes. J Appl Polym Sci 130:1959–1974
103. Lecouvet B, Sclavons M, Bourbigot S, Bailly C (2014) Towards scalable production of polyamide 12/halloysite nanocomposites via water-assisted extrusion: mechanical modeling, thermal and fire properties. Polym Adv Technol 25:137–151
104. Fedullo N, Sorlier E, Sclavons M, Bailly C, Lefebvre J-M, Devaux J (2007) Polymer-based nanocomposites: overview, applications and perspectives. Prog Org Coat 58:87–95
105. Dini M, Mousavand T, Carreau PJ, Kamal MR, Ton-That MT (2014) Microstructure and properties of poly (ethylene terephthalate)/organoclay nanocomposites prepared by water-assisted extrusion: effect of organoclay concentration. Polym Eng Sci 54:1879–1892
106. Dini M, Mousavand T, Carreau PJ, Kamal MR, Ton-That MT (2014) Effect of water-assisted extrusion and solid-state polymerization on the microstructure of PET/Clay nanocomposites. Polym Eng Sci 54:1723–1736
107. Lecouvet B, Sclavons M, Bourbigot S, Bailly C (2013) Thermal and flammability properties of polyethersulfone/halloysite nanocomposites prepared by melt compounding. Polym Degrad Stab 98:1993–2004
108. Korbee RA, Van Geenen AA (2002) Process for the preparation of a polyamide nanocomposite composition. Google Patents
109. Peng J, Walsh PJ, Sabo RC, Turng L-S, Clemons CM (2016) Water-assisted compounding of cellulose nanocrystals into polyamide 6 for use as a nucleating agent for microcellular foaming. Polymer 84:158–166
110. Hasegawa N, Okamoto H, Kato M, Usuki A, Sato N (2003) Nylon 6/Na–montmorillonite nanocomposites prepared by compounding Nylon 6 with Na–montmorillonite slurry. Polymer 44:2933–2937
111. Fedullo N, Sclavons M, Bailly C, Lefebvre JM, Devaux J (2006) Nanocomposites from untreated clay: a myth? Macromolecular symposia: Wiley Online Library, pp 235–245
112. Wevers MGM, Pijpers TFJ, Mathot VBF (2007) The way to measure quantitatively full dissolution and crystallization of polyamides in water up to 200°C and above by DSC. Thermochim Acta 453:67–71
113. Charlet K, Mathot V, Devaux J (2011) Crystallization and dissolution behaviour of polyamide 6-water systems under pressure. Polym Int 60:119–125
114. Shin YM, Hohman MM, Brenner MP, Rutledge GC (2001) Electrospinning: a whipping fluid jet generates submicron polymer fibers. Appl Phys Lett 78:1149–1151
115. Li D, Xia Y (2004) Electrospinning of nanofibers: reinventing the wheel? Adv Mater 16:1151–1170
116. Frenot A, Chronakis IS (2003) Polymer nanofibers assembled by electrospinning. Curr Opin Colloid Interface Sci 8:64–75
117. Liao Y, Loh C-H, Tian M, Wang R, Fane AG (2018) Progress in electrospun polymeric nanofibrous membranes for water treatment: fabrication, modification and applications. Prog Polym Sci 77:69–94

118. Wang G, Yu D, Kelkar AD, Zhang L (2017) Electrospun nanofiber: emerging reinforcing filler in polymer matrix composite materials. Prog Polym Sci 75:73–107
119. Anton F (1944) Method and apparatus for spinning. Anton, Formhals, United States
120. Formhals A (1934) Process and apparatus for preparaing artificial threads. Anton Formhals, United States
121. Doshi J, Reneker DH (1995) Electrospinning process and applications of electrospun fibers. J Electrostat 35:151–160
122. Reneker DH, Chun I (1996) Nanometre diameter fibres of polymer, produced by electrospinning. Nanotechnology 7:216–223
123. Reneker DH, Yarin AL, Fong H, Koombhongse S (2000) Bending instability of electrically charged liquid jets of polymer solutions in electrospinning. J Appl Phys 87:4531–4547
124. Haider A, Haider S, Kang I-K (2015) A comprehensive review summarizing the effect of electrospinning parameters and potential applications of nanofibers in biomedical and biotechnology. Arab J Chem
125. Brown TD, Dalton PD, Hutmacher DW (2016) Melt electrospinning today: an opportune time for an emerging polymer process. Prog Polym Sci 56:116–166
126. Taylor G (1964) Disintegration of water drops in an electric field. Proc R Soc Lond A 280:383–397
127. Bhardwaj N, Kundu SC (2010) Electrospinning: a fascinating fiber fabrication technique. Biotechnol Adv 28:325–47
128. Ding Y, Hou H, Zhao Y, Zhu Z, Fong H (2016) Electrospun polyimide nanofibers and their applications. Prog Polym Sci 61:67–103
129. Bergshoef MM, Vancso GJ (1999) Transparent nanocomposites with ultrathin, electrospun nylon-4,6 fiber reinforcement. Adv Mater 11:1362
130. Chang Z (2011) "Firecracker-shaped" ZnO/polyimide hybrid nanofibers via electrospinning and hydrothermal process. Chem Commun 2011:4427–4429
131. Schiffman JD, Elimelech M (2011) Antibacterial activity of electrospun polymer mats with incorporated narrow diameter single-walled carbon nanotubes. ACS Appl Mater Interfaces 3:462–468
132. Huang ZB, Li G, Sui G, Yang XP (2009) Synergy effects of electrospun polysulfone/CNTs hybrid nanofibers toughened and reinforced epoxy. Trans Tech Publ, Adv Mater Res, pp 517–520
133. Xu W, Ding Y, Jiang S, Zhu J, Ye W, Shen Y et al (2014) Mechanical flexible PI/MWCNTs nanocomposites with high dielectric permittivity by electrospinning. Eur Polymer J 59:129–135
134. Chen D, Wang R, Tjiu WW, Liu T (2011) High performance polyimide composite films prepared by homogeneity reinforcement of electrospun nanofibers. Compos Sci Technol 71:1556–1562
135. Chen D, Liu T, Zhou X, Tjiu WC, Hou H (2009) Electrospinning fabrication of high strength and toughness polyimide nanofiber membranes containing multiwalled carbon nanotubes. J Phys Chem B 113:9741–9748
136. Ko F, Gogotsi Y, Ali A, Naguib N, Ye H, Yang G et al (2003) Electrospinning of continuous carbon nanotube-filled nanofiber yarns. Adv Mater 15:1161–1165
137. Heikkilä P, Harlin A (2009) Electrospinning of polyacrylonitrile (PAN) solution: effect of conductive additive and filler on the process. Express Polymer Lett 3:437–445
138. Prilutsky S, Zussman E, Cohen Y (2008) The effect of embedded carbon nanotubes on the morphological evolution during the carbonization of poly (acrylonitrile) nanofibers. Nanotechnology 19:165603
139. Pilehrood MK, Heikkilä P, Harlin A (2012) Preparation of carbon nanotube embedded in polyacrylonitrile (pan) nanofibre composites by electrospinning process. AUTEX Res J 12:1–6
140. Kaur N, Kumar V, Dhakate SR (2016) Synthesis and characterization of multiwalled CNT—PAN based composite carbon nanofibers via electrospinning. SpringerPlus 5:483
141. Song Y, Sun Z, Xu L, Shao Z (2017) Preparation and characterization of highly aligned carbon nanotubes/polyacrylonitrile composite nanofibers. Polymers 9:1

142. Mathew G, Hong J, Rhee J, Lee H, Nah C (2005) Preparation and characterization of properties of electrospun poly (butylene terephthalate) nanofibers filled with carbon nanotubes. Polym Testing 24:712–717
143. Sundaray B, Subramanian V, Natarajan TS, Krishnamurthy K (2006) Electrical conductivity of a single electrospun fiber of poly(methyl methacrylate) and multiwalled carbon nanotube nanocomposite. Appl Phys Lett 88:143114
144. Sung JH, Kim HS, Jin H-J, Choi HJ, Chin I-J (2004) Nanofibrous membranes prepared by multiwalled carbon nanotube/poly (methyl methacrylate) composites. Macromolecules 37:9899–9902
145. Zhu J, Wei S, Chen X, Karki AB, Rutman D, Young DP et al (2010) Electrospun polyimide nanocomposite fibers reinforced with core—shell Fe–FeO nanoparticles. J Phys Chem C 114:8844–8850
146. Xu W, Ding Y, Jiang S, Chen L, Liao X, Hou H (2014) Polyimide/BaTiO3/MWCNTs three-phase nanocomposites fabricated by electrospinning with enhanced dielectric properties. Mater Lett 135:158–161
147. Li X, Wang N, Fan G, Yu J, Gao J, Sun G et al (2015) Electreted polyetherimide–silica fibrous membranes for enhanced filtration of fine particles. J Colloid Interface Sci 439:12–20
148. Wan H, Wang N, Yang J, Si Y, Chen K, Ding B et al (2014) Hierarchically structured polysulfone/titania fibrous membranes with enhanced air filtration performance. J Colloid Interface Sci 417:18–26
149. Yang T, Yang H, Zhen SJ, Huang CZ (2015) Hydrogen-bond-mediated in situ fabrication of AgNPs/agar/PAN electrospun nanofibers as reproducible SERS substrates. ACS Appl Mater Interfaces 7:1586–1594
150. Carlberg B, Ye LL, Liu J (2011) Surface-confined synthesis of silver nanoparticle composite coating on electrospun polyimide nanofibers. Small (Weinheim an der Bergstrasse, Germany) 7:3057–3066
151. Bai L, Jia L, Yan Z, Liu Z, Liu Y (2018) Plasma-assisted fabrication of nanoparticle-decorated electrospun nanofibers. J Taiwan Inst Chem Eng 82:360–366
152. Gibson I (2017) The changing face of additive manufacturing. J Manuf Technol Manage 28:10–17
153. Lee J-Y, An J, Chua CK (2017) Fundamentals and applications of 3D printing for novel materials. Appl Mater Today 7:120–133
154. Stansbury JW, Idacavage MJ (2016) 3D printing with polymers: challenges among expanding options and opportunities. Dent Mater 32:54–64
155. Wang X, Jiang M, Zhou Z, Gou J, Hui D (2017) 3D printing of polymer matrix composites: a review and prospective. Compos B Eng 110:442–458
156. Berman B (2012) 3-D printing: the new industrial revolution. Bus Horiz 55:155–162
157. Gibson I, Rosen D, Stucker B (2010) Additive manufacturing technologies, 3D printing, rapid prototyping, and direct digital manufacturing. Springer, New York
158. Campbell I, Bourell D, Gibson I (2012) Additive manufacturing: rapid prototyping comes of age. Rapid Prototyp J 18:255–258
159. de Leon AC, Chen Q, Palaganas NB, Palaganas JO, Manapat J, Advincula RC (2016) High performance polymer nanocomposites for additive manufacturing applications. React Funct Polym 103:141–155
160. Okada A, Kawasumi M, Usuki A, Kojima Y, Kurauchi T, Kamigaito O (1989) Nylon 6–clay hybrid. MRS Proceed 171:45
161. Kojima Y, Usuki A, Kawasumi M, Okada A, Kurauchi T, Kamigaito O (1993) Synthesis of nylon 6-clay hybrid by montmorillonite intercalated with ϵ-caprolactam. J Polym Sci Part A Polym Chem 31:983–986
162. Kojima Y, Usuki A, Kawasumi M, Okada A, Kurauchi T, Kamigaito O (1993) One-pot synthesis of nylon 6–clay hybrid. J Polym Sci Part A Polym Chem 31:1755–1758
163. Usuki A, Koiwai A, Kojima Y, Kawasumi M, Okada A, Kurauchi T et al (1995) Interaction of nylon 6-clay surface and mechanical properties of nylon 6-clay hybrid. J Appl Polym Sci 55:119–123

164. Kojima Y, Usuki A, Kawasumi M, Okada A, Kurauchi T, Kamigaito O (1993) Sorption of water in nylon 6-clay hybrid. J Appl Polym Sci 49:1259–1264
165. Reichert P, Kressler J, Thomann R, Müllhaupt R, Stöppelmann G (1998) Nanocomposites based on a synthetic layer silicate and polyamide-12. Acta Polym 49:116–123
166. Maxfield M, Christiani BR, Murthy SN, Tuller H (1995) Nanocomposites of gamma phase polymers containing inorganic particulate material. Google Patents
167. Liu L, Qi Z, Zhu X (1999) Studies on nylon 6/clay nanocomposites by melt-intercalation process. J Appl Polym Sci 71:1133–1138
168. Cho JW, Paul DR (2001) Nylon 6 nanocomposites by melt compounding. Polymer 42:1083–1094
169. Phang IY, Liu T, Mohamed A, Pramoda KP, Chen L, Shen L et al (2005) Morphology, thermal and mechanical properties of nylon 12/organoclay nanocomposites prepared by melt compounding. Polym Int 54:456–464
170. Zhang Y, Yang JH, Ellis TS, Shi J (2006) Crystal structures and their effects on the properties of polyamide 12/clay and polyamide 6–polyamide 66/clay nanocomposites. J Appl Polym Sci 100:4782–4794
171. Mohanty S, Nayak SK (2007) Effect of clay exfoliation and organic modification on morphological, dynamic mechanical, and thermal behavior of melt-compounded polyamide-6 nanocomposites. Polym Compos 28:153–162
172. Alexandre B, Marais S, Langevin D, Médéric P, Aubry T (2006) Nanocomposite-based polyamide 12/montmorillonite: relationships between structures and transport properties. Desalination 199:164–166
173. Hocine NA, Médéric P, Aubry T (2008) Mechanical properties of polyamide-12 layered silicate nanocomposites and their relations with structure. Polym Testing 27:330–339
174. Kim GM, Lee DH, Hoffmann B, Kressler J, Stöppelmann G (2001) Influence of nanofillers on the deformation process in layered silicate/polyamide-12 nanocomposites. Polymer 42:1095–1100
175. Liu X, Wu Q (2002) Polyamide 66/clay nanocomposites via melt intercalation. Macromol Mater Eng 287:180–186
176. Beatrice CAGa, Santos CRd, Branciforti MC, Bretas RES (2012) Nanocomposites of polyamide 6/residual monomer with organic-modified montmorillonite and their nanofibers produced by electrospinning. Mater Res 15:611–621
177. Wu H, Krifa M, Koo JH (2014) Flame retardant polyamide 6/nanoclay/intumescent nanocomposite fibers through electrospinning. Text Res J 84:1106–1118
178. Kim G-M, Michler GH, Ania F, Calleja FJB (2007) Temperature dependence of polymorphism in electrospun nanofibres of PA6 and PA6/clay nanocomposite. Polymer 48:4814–4823
179. Saeed K, Park SY (2012) Effect of nanoclay on the thermal, mechanical, and crystallization behavior of nanofiber webs of nylon-6. Polym Compos 33:192–195
180. Ristolainen N, Heikkilä P, Harlin A, Seppälä J (2006) Poly(vinyl alcohol) and polyamide-66 nanocomposites prepared by electrospinning. Macromol Mater Eng 291:114–122
181. Cai Y, Tao D, Wei Q, Gao W (2008) Preparation, surface morphology, and thermal stability of polyamide 6 composite nanofibres by electrospinning. Polym Polym Compos 16:605–610
182. Li Q, Gao D, Wei Q, Ge M, Liu W, Wang L et al (2010) Thermal stability and crystalline of electrospun polyamide 6/organo-montmorillonite nanofibers. J Appl Polym Sci 117:1572–1577
183. Li Q, Wei Q, Wu N, Cai Y, Gao W (2008) Structural characterization and dynamic water adsorption of electrospun polyamide6/montmorillonite nanofibers. J Appl Polym Sci 107:3535–3540
184. Fong H, Liu W, Wang C-S, Vaia RA (2002) Generation of electrospun fibers of nylon 6 and nylon 6-montmorillonite nanocomposite. Polymer 43:775–780
185. Wang Y, Shi Y, Huang S (2005) Selective laser sintering of polyamide-rectorite composite. Proceed Inst Mech Eng, Part L: J Mater Design Appl 219:11–15
186. Jain PK, Pandey PM, Rao P (2010) Selective laser sintering of clay-reinforced polyamide. Polym Compos 31:732–743

187. Koo J, Lao S, Ho W, Ngyuen K, Cheng J, Pilato L et al (2006) Polyamide nanocomposites for selective laser sintering. In: Proceedings of the SFF Symposium, Austin, pp 392–409
188. Kim J, Creasy TS (2004) Selective laser sintering characteristics of nylon 6/clay-reinforced nanocomposite. Polym Testing 23:629–636
189. Yan CZCCZ (2011) An organically modified montmorillonite/nylon-12 composite powder for selective laser sintering. Rapid Prototyp J 17:28–36
190. Vaia RA, Ishii H, Giannelis EP (1993) Synthesis and properties of two-dimensional nanostructures by direct intercalation of polymer melts in layered silicates. Chem Mater 5:1694–1696
191. Follain N, Alexandre B, Chappey C, Colasse L, Médéric P, Marais S (2016) Barrier properties of polyamide 12/montmorillonite nanocomposites: effect of clay structure and mixing conditions. Compos Sci Technol 136:18–28
192. Bourbigot S, Bras ML, Dabrowski F, Gilman JW, Kashiwagi T (2000) PA-6 clay nanocomposite hybrid as char forming agent in intumescent formulations. Fire Mater 24:201–208
193. Devaux E, Bourbigot S, Achari AE (2002) Crystallization behavior of PA-6 clay nanocomposite hybrid. J Appl Polym Sci 86:2416–2423
194. Fornes T, Paul D (2004) Structure and properties of nanocomposites based on nylon-11 and-12 compared with those based on nylon-6. Macromolecules 37:7698–7709
195. Mehrabzadeh M, Kamal MR (2004) Melt processing of PA-66/clay, HDPE/clay and HDPE/PA-66/clay nanocomposites. Polym Eng Sci 44:1152–1161
196. Shen L, Phang IY, Chen L, Liu T, Zeng K (2004) Nanoindentation and morphological studies on nylon 66 nanocomposites, I: effect of clay loading. Polymer 45:3341–3349
197. Qin H, Su Q, Zhang S, Zhao B, Yang M (2003) Thermal stability and flammability of polyamide 66/montmorillonite nanocomposites. Polymer 44:7533–7538
198. Risite H, Mabrouk KE, Bousmina M, Fassi-Fehri O (2016) Role of polyamide 11 interaction with clay and modifier on thermal, rheological and mechanical properties in polymer clay nanocomposites. J Nanosci Nanotechnol 16:7584–7593
199. He X, Yang J, Zhu L, Wang B, Sun G, Lv P et al (2006) Morphology and melt rheology of nylon 11/clay nanocomposites. J Appl Polym Sci 102:542–549
200. Lecouvet B, Gutierrez JG, Sclavons M, Bailly C (2011) Structure–property relationships in polyamide 12/halloysite nanotube nanocomposites. Polym Degrad Stab 96:226–235
201. Médéric P, Razafinimaro T, Aubry T (2006) Influence of melt-blending conditions on structural, rheological, and interfacial properties of polyamide-12 layered silicate nanocomposites. Polym Eng Sci 46:986–994
202. Hu Y, Wang S, Ling Z, Zhuang Y, Chen Z, Fan W (2003) Preparation and combustion properties of flame retardant nylon 6/montmorillonite nanocomposite. Macromol Mater Eng 288:272–276
203. Gilman J (1999) Flammability and thermal stability studies of polymer layered-silicate (clay) nanocomposites. Appl Clay Sci 15:31–49
204. Bourbigot S, Samyn F, Turf T, Duquesne S (2010) Nanomorphology and reaction to fire of polyurethane and polyamide nanocomposites containing flame retardants. Polym Degrad Stab 95:320–326
205. Agarwal A, Raheja A, Natarajan TS, Chandra TS (2014) Effect of electrospun montmorillonite-nylon 6 nanofibrous membrane coated packaging on potato chips and bread. Innov Food Sci Emerg Technol 26:424–430
206. Bhattacharyya AR, Pötschke P, Abdel-Goad M, Fischer D (2004) Effect of encapsulated SWNT on the mechanical properties of melt mixed PA12/SWNT composites. Chem Phys Lett 392:28–33
207. Nie M, Xia H, Wu J (2013) Preparation and characterization of poly(styrene-*co*-butyl acrylate)-encapsulated single-walled carbon nanotubes under ultrasonic irradiation. Iran Polym J 22:409–416
208. Xia H, Wang Q, Qiu G (2003) Polymer-encapsulated carbon nanotubes prepared through ultrasonically initiated in situ emulsion polymerization. Chem Mater 15:3879–3886
209. Zhao C, Hu G, Justice R, Schaefer DW, Zhang S, Yang M et al (2005) Synthesis and characterization of multi-walled carbon nanotubes reinforced polyamide 6 via in situ polymerization. Polymer 46:5125–5132

210. Xu C, Jia Z, Wu D, Han Q, Meek T (2006) Fabrication of nylon-6/carbon nanotube composites. J Electron Mater 35:954–957
211. Scaffaro R, Maio A, Tito AC (2012) High performance PA6/CNTs nanohybrid fibers prepared in the melt. Compos Sci Technol 72:1918–923
212. Zhu X-D, Zang C-G, Jiao Q-J (2014) High electrical conductivity of nylon 6 composites obtained with hybrid multiwalled carbon nanotube/carbon fiber fillers. J Appl Polym Sci 131
213. Zeng H, Gao C, Wang Y, Watts PC, Kong H, Cui X et al (2006) In situ polymerization approach to multiwalled carbon nanotubes-reinforced nylon 1010 composites: mechanical properties and crystallization behavior. Polymer 47:113–122
214. Krause B, Pötschke P, Häußler L (2009) Influence of small scale melt mixing conditions on electrical resistivity of carbon nanotube-polyamide composites. Compos Sci Technol 69:1505–1515
215. Ha H, Kim SC, Ha K (2010) Morphology and properties of polyamide/multi-walled carbon nanotube composites. Macromol Res 18:660–667
216. Wang M, Wang W, Liu T, Zhang W-D (2008) Melt rheological properties of nylon 6/multi-walled carbon nanotube composites. Compos Sci Technol 68:2498–2502
217. Faghihi M, Shojaei A, Bagheri R (2015) Characterization of polyamide 6/carbon nanotube composites prepared by melt mixing-effect of matrix molecular weight and structure. Compos B Eng 78:50–64
218. Sahoo NG, Cheng HKF, Cai J, Li L, Chan SH, Zhao J et al (2009) Improvement of mechanical and thermal properties of carbon nanotube composites through nanotube functionalization and processing methods. Mater Chem Phys 117:313–320
219. Chatterjee S, Nüesch F, Chu B (2013) Crystalline and tensile properties of carbon nanotube and graphene reinforced polyamide 12 fibers. Chem Phys Lett 557:92–96
220. Perrot C, Piccione PM, Zakri C, Gaillard P, Poulin P (2009) Influence of the spinning conditions on the structure and properties of polyamide 12/carbon nanotube composite fibers. J Appl Polym Sci 114:3515–3523
221. Latko P, Kolbuk D, Kozera R, Boczkowska A (2016) Microstructural characterization and mechanical properties of PA11 nanocomposite fibers. J Mater Eng Perform 25:68–75
222. Sandler J, Pegel S, Cadek M, Gojny F, Van Es M, Lohmar J et al (2004) A comparative study of melt spun polyamide-12 fibres reinforced with carbon nanotubes and nanofibres. Polymer 45:2001–2015
223. Havel M, Behler K, Korneva G, Gogotsi Y (2008) Transparent thin films of multiwalled carbon nanotubes self-assembled on polyamide 11 nanofibers. Adv Func Mater 18:2322–2327
224. Bai J, Goodridge RD, Hague RJ, Song M (2013) Improving the mechanical properties of laser-sintered polyamide 12 through incorporation of carbon nanotubes. Polym Eng Sci 53:1937–1946
225. Bai J, Goodridge RD, Hague RJ, Song M, Okamoto M (2014) Influence of carbon nanotubes on the rheology and dynamic mechanical properties of polyamide-12 for laser sintering. Polym Testing 36:95–100
226. Bai J, Goodridge RD, Yuan S, Zhou K, Chua CK, Wei J (2015) Thermal influence of CNT on the polyamide 12 nanocomposite for selective laser sintering. Molecules 20:19041–19050
227. Zhou L, Liu H, Zhang X (2015) Graphene and carbon nanotubes for the synergistic reinforcement of polyamide 6 fibers. J Mater Sci Full Set Incl J Mater Sci Lett 50:2797–2805
228. Socher R, Krause B, Hermasch S, Wursche R, Pötschke P (2011) Electrical and thermal properties of polyamide 12 composites with hybrid fillers systems of multiwalled carbon nanotubes and carbon black. Compos Sci Technol 71:1053–1059
229. Socher R, Krause B, Boldt R, Hermasch S, Wursche R, Pötschke P (2011) Melt mixed nano composites of PA12 with MWNTs: influence of MWNT and matrix properties on macrodispersion and electrical properties. Compos Sci Technol 71:306–314
230. Salmoria GV, Paggi RA, Lago A, Beal VE (2011) Microstructural and mechanical characterization of PA12/MWCNTs nanocomposite manufactured by selective laser sintering. Polym Testing 30:611–615

231. Goodridge R, Shofner M, Hague R, McClelland M, Schlea M, Johnson R et al (2011) Processing of a polyamide-12/carbon nanofibre composite by laser sintering. Polym Testing 30:94–100
232. Chatterjee S, Nüesch F, Chu BT (2011) Comparing carbon nanotubes and graphene nanoplatelets as reinforcements in polyamide 12 composites. Nanotechnology 22:275714
233. Versavaud S, Regnier G, Gouadec G, Vincent M (2014) Influence of injection molding on the electrical properties of polyamide 12 filled with multi-walled carbon nanotubes. Polymer 55:6811–6818
234. Song KH, Choi CH, Choi C, Lee Mh, Lee ST (2013) Method of manufacturing polyamide and carbon nanotube composite using high shearing process. Google Patents
235. Huang S, Wang M, Liu T, Zhang WD, Tjiu WC, He C et al (2009) Morphology, thermal, and rheological behavior of nylon 11/multi-walled carbon nanotube nanocomposites prepared by melt compounding. Polym Eng Sci 49:1063–1068
236. Yang Z, Huang S, Liu T (2011) Crystallization behavior of polyamide 11/multiwalled carbon nanotube composites. J Appl Polym Sci 122:551–560
237. Bhattacharyya AR, Bose S, Kulkarni AR, Pötschke P, Häußler L, Fischer D et al (2007) Styrene maleic anhydride copolymer mediated dispersion of single wall carbon nanotubes in polyamide 12: crystallization behavior and morphology. J Appl Polym Sci 106:345–353
238. Chiu F-C, Kao G-F (2012) Polyamide 46/multi-walled carbon nanotube nanocomposites with enhanced thermal, electrical, and mechanical properties. Compos A Appl Sci Manuf 43:208–218
239. Mago G, Kalyon DM, Fisher FT (2011) Nanocomposites of polyamide-11 and carbon nanostructures: Development of microstructure and ultimate properties following solution processing. J Polym Sci Part B Polym Phys 49:1311–1321
240. Fu X, Yao C, Yang G (2015) Recent advances in graphene/polyamide 6 composites: a review. RSC Adv 5:61688–61702
241. O'Neill A, Bakirtzis D, Dixon D (2014) Polyamide 6/Graphene composites: the effect of in situ polymerisation on the structure and properties of graphene oxide and reduced graphene oxide. Eur Polymer J 59:353–362
242. Zhang X, Fan X, Li H, Yan C (2012) Facile preparation route for graphene oxide reinforced polyamide 6 composites via in situ anionic ring-opening polymerization. J Mater Chem 22:24081–24091
243. Xu Z, Gao C (2010) In situ polymerization approach to graphene-reinforced nylon-6 composites. Macromolecules 43:6716–6723
244. Ding P, Su S, Song N, Tang S, Liu Y, Shi L (2014) Highly thermal conductive composites with polyamide-6 covalently-grafted graphene by an in situ polymerization and thermal reduction process. Carbon 66:576–584
245. Nguyễn L, Choi S-M, Kim D-H, Kong N-K, Jung P-J, Park S-Y (2014) Preparation and characterization of nylon 6 compounds using the nylon 6-grafted GO. Macromol Res 22:257–263
246. Yan D, Zhang H-B, Jia Y, Hu J, Qi X-Y, Zhang Z et al (2012) Improved electrical conductivity of polyamide 12/graphene nanocomposites with maleated polyethylene-octene rubber prepared by melt compounding. ACS Appl Mater Interfaces 4:4740–4745
247. Chiu F-C, Huang IN (2012) Phase morphology and enhanced thermal/mechanical properties of polyamide 46/graphene oxide nanocomposites. Polym Testing 31:953–962
248. Fabiola NP, Ana LMH, Carlos VS (2016) Carbon nanotube and graphene based polyamide electrospun nanocomposites: a review. J Nanomater
249. Ji X, Xu Y, Zhang W, Cui L, Liu J (2016) Review of functionalization, structure and properties of graphene/polymer composite fibers. Compos A 87:29–45
250. Tyan H-L, Liu Y-C, Wei K-H (1999) Thermally and mechanically enhanced clay/polyimide nanocomposite via reactive organoclay. Chem Mater 11:1942–1947
251. Yano K, Usuki A, Okada A (1997) Synthesis and properties of polyimide-clay hybrid films. J Polym Sci Part A Polym Chem 35:2289–2294
252. Leu C-M, Wu Z-W, Wei K-H (2002) Synthesis and properties of covalently bonded layered silicates/polyimide (BTDA-ODA) nanocomposites. Chem Mater 14:3016–3021

253. Yano K, Usuki A, Okada A, Kurauchi T, Kamigaito O (1993) Synthesis and properties of polyimide–clay hybrid. J Polym Sci Part A Polym Chem 31:2493–2498
254. Zhang Y-H, Dang Z-M, Fu S-Y, Xin JH, Deng J-G, Wu J et al (2005) Dielectric and dynamic mechanical properties of polyimide–clay nanocomposite films. Chem Phys Lett 401:553–557
255. Kakiage M, Ando S (2011) Effects of dispersion and arrangement of clay on thermal diffusivity of polyimide-clay nanocomposite film. J Appl Polym Sci 119:3010–3018
256. Park JS, Chang JH (2009) Colorless polyimide nanocomposite films with pristine clay: thermal behavior, mechanical property, morphology, and optical transparency. Polym Eng Sci 49:1357–1365
257. Tang Q-Y, Chan YC, Zhang K (2011) Fast response resistive humidity sensitivity of polyimide/multiwall carbon nanotube composite films. Sens Actuators B: Chem 152:99–106
258. Jiang Y, Yu S, Li J, Jia L, Wang C (2013) Improvement of sensitive Ni(OH)2 nonenzymatic glucose sensor based on carbon nanotube/polyimide membrane. Carbon 63:367–375
259. Yoo K-P, Lim L-T, Min N-K, Lee MJ, Lee CJ, Park C-W (2010) Novel resistive-type humidity sensor based on multiwall carbon nanotube/polyimide composite films. Sens Actuators B: Chem 145:120–125
260. Kim BS, Bae SH, Park Y-H, Kim J-H (2007) Preparation and characterization of polyimide/carbon-nanotube composites. Macromol Res 15:357–362
261. Guo Y, Xu G, Yang X, Ruan K, Ma T, Zhang Q, et al (2018) Significantly enhanced and precisely modeled thermal conductivity in polyimide nanocomposites by chemically modified graphene via in-situ polymerization and electrospinning-hot press technology. J Mater Chem C
262. Wu G, Cheng Y, Wang Z, Wang K, Feng A (2017) In situ polymerization of modified graphene/polyimide composite with improved mechanical and thermal properties. J Mater Sci Mater Electron 28:576–581
263. Liu J, Tian G, Qi S, Wu Z, Wu D (2014) Enhanced dielectric permittivity of a flexible three-phase polyimide–graphene–BaTiO3 composite material. Mater Lett 124:117–119
264. Raju MP, Alam S (2009) Synthesis of polyimide nanocomposites. J Macromol Sci Part A 46:1136–1141
265. Liu P, Tran TQ, Fan Z, Duong HM (2015) Formation mechanisms and morphological effects on multi-properties of carbon nanotube fibers and their polyimide aerogel-coated composites. Compos Sci Technol 117:114–120
266. Fan W, Zuo L, Zhang Y, Chen Y, Liu T (2018) Mechanically strong polyimide/carbon nanotube composite aerogels with controllable porous structure. Compos Sci Technol 156:186–191
267. Wang K, Chang Y-H, Zhang C, Wang B (2016) Conductive-on-demand: tailorable polyimide/carbon nanotube nanocomposite thin film by dual-material aerosol jet printing. Carbon 98:397–403
268. Wu DC, Shen L, Low JE, Wong SY, Li X, Tjiu WC et al (2010) Multi-walled carbon nanotube/polyimide composite film fabricated through electrophoretic deposition. Polymer 51:2155–2160
269. Wang HW, Dong RX, Liu CL, Chang HY (2007) Effect of clay on properties of polyimide-clay nanocomposites. J Appl Polym Sci 104:318–324
270. Agag T, Koga T, Takeichi T (2001) Studies on thermal and mechanical properties of polyimide–clay nanocomposites. Polymer 42:3399–3408
271. Mya KY, Wang K, Chen L, Lin TT, Pallathadka PK, Pan J et al (2008) The effect of nanofiller on the thermomechanical properties of polyimide/clay nanocomposites. Macromol Chem Phys 209:643–650
272. Yu YH, Yeh JM, Liou SJ, Chen CL, Liaw DJ, Lu HY (2004) Preparation and properties of polyimide–clay nanocomposite materials for anticorrosion application. J Appl Polym Sci 92:3573–3582
273. Tyan HL, Wei KH, Hsieh TE (2000) Mechanical properties of clay–polyimide (BTDA–ODA) nanocomposites via ODA-modified organoclay. J Polym Sci Part B Polym Phys 38:2873–2878
274. Tyan HL, Wu CY, Wei KH (2001) Effect of montmorillonite on thermal and moisture absorption properties of polyimide of different chemical structures. J Appl Polym Sci 81:1742–1747

275. Tyan H-L, Liu Y-C, Wei K-H (1999) Enhancement of imidization of poly (amic acid) through forming poly (amic acid)/organoclay nanocomposites. Polymer 40:4877–4886
276. Tyan H-L, Leu C-M, Wei K-H (2001) Effect of reactivity of organics-modified montmorillonite on the thermal and mechanical properties of montmorillonite/polyimide nanocomposites. Chem Mater 13:222–226
277. Mo TC, Wang HW, Chen SY, Yeh YC (2008) Synthesis and characterization of polyimide/multi-walled carbon nanotube nanocomposites. Polym Compos 29:451–457
278. Cai H, Yan F, Xue Q (2004) Investigation of tribological properties of polyimide/carbon nanotube nanocomposites. Mater Sci Eng, A 364:94–100
279. Deshmukh S, Ounaies Z (2009) Single walled carbon nanotube (SWNT)–polyimide nanocomposites as electrostrictive materials. Sens Actuators A 155:246–252
280. Ye X, Liu X, Yang Z, Wang Z, Wang H, Wang J et al (2016) Tribological properties of fluorinated graphene reinforced polyimide composite coatings under different lubricated conditions. Compos A Appl Sci Manuf 81:282–288
281. Min C, Nie P, Song H-J, Zhang Z, Zhao K (2014) Study of tribological properties of polyimide/graphene oxide nanocomposite films under seawater-lubricated condition. Tribol Int 80:131–140
282. Zuo L, Fan W, Zhang Y, Zhang L, Gao W, Huang Y et al (2017) Graphene/montmorillonite hybrid synergistically reinforced polyimide composite aerogels with enhanced flame-retardant performance. Compos Sci Technol 139:57–63
283. Zhang L-B, Wang J-Q, Wang H-G, Xu Y, Wang Z-F, Li Z-P et al (2012) Preparation, mechanical and thermal properties of functionalized graphene/polyimide nanocomposites. Compos A Appl Sci Manuf 43:1537–1545
284. Luong ND, Hippi U, Korhonen JT, Soininen AJ, Ruokolainen J, Johansson L-S et al (2011) Enhanced mechanical and electrical properties of polyimide film by graphene sheets via in situ polymerization. Polymer 52:5237–5242
285. Tseng IH, Liao Y-F, Chiang J-C, Tsai M-H (2012) Transparent polyimide/graphene oxide nanocomposite with improved moisture barrier property. Mater Chem Phys 136:247–253
286. Tsai M-H, Chang C-J, Lu H-H, Liao Y-F, Tseng IH (2013) Properties of magnetron-sputtered moisture barrier layer on transparent polyimide/graphene nanocomposite film. Thin Solid Films 544:324–330
287. Liu M, Du Y, Miao Y-E, Ding Q, He S, Tjiu WW et al (2015) Anisotropic conductive films based on highly aligned polyimide fibers containing hybrid materials of graphene nanoribbons and carbon nanotubes. Nanoscale 7:1037–1046
288. Li H, Dai S, Miao J, Wu X, Chandrasekharan N, Qiu H et al (2018) Enhanced thermal conductivity of graphene/polyimide hybrid film via a novel "molecular welding" strategy. Carbon 126:319–327

Chapter 3
Rubber Nanocomposites: Processing, Structure–Property Relationships, Applications, Challenges, and Future Trends

Reza Salehiyan and Suprakas Sinha Ray

Abstract This chapter discusses the roles of different nanoparticle types such as clays, CNTs, and graphene-based materials in the rubber manufacturing processes. It is shown that nanoparticles not only reinforce rubber matrices, but they can also accelerate cross-linking reactions during vulcanization/curing and save energy. Further, the degree of reinforcement depends strongly on the dispersion of the nanoparticles within the nanocomposites. Accordingly, different rubber fabrication technologies can give rise to different dispersion states, and, hence, different final properties. Often, nanocomposites prepared via solution mixing or in situ polymerization exhibit better dispersion than those prepared via the melt-intercalation method. However, the environmental and cost issues associated with the solvents used in these methods limit their widespread and large-scale use. Finally, this chapter shows that the morphology of the nanoparticles (i.e., segregated structures) within the matrix can enhance properties such as electrical conductivity and permeability more effectively than dispersion itself (i.e., non-segregated structures).

R. Salehiyan · S. Sinha Ray (✉)
DST-CSIR National Centre for Nanostructured Materials, Council for Scientific and Industrial Research, Pretoria 0001, South Africa
e-mail: rsuprakas@csir.co.za

R. Salehiyan
e-mail: rsalehiyan@csir.co.za

S. Sinha Ray
Department of Applied Chemistry, University of Johannesburg, Doornfontein 2028, Johannesburg, South Africa
e-mail: ssinharay@uj.ac.za

S. Sinha Ray (ed.), *Processing of Polymer-based Nanocomposites*, Springer Series in Materials Science 278, https://doi.org/10.1007/978-3-319-97792-8_3

3.1 Introduction

From the industrial viewpoint, polymer nanocomposites are of significant interest owing to their valuable characteristics that are related to their large specific area per unit volume, and make them attractive building blocks for the development of new materials with outstanding properties. Nanoparticles incorporated into a polymeric medium can act as reinforcement agents to enhance the performance of the final polymer in terms of mechanical, thermal, barrier, and other properties. The enhancement effect, however, is proportional to the concentration, dispersion, and distribution of the nanoparticle fillers within the polymer. Elastomer/rubber nanocomposites have been used previously in rubber industries to achieve similar beneficial effects. By definition, rubbers are massive macromolecules with negative glass transition temperatures ($T_g << 0$) and randomly coiled long chains that can undergo uncoiling–recoiling in response to stretching–release force cycles [1, 2]. Advances in nanotechnology are expected to give manufacturers the opportunity to develop rubber nanocomposites with enhanced properties, e.g., tear strength, electrical conductivity, and anti-abrasion characteristics, and in which load transfer from the matrix is enhanced and crack propagation is hampered. Such improvements, however, require good nanoparticle dispersion and adhesion to the rubber matrix [2–5]. To date, nanoparticles of different sizes and shapes (clay, carbon black, nanotubes, silica, etc.) have been incorporated into rubbers in order to attain such enhancements [2, 6]. The key parameters influencing the properties of the resulting nanocomposites are the degree of dispersion/distribution and the type and degree of surface modification of the particles. In this chapter, the structure–property relationships of rubber nanocomposites filled with different nanoparticle types will be discussed with respect to the different methods of fabrication, as well as the associated challenges and the end-use applicability of the materials. Finally, this chapter will conclude with a brief statement of future trends.

3.2 Rubber Fabrication

Generally, the process of rubber manufacturing involves three steps, namely, compounding, molding, and vulcanization. In the compounding step, rubber and other fillers such as anti-oxidants, accelerants, and stabilizers are mixed together using a two-roll mill or internal mixers and the mixture is transferred to a mold that determines the shape of the final product. The material is subsequently cured, and vulcanization occurs as the mixture is maintained in the mold at high temperatures [2]. Vulcanization is an inevitable step in the rubber industry as rubber macromolecules undergo cross-linking reactions in the presence of reactive agents such as sulfurs or peroxides to create elastomers with enhanced elasticity [1]. In order to produce rubber materials with efficient properties, it is necessary to optimize the mixing time, curing time and temperature, and the order of filler incorporations.

3.3 Rubber Nanocomposite Manufacturing

Taking into consideration the three steps involved in the production of rubbers, nanoparticles can be incorporated into the resulting material as reinforcement agents using four main techniques.

3.3.1 *Melt Compounding*

It can be said that the melt mixing of polymer nanocomposites is one of the most popular, commercially available techniques in the rubber manufacturing industry owing to its solvent-free and cost-effective characteristics [2, 6–8]. In this method, nanoparticles are added to the rubber matrix while the melt is being mixed inside a mixer. The final quality of the rubber nanocomposite is related to the extent of dispersion of the nanoparticles, and, in the case of nanoclays, to the degree of intercalation/exfoliation. In other words, ensuring that the nanoparticles are well dispersed and the nanoclays highly exfoliated can enhance the properties of the resulting rubber nanocomposites remarkably [4, 5, 8]. However, attaining such ideal dispersion/degree of exfoliation is rather challenging. The surface energies and polymer-particle adhesion must both be sufficient to facilitate nanoparticle dispersion and nanoclay delamination. For that reason, nanoparticles are compatibilized with the matrix by means of specific surface modifications in order to facilitate improved polymer–particle interactions. In addition, application of force can be used to break down nanoparticle aggregates into smaller bundles, enhancing thus polymer adsorption onto the surface of the nanoparticles and resulting in their disassociation via a so-called "adsorption–shear" mechanism [5]. Moreover, the applied shear force can break down anisotropic particles (long aspect ratio) into particles with shorter lengths, thereby facilitating increased polymer adsorption onto the nanoparticles and simultaneously leading to improved stress dissipation. It should be noted, however, that cross-linking reactions occurring during the vulcanizing process can also affect the dispersion of nanoparticles [6]. This curing reaction would increase the molecular weights and in turn the viscosity of the rubber. Hence, further movement of nanoparticles within the cross-linked rubber could be limited. More details of the effects of nanoparticles on the vulcanization process of the rubbers is discussed in Sect. 3.4.

3.3.2 *Solution Mixing*

Another method used for the production of rubber nanocomposites is solvent mixing (Fig. 3.1). Generally, in this method, nanoparticles are dispersed in a particular solvent such as THF or DMF and stirred for a certain period of time via ultrasonication or a magnetic bar to produce a stable and uniform suspension of nanoparticles. At

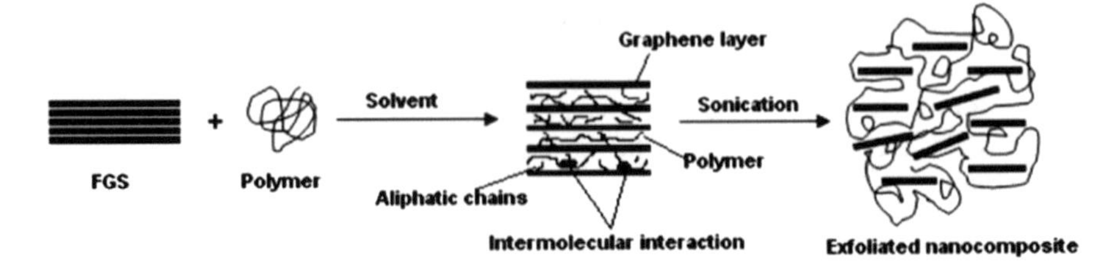

Fig. 3.1 Schematic diagram illustrating the nanocomposite solvent mixing preparation method. FGS represents the foliated graphene sheets. Reprinted with permission from [13]. Copyright 2011 Elsevier Science Ltd.

the same time, rubber is dissolved in another solvent and stirred vigorously. Once the dissolution process is completed, the nanoparticle suspension can gradually be added to the rubber solution while it continues to be stirred vigorously. Subsequently, after sufficient stirring, the solvent is removed in a vacuum oven to yield the desired nanocomposites. Finally, the extracted rubber nanocomposite is cured in a hot press in the presence of vulcanizing agents. In this procedure, the dispersion quality of the solvent-mixed nanocomposites depends on the mixing conditions, polymer-solvent compatibility, and surface functionality of the nanoparticles [9–11]. It has been shown that nanocomposites prepared by the solvent mixing method typically exhibit better nanoparticle dispersions and distributions than those produced by melt compounding [12]. However, this method is limited by the solvents used—in particular, the often slow evaporation rate, which might lead to particle re-aggregation, and the associated environmental issues [6, 12].

3.3.3 *In Situ Polymerization*

In this technique, monomer or oligomer is added to the nanoparticle suspension and in situ polymerization occurs in the presence of nanoparticles [12, 14]. Although certain studies suggest that the nanocomposites prepared using the in situ polymerization method can have better dispersion qualities, this method is not used frequently owing to the limitations related to the need for pre-synthesized monomers, particle purification, and the use of solvents [12, 15].

3.3.4 *Latex Compounding*

Undoubtedly, latex compounding is one of the most well-known and most practiced methods of rubber production [16–20]. The cost–performance ratio has been stated as the most important advantage of the latex compounding method [6]. Some rubbers exist in a liquid state as particles dispersed in an aqueous medium, and such rubbers

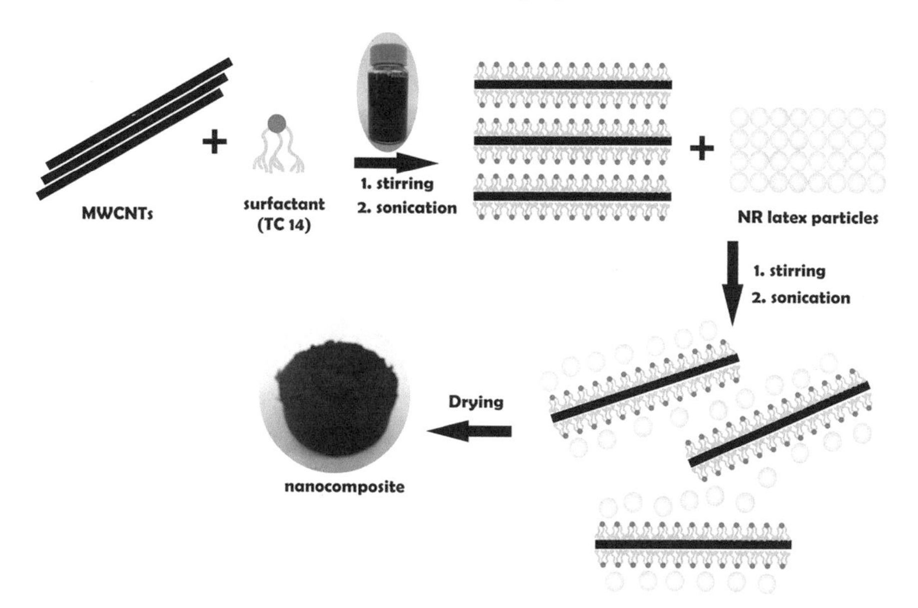

Fig. 3.2 Diagram illustrating the manufacturing of a rubber latex nanocomposite based on NR and multiwalled carbon nanotube (MWCNT). Reprinted with permission from [19]. Copyright 2014 Springer Verlag

are referred to as latex. Latex technology is particularly beneficial for manufacturers, as it can allow facile fabrication of environmentally friendly rubber nanocomposites [20, 21]. For latex based nanocomposites, aqueous solution of dispersed nanoparticles and surfactants are added to the rubber latex and stirred for an appropriate period. The fabrication process of NR/MWCNT nanocomposite is illustrated in Fig. 3.2. The efficiencies of rubber latex nanocomposites are strongly dependent on the concentration of the nanoparticles, size of the rubber latex, interactions and adhesion between the particles and latex, dispersion of the nanoparticles, and the nanoparticle/surfactant ratio [19, 22, 23].

3.4 Vulcanization Characteristics

In the rubber manufacturing process, vulcanization or curing is a crucial step for all materials with the exception of thermoplastic elastomers. Uncured rubbers usually suffer from poor physical and chemical properties. Therefore, their end-use application could be limited [1]. Thus, vulcanization mediated by the introduction of curing agents was introduced to induce cross-linking reactions between the rubber molecules and to improve the properties of the resulting material. The vulcanization process was reported first by Goodyear [24], who applied it to the tire industry, facilitating thus the widespread use of rubbers. The irreversible cross-linking reactions increase the molecular weight of rubbers. This increase, in turn, increases their elasticity and endows them with properties that make them suitable for use in different applications such as tires, seals, gaskets, insulating wire coatings, dampers, and many more. Over time, further improvements in the mechanical, electrical, and barrier properties of rubber nanocomposites have been accomplished by the introduction of nanofillers such as carbon black [25, 26], clay [6], silica [27], nanotubes [28] and others as nano reinforcements [2]. The extent of cross-linking reactions (cross-link density) defines the final performance of the rubber. The target properties may not be achieved if the curing temperatures are too low or too high—under such conditions the curing can yield pre-mature or over-cured (burnt) rubbers, respectively. As a consequence, the knowledge of curing kinetics is of critical importance in the rubber manufacturing process. A very common way of understanding the curing kinetics is to monitor the torque values during the vulcanization process using an oscillating disc rheometer (ODR) [8, 29] or rubber process analyzer (RPA) [30]. It is known that the torque values are proportional to the viscosity and thus also to the cross-linking density of the materials. Therefore, the curing characteristics can be explained according to the torque values at different times using the following definitions [6, 8, 31]. The scorch time (s_t) or induction period, during which the majority of accelerator reactions occur, reflects the time required to increase the torque by 2.26×10^{-2} N m over the minimum [8]. Another very important parameter is the cure time (t_{90}), i.e., the time point at which 90% of the cross-linking reactions have taken place. This parameter also reflects the time required to reach 90% of the maximum possible difference between the maximum and minimum torque [31]. Accordingly, the cure rate index (CRI) can be obtained using (3.1) [30].

$$CRI = 100/(t_{90} - t_s) \tag{3.1}$$

The knowledge of these parameters allows the curing kinetics to be studied at a given temperature and over a specific time. It is also worth noting that, apart from the physical reinforcements associated with nanofiller incorporations, which will be discussed in more detail in Sects. 3.6, 3.7, and 3.8, the nano-sized fillers can also play an important role during the curing process of rubber nanocomposites. It has been shown that the presence of ammonium functional groups on the surface of organo-modified clays facilitates the curing action by their reaction with sulfur. Therefore, these functional groups can act as cross-linking accelerators and increase the *CRI* of the rubber compounds [8, 29–31]. Moreover, the combination of amine groups with benzothiazyl accelerant can have a synergetic effect on the acceleration of the curing process [32]. By contrast, it has been reported that acidic pH and zinc oxide adsorption on the surface of silica particles can impede vulcanization reactions with sulfur [29]. Nanokaolin was found to reduce the optimum curing time of rubbers (NR, BR, SBR and EPDM) more significantly than nano silica particles [27]. Similarly, CNTs can also increase the curing time and impede vulcanization by absorbing accelerator agents during curing. Their high surface area per unit volume provides sufficient physical interaction with rubber molecules, resulting thus in their immobilization. This disruption in the mobility of the rubber chains also contributes to reduced reactivity, as in the case of NR/CNT nanocomposites [33]. Similarly, the addition of a coupling agent to NR/OMMT nanocomposites increases the curing time. Typically, the strong interactions between amine groups present in the clays and the silane groups of the coupling agent hinder the vulcanization process [32]. Despite these results, a decrease in the scorch time upon addition of CNTs into NR has also been observed [34]. The authors of this work related the observed behavior to the excellent thermal conductivity of the CNTs, which could accelerate reactions within an induction period. Previous work has also revealed that CNTs are able to overcome the over-curing reversion associated with carbon black CB filled rubbers [34]. Liu et al. [35] proposed that the curing process of SBR/graphene oxide nanocomposites can occur in two stages, driven by chemical and diffusion controlled reactions. At low content of graphene, oxides can accelerate the curing reactions by reducing the activation energy of the curing process, whereas at higher concentrations of graphene, they can interrupt the diffusion of the curing agents and decrease the cross-link density of the nanocomposites [35, 36]. However, the cross-link density was still higher than that of neat SBR. This result was suggested to stem from the plasticizing effect of alkylamine molecules of the modified GO on the rubber molecules. The copolymerization mechanism proposed by the authors for the reaction between oleylamine double bonds of the modified GO and sulfur in SBR/GO nanocomposites is shown in Fig. 3.3. An alternative mechanism could be envisaged, which involves the reaction between previously reacted hydroxyl and carboxyl groups present in the graphene oxide sheets with sulfur derived from with benzothiazyl accelerants. However, it was shown that excessive amounts of graphene oxide (>1 phr) reduce the rate of vulcanization of NR/graphene oxide nanocomposites to levels below that observed for unfilled NR, and this effect is similar to the hindering effect exerted by high-surface-area graphene oxide sheets on the diffusion of curing agents at high GO concentrations [36]. From these results, it is apparent that fillers can influence the curing process of rubbers in different ways that depend on their surface chemistry, concentration, dispersion, and characteristics of the rubber. In order to fabricate a flawless rubber nanocomposite, all of these parameters must be carefully considered and taken into account. Finally, it can be said that the incorporation of nano-particles (fillers) not only reinforces the nanocomposites by virtue of their inherent properties and improves the final performance of rubber nanocomposites, but it can also accelerate the curing process and prevent over-curing, which can reduce the processing energy required. Some of the curing characteristics of filler rubber nanocomposites are described in Table 3.1 based on the different types of fillers used.

3.5 Dynamic Properties of Filled Rubbers

Undoubtedly, two of the most important and oldest features of filled rubbers are known as the Mullins effect [37–41] and the Payne effect [42–44]. The Mullins

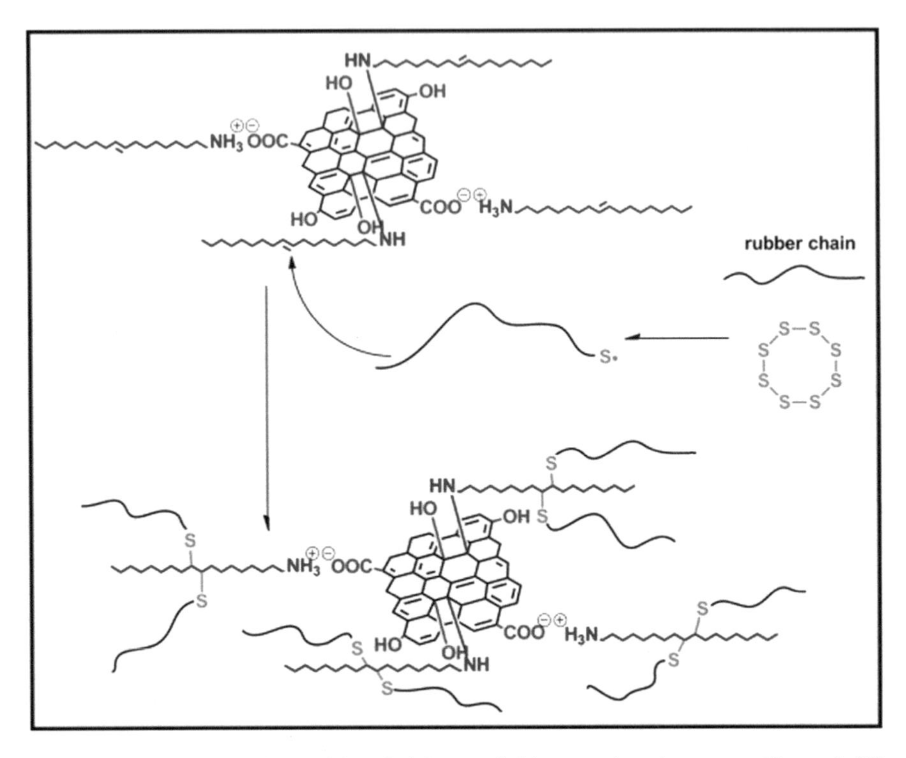

Fig. 3.3 Proposed mechanism of interfacial cross-linking reactions between sulfur and GO. Reprinted with permission from [35]. Copyright 2015 Elsevier Science Ltd.

effect explains the stress softening behavior of rubbers. When a rubber is subjected to a load, it deforms to a particular length that depends on the applied stress; after a certain relaxation time, further stretching of the rubber requires less force to reach to the same deformation [45]. Consequently, according to the Mullins effect, a filled rubber nanocomposite exhibits a larger tensile modulus on the first loading than in the subsequent ones. This phenomenological non-linear strain-induced stress softening behavior leads to the appearance of hysteretic behavior in loading–release cycles [39]. The residual deformation that remains after releasing the stretched rubber is often referred to as the permanent set [46, 47]. After a large number of loading–release cycles, the stress reaches a plateau region where further stretching will not affect the stress response anymore. A number of explanations for this behavior of filled rubber nanocomposites have been proposed to date. For example, Harwood and co-workers [39, 47] reported that such stress softening behavior in filled rubber systems is caused by the breakdown of carbon black agglomerates, which occurs at lower levels of stress. This effect also manifests in the dependency of the shear modulus of filled rubbers on the oscillating strain (Fig. 3.4), and is known as the Payne effect [44].

Table 3.1 The curing characteristics of selected rubber nanocomposites

Rubber	Filler	Φ (phr)	Preparation method	Vulcanization Temperature (°C)	ts (min)	t90 (min)	References
SBR	Nanokaolin	50	Mechanically mixed	153	9.23	18	[27]
SBR	Nanosilica	50	Mechanically mixed	153	7.45	48	
NR	Nanokaolin	45	Mechanically mixed	153	1.30	5	
NR	Nanosilica	45	Mechanically mixed	153	2.9	19	
BR	Nanokaolin	60	Mechanically mixed	153	9.50	16	
BR	Nanosilica	60	Mechanically mixed	153	2.37	42	
EPDM	Nanokaolin	60	Mechanically mixed	153	2.05	26	
EPDM	Nanosilica	60	Mechanically mixed	153	1.02	44	
BR	Cloisite® 15A	3	Mechanically mixed	155	2.11	3.55	[31]
BR	Cloisite® 20A	3	Mechanically mixed	155	4.08	7.50	
BR	Carbon Black	10	Mechanically mixed	155	16.33	23.87	
NR	OMMT	10	Mechanically mixed	150	3.24	6.87	[32]
NR	OMMT	10	Solution mixed	150	1.32	3.71	
NR	OMMT-Si69[a]	10	Solution mixed	150	1.27	7.26	
NR	Zeosil Silica	50	Mechanically mixed	160	2.1	3.4	[29]
NR	Cloisite® 15A	2	Mechanically mixed	160	1.3	2.1	
NR	Cloisite® 15A	4	Mechanically mixed	160	1.1	1.9	
NR	Cloisite® 15A	8	Mechanically mixed	160	1.1	2.1	
NR	Na-MMT	5	Mechanically mixed	143	4.70	7.63	[30]
NR	OMMT	5	Mechanically mixed	143	1.51	2.95	
NR	OMMT	8	Mechanically mixed	143	1.17	2.66	
NR	P-CNT[b]	25	Solution mixed	150	3.65	7.13	[33]
NR	B-CNT[c]	25	Solution mixed	150	3.76	9.23	

[a] Bis(triethoxysilylpropyl)tetrasulfan (TESPT) (Si69) was used as coupling agent
[b] Purified CNT
[c] Ball-milled CNT

Figure 3.4b demonstrates the different parameters that contribute to the dynamic oscillatory shear behavior of filled rubbers at low and high deformations. The full description of each contribution can be explained as follows [44, 48]:

- The low strain regions reflect the contributions of filler–filler interactions, in which the early softening behavior can be attributed to the breakdown of the filler structures. That is, higher concentrations of fillers result in increased levels of aggregations, and hence breakdown occurs at an earlier point in the lower deformation regions;
- The hydrodynamic contribution explains the effect of un-deformable rigid particles (e.g., silica and CB). Such rigid particles have an inherently high modulus, and,

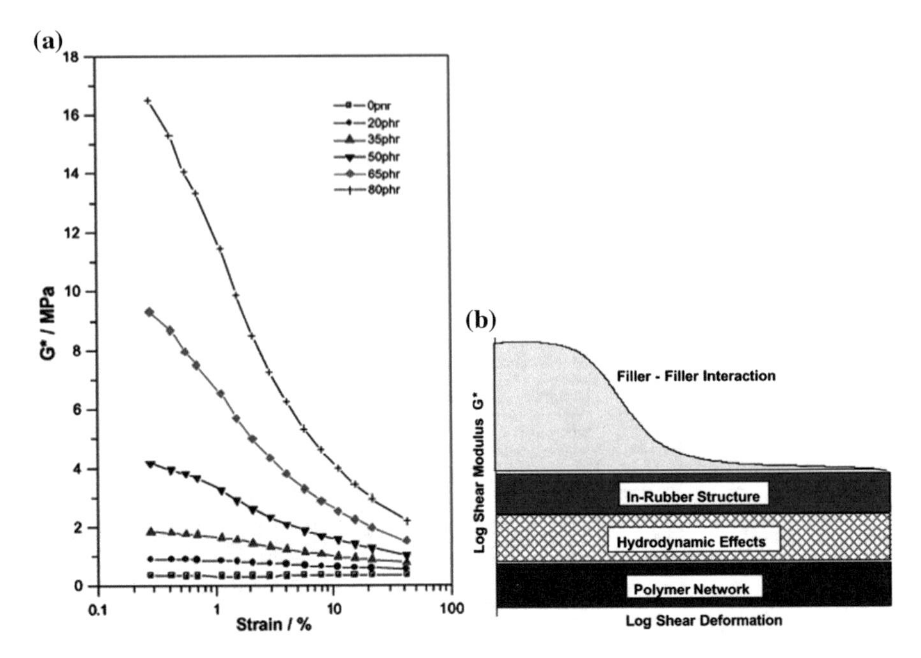

Fig. 3.4 **a** Complex modulus G* of vulcanized carbon black filled-SBR measured as a function of strain amplitude at different concentrations. **b** Different contributions defining the typical strain dependency of the oscillatory shear modulus of a filled rubber system. Reprinted with permission from [48]. Copyright 2005 Elsevier Science Ltd.

consequently, the higher the concentration of these fillers in a system, the higher the modulus;

- Rubber–particle interactions (adhesions) denoted as "In-Rubber" structures, which mostly define the modulus at large deformations, contribute to the strain-independent behaviors;
- Finally, the polymer network, which depends on the nature of the rubber itself and the cross-link density of the rubber, also contributes to the stress-strain curves.

There are additional mechanisms that are involved in the Mullins effect, including, for example, the disentanglement of rubber chains, chain scission, and formation of defects related to weak rubber–filler interactions, each of which has been discussed in detail elsewhere [46, 49]. It is clear that filler–filler and rubber–filler interactions play a critical role in determining the deformation behavior of filled rubber systems. This means that parameters such as the filler geometry, modification, and coupling agents can also have a remarkable influence on such behaviors. It should be noted that thus far, the majority of studies reported in the literature have dealt with spherical particles such as carbon black and silica particles. Consequently, it would be of considerable interest to investigate the degree of intercalation/exfoliation or the strength of a coiled nanotube bundle as a function of the applied deformation by means of such dynamic oscillatory shear analyses. In the following sections, the relationship

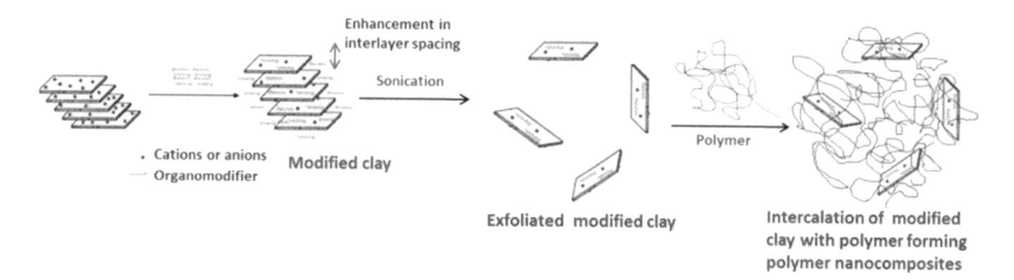

Fig. 3.5 Schematic diagram illustrating the production of clays with enhanced exfoliated by means of surface modification. Reprinted with permission from [51]. Copyright 2015 Elsevier Science Ltd.

between the properties of rubber nanocomposites and the type of reinforcement agent used will be discussed. Since most of the early discoveries and breakthroughs in rubber nanocomposites involved spherical fillers such as carbon black, in the following section, we will focus on the more recently reported rubber nanocomposite systems filled with large aspect ratio particles. It will be discussed that how the particles with higher aspect ratio can effectively contribute in enhancing physical, barrier and electrical properties of the rubber nanocomposites at low loadings.

3.6 Rubber/Clay Nanocomposites

In clay-based nanocomposites, the extent of intercalation and exfoliation of the silicate layers determines the effectiveness of the nanocomposites. In order to produce nanoclays that are delaminated in the polymer matrix, the silicate nanoclays need to be organically modified in order to strengthen their interactions with the polymer chains. This, in turn, will allow them to diffuse between the nanoclay galleries and break down nanoclay aggregates into intercalated structures (Fig. 3.5) [50, 51]. Moreover, surface modifications can enhance the exfoliation during the vulcanization process by reducing the van der Waals forces between the clay layers [8].

3.6.1 Mechanical Properties

One of the greatest advantages of nanocomposites is their enhanced mechanical strength, which makes them useful for a wide range of applications in which higher stiffness and tear strength are required. As mentioned earlier, however, achieving enhanced mechanical strength requires optimized intercalation/exfoliation of nanoclay particles within a polymer matrix. Thus, the interactions between the polymer and the nanoclays need to be maximized. Wu et al. [52] revealed that polar organo-modified clays will disperse more efficiently if the polarity of the rubber is increased, leading thus to a higher degree of exfoliation. The authors found that organo-modified clays exfoliate successfully in a polar NR. By contrast, the same clays exhibited intercalated structures in SBR and EPDM; in the case of EPDM, the structures displayed characteristics small interlayer distances. The differences in interactions, and, as a result, also in the intercalation/exfoliation structures formed, resulted in different degrees of enhancements and the moduli of the EPDM/clay and SBR/clay nanocomposites did not vary significantly from those of the pure rubbers. By contrast, the modulus of the NR/clay was shown to be higher than that of pure NR as a result of the exfoliation effect. This result suggests that compatibility between organoclays and polymer chains can facilitate the penetration of polymer chains in-between clay galleries, thereby facilitating higher level of intercalation [53]. Wang et al. [54] reported that SBR/clay nanocomposites prepared via the latex method typically have better mechanical properties than the nanocomposites prepared via the solution method. The authors reported that the tear strength of the nanocomposite prepared via the latex method was 23.8 kN/m at 20 phr clay content, while the nanocomposite produced using the solvent mixing method exhibited a markedly lower tear strength of 12.8 kN/m. This difference in tear strength was attributed to the enhanced dispersion of nanoclays in the nanocomposites derived from the latex method. The enhancement in properties of nanocomposites also depends on the concentration of the filler material—above a certain critical concentration, filler agglomerations reduce the mechanical strength of nanocomposites as a result of uneven stress distribution [55]. Manchado et al. [32] revealed that NR/clay nanocomposites prepared by means of solution intercalation show comparatively better mechanical properties, e.g., hardness, tensile modulus, and strength, than nanocomposites mixed mechanically with a two-roll mill. This outcome was attributed to the formation of improved intercalation structures in the nanocomposites prepared via the solution mixing method, which was mediated by the solvent-driven increase in the space between clay galleries and the swelling of rubber chains. The tensile properties of selected rubber/clay nanocomposites are summarized in Table 3.2.

Generally, platelet-like particles can enhance the tensile properties of rubber nanocomposites since the planar surfaces of the particles can interact with a large number of rubber chains, leading thus to the immobilization of the motion of rubber molecules [27].

3.6.2 Barrier Properties

Clay platelets can improve the barrier properties by creating a "torturous path" in the nanocomposites, which increases the distance that a gas molecule must diffuse through the width of a material or by decreasing the volume of the permeable amorphous phase relative to the content of the impermeable phase (clays), and, hence, reducing the diffusion rate [58]. Therefore, the higher the surface area of the fillers, the less are the gas molecules able to pass through the medium. That is, high aspect ratio of clays is expected to be advantageous in enhancing the barrier properties of

Table 3.2 Tensile properties of selected rubber/clay nanocomposites prepared using different methods

Rubber	Clay type	Φ wt% or Phr	Preparation method	Tensile strength (MPa)	Tensile modulus (MPa)	Elongation at break (%)	References
EVA	OMMT	2	Solution	5.7	0.79[a]	1471	[56]
EVA	OMMT	4	Solution	8.1	1.26[a]	1135	
EVA	OMMT	6	Solution	9.6	1.27[a]	1397	
SBR	OMMT	10	Latex	5.3	–	318	[54]
SBR	OMMT	20	Latex	7.5	–	454	
SBR	OMMT	30	Latex	12.4	–	504	
SBR	OMMT	40	Latex	15.8	–	568	
SBR	OMMT	60	Latex	13.9	–	732	
SBR	OMMT	5	Solution	2.1	–	505	[54]
SBR	OMMT	10	Solution	4.3	–	524	
SBR	OMMT	20	Solution	7.2	–	530	
SBR	OMMT	40	Solution	11.3	–	750	
BR	OMMT	5	Solution	1.6	–	225	[54]
BR	OMMT	10	Solution	3.1	–	360	
BR	OMMT	20	Solution	6.4	–	724	
BR	OMMT	40	Solution	8.9	–	670	
NR	OMMT	10	Two-roll mill	20.6 ± 0.75	4.31 ± 0.2[a]	1012 ± 52	[32]
NR	OMMT	10	Solution	22.2 ± 0.68	5.58 ± 25[a]	919 ± 33	
BR	Cloisite® 15A	3	Melt	5.68 ± 0.14	0.95 ± 0.06[a]	997 ± 21	[31]
BR	Cloisite® 20A	3	Melt	5.73 ± 0.16	1.28 ± 0.05[a]	783 ± 29	
PU	Cloisite® 30B	1	In situ	8.53	0.527	211.62	[57]
PU	Cloisite® 30B	2	In situ	9.26	0.517	225.21	
PU	Cloisite® 30B	3	In situ	10.21	1.102	196.38	
PU	Cloisite® 30B	4	In situ	13.23	1.074	177.60	
PU	Cloisite® 30B	5	In situ	17.10	1.434	127.35	

[a]Tensile modulus at 300%

nanocomposites. The permeability of nanocomposites with respect to the aspect ratio (AR) of the fillers can be predicted using the Nielsen model (3.2) [59]:

$$P = P_0\left(\frac{1-\emptyset}{1+\frac{(AR)}{2}\emptyset}\right) \tag{3.2}$$

where P and P_0 are the permeability of the filled and unfilled systems, respectively, and $\emptyset$ is the volume fraction of the filler. It has been shown that the addition of clay platelets to rubber reduces the N_2 permeability of rubber nanocomposites significantly [60, 61]. In addition, the incorporation of 7% volume fraction of rectorite and

30 phr of clay was shown to reduce the nitrogen permeability of NR and NBR by 50% and 48%, respectively. However, it should be noted that the orientation of the fillers along the diffusion path can influence significantly the performance. In particular, a planar orientation would be expected to result in enhanced barrier characteristics.

3.7 Rubber/Carbon Nanotube Nanocomposites

CNTs represent another class of high-aspect-ratio one-dimensional nanoparticles with attractive characteristics such as high mechanical strength, and good thermal and electrical conductivities, arising from their inherently unique properties [62]. However, in most solvents, CNTs usually exist as coiled-structure aggregates, which causes poor dispersity [63, 64]. Therefore, weak reinforcement provided by the slippage of nanotubes inside the aggregates can diminish mechanical properties and even electrical conductivities. Thus, it can be said that the key issue in CNT-reinforced rubber nanocomposites is the effective dispersal of nanotubes. Significant research efforts have been made to increase the dispersity of nanotubes inside the matrix by improving the interfacial interactions between the nanotubes and the rubber matrix. Surface modification [65] and ultrasonication of nanotubes [19] are two example techniques that can be exploited to improve nanotube dispersion, as illustrated in Fig. 3.6. CNTs can be modified with amphiphilic molecules via three different modes of action, including cylindrical micelle encapsulation and hemi-micellar and random adsorption of molecules onto the surface of CNTs.

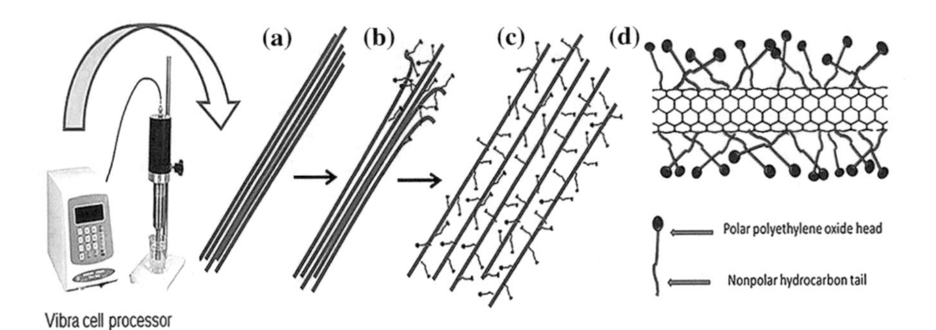

Fig. 3.6 Mechanisms of dispersing by means of ultrasonication and Vulcastab (Polyethylene oxide condensate) VL surfactant. **a** Separation of nanotubes aggregates via sonication; **b** "unzipping" of nanotube bundles by adsorption of surfactants; **c** complete coverage of nanotubes with surfactants, and d) proposed route of surfactant adsorption. Reprinted with permission from [22]. Copyright 2017 Elsevier Science Ltd.

3.7.1 Mechanical Properties

The modulus of rubber nanocomposites reinforced with tubular particles can be expressed using (3.3), which indicates that the reinforcement of such nanocomposites depends only on the *AR* and volume fraction of the particles (Ø) [41]:

$$E = E_0\left[1 + 0.67(AR)\text{Ø} + 1.62(AR)^2\text{Ø}^2\right] \tag{3.3}$$

where E and E_0 are the moduli of the reinforced and virgin rubber, respectively. The Halpin–Tsai model was also shown to be useful in the prediction of modulus of reinforced nanocomposites parallel to the fiber direction (3.4–3.6) [66]:

$$E_0/E = \left(1 + \eta\nu_f\right)/\left(1 - \eta\nu_f\right) \tag{3.4}$$

$$\eta = \left(E_f/E - 1\right)/\left(E_f/E + \xi\right) \tag{3.5}$$

$$\xi = 2(l/d) \tag{3.6}$$

where ν_f is the Poisson's ratio and l/d is the length to dimeter (AR, aspect ratio) ratio. For the estimation of the modulus transverse to the fiber direction, the value of the AR needs to be 1 ($l = d$). It is also known that at specific volume fractions with respect to different ARs, the formation of particle networks contributes significantly to the reinforcement of nanocomposites. As previously discussed in Sect. 3.5, in filled rubber systems, filler–filler and polymer–filler interactions dictate the magnitude of the modulus in nanocomposites. Therefore, the treatment of particle surface can be beneficial for enhancing the network rigidity as well as particle–polymer interactions [23, 67]. Jiang et al. [62, 63] revealed that MWNTs chemically modified with a silane coupling agent display improved bonding between nanotubes and methylvinyl silicon rubber. This improved bonding resulted in the enhancements in tensile modulus and elongation at break of the rubber nanocomposites. Specifically, the addition of the coupling agent at 1 wt% enhanced the elongation at break and tensile modulus to 210% and 3.8 MPa, respectively, from the 167% and 3 MPa obtained if no coupling agent was used. However, the authors found that there is a critical coupling agent concentration (5 wt%) above which a further increase in concentration to 7 wt% produces a reduction in mechanical properties. Peng et al. [17] utilized a self-assembling mechanism to enhance the dispersion of MWNTs in NR latex. In their method, nanotubes were first positively charged using an electrostatic adsorption technique to suppress the self-aggregation of the nanotubes. Subsequently, negatively charged rubber latex particles were added to enhance the interactions between the MWNTs and rubber particles, thereby increasing the dispersion. This work revealed that nanotube-reinforced rubbers displayed concurrently enhanced tensile stiffness and toughness, and those employing modified nanotubes exhibited particularly high values (Table 3.3) owing to the improved dispersion. Generally, in rubber latex nanocomposites, it is the hydrophilic/hydrophobic interactions that govern the homogeneity of the dispersions—specifically, the hydrophilic heads

Table 3.3 Tensile properties of NR reinforced with MWNTs via latex compounding

Rubber	CNT	ϕ	Preparation method	Tensile strength (MPa)	Tensile modulusa (MPa)	Elongation at break (%)	References
NR	p-MWNT[b]	1	Latex	24.9 ± 1.2	1.2	932	[17]
NR	m-MWNT[c]	1	Latex	30.0 ± 1.2	1.3	962	
NR	m-MWNT	0.05	Latex	30.0 ± 0.52	1.68 ± 0.05	1812 ± 61	[22]
NR	m-MWNT	0.1	Latex	35.2 ± 0.70	1.91 ± 0.03	1788 ± 43	
NR	m-MWNT	0.3	Latex	33.7 ± 0.65	1.86 ± 0.05	1780 ± 58	
NR	m-MWNT	0.5	Latex	32.2 ± 0.31	1.71 ± 0.02	1769 ± 78	
NR	m-MWNT	1.0	Latex	29.6 ± 0.57	1.44 ± 0.03	1728 ± 63	

[a]Tensile modulus at 300%
[b] Pure MWNTs
[c]Modified MWNTs

interact with the rubber latex and the hydrophobic tails interact with the surfaces of the nanotubes [19]. The nanotubes dispersed uniformly in the rubber matrix can form a 3D network, confining thus the rubber matrix into small unit cells [22, 67, 68]. This so-called 3D network can therefore transfer load evenly and enhance the mechanical properties of the rubber. The mechanical properties of selected NR/MWNT nanocomposites prepared via the latex method are summarized in Table 3.3.

3.7.2 Electrical Properties

As explained earlier on in the discussion of mechanical properties, the dispersion of nanotubes is a prominent factor influencing the properties of rubber nanocomposites. The same concept can also be applied to the electrical characteristics of nanocomposites, i.e., the higher the dispersion degree, the higher the electrical conductivity displayed by a nanocomposite. Carbon nanotubes can enhance the electrical conductivities by facilitating electron passage across the nanocomposite via the tunneling [69] and hopping (jumping) [3] effects. Such phenomena require the formation of a well-distributed and dispersed network of fillers. As was noted for mechanical properties, enhancements in electrical conductivities are affected by surface modification [19, 20] and the AR of the nanotubes [23]. Normally, nanotubes with a high AR would be expected to produce a more efficient tunneling effect, and hence, result in higher conductivities. However, Jiang et al. [63, 64] revealed the opposite effect. The authors showed that MWNTs with lower ARs are more efficient in promoting electron conduction in silicon rubber/MWNT nanocomposites than MWNTs with higher ARs (Fig. 3.7).

This disparity was attributed to the dispersion of the corresponding nanotubes within the nanocomposites. The nanotubes with higher AR were entangled, and, consequently, the bundles were hard to unwind. Therefore, poorly dispersed nanotubes

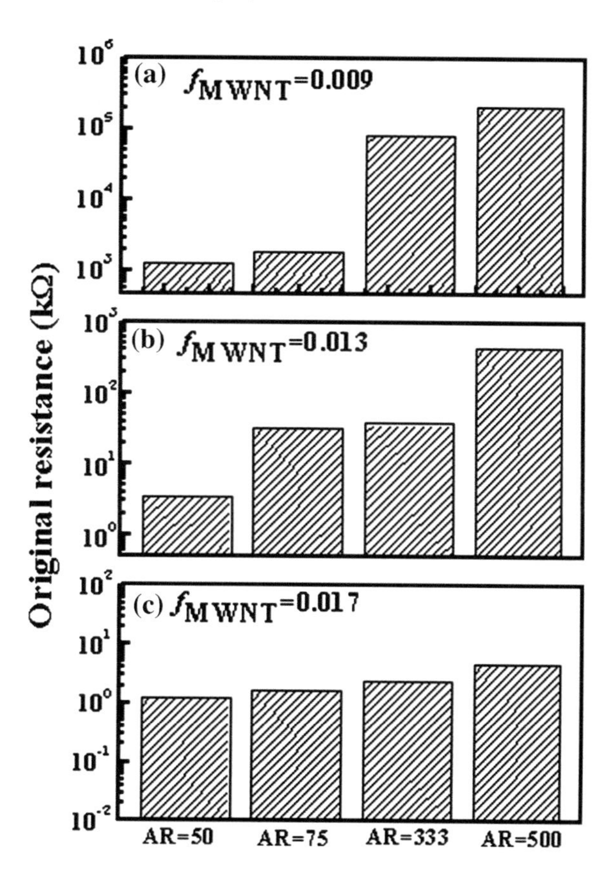

Fig. 3.7 Electrical resistance (kΩ) of silicon rubber/MWNT nanocomposites with different aspect ratios (AR = 50, 75, 333, and 500) and different volume fractions of nanotubes (f_{MWNT}). Reprinted with permission from [64]. Copyright 2007 American Institute of Physics

with high ARs resulted in nanocomposites with conductivities that were comparatively lower than those produced with nanotubes with lower ARs. Another parameter influencing the dispersion is the surface modification of the nanotubes. As discussed previously, surface modification can enhance the rubber–nanotube interactions, and therefore also the dispersion of nanotubes. However, it has been reported that the improvements in dispersion are dependent on the concentrations of the introduced modifications, and beyond a critical amount, the tunneling effect can be hindered [62]. Jiang et al. [63] reported that in cases utilizing an excessive amount of surfactant, a thick layer of rubber on the surfaces of the nanotubes interfered with effective tunneling by preventing direct contacts between the nanotubes. Azmi et al. [19] also found that the addition of non-aromatic methylated chains to sodium-based surfactants lowered the surface energies significantly and increased their interactions with MWNTs in NR/MWNT nanocomposites. It was later reported that surfactants functionalized with phenyl groups are even more efficient in stabilizing nanotubes in NR, yielding thus higher electrical conductivities than those of nanocomposites utilizing di-chain and tri-chain surfactants [20]. In a different study, Thomas et al. [3] found that functionalization of acid-treated CNTs with 1-octadecanol as the surfactant increased the resistivity of the NR/CNT nanocomposites by opening the double bond on the surface of the CNTs, which were required initially for electron passage. These results therefore suggest that the fabrication of a rubber nanocomposite with an efficient electrical conductivity requires the formation of a 3D percolated network. Such homogenous dispersion of a conductive network can be accomplished by enhancing the interactions between rubber chains and nanotube surfaces. Studies reported to date have demonstrated that parameters such as the choice of the surfactant, surfactant concentration, and the aspect ratio of the nanotubes could affect the formation of the said 3D network.

3.7.3 Barrier Properties

Jose et al. [23] described permeability as a combination of sorption and diffusion processes, where sorption explains the degree of gas molecules absorbed during the penetration, while diffusion determines the diffusivity of gas molecules depending on their diffusion coefficient. It was shown that increasing the concentration of MWNTs decreases the solvent uptake as a result of the increased cross-link density, thereby lowering the free volume available for the absorption of solvents. By contrast, increasing the fraction of nanotubes created a torturous path and reduced the diffusion of the solvent molecules through the NR/MWNT nanocomposites. Jo et al. [70] showed that the O_2 gas permeability of chlorobutyl rubber/epoxidized natural rubber/CNT nanocomposites (CIIR/ENR/CNT) was reduced significantly when the CNTs were functionalized with TiO_2. It was shown that metal oxides block the passage of gas molecules through the rubber matrix in a manner illustrated in Fig. 3.8. It is worth noting that when it comes to enhancing barrier properties, nanotubes may not be as efficient as platelet-like and disk-like particles. Platelet-like particles such as clays and graphene oxides possess larger planar surfaces that enable them to block the passage of gas molecule through the matrix more effectively [9, 71].

3.8 Rubber/Graphene-Based Nanocomposites

Graphite-based nano platelets possess excellent mechanical, thermal, and electrical properties, and represent another type of attractive fillers used as nano reinforcement agents [72]. Graphite exists naturally in the form of stacked aggregates, which need to be expanded in order to permit efficient intercalation. Thus, graphite is heated at high temperatures (above 800 °C) in order to furnish expanded (fluffy shaped) graphite structures with large surface areas. Moreover, sonication can be applied to the thermally shocked graphite to attain exfoliated graphite nanosheets. These nanosheets can be subsequently functionalized with chemical treatments in order to facilitate interactions between particles and the polymer matrix and achieve better

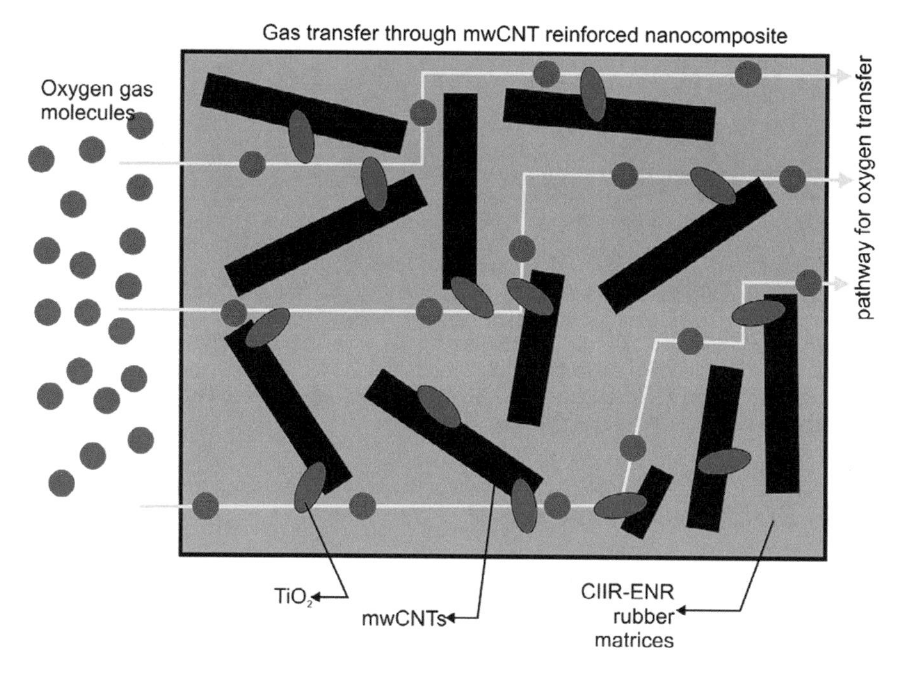

Fig. 3.8 Diagram illustrating the barrier mechanism in CIIR/ENR/CNT nanocomposites functionalized with TiO_2. Reprinted with permission from [70]. Copyright 2015 Elsevier Science Ltd.

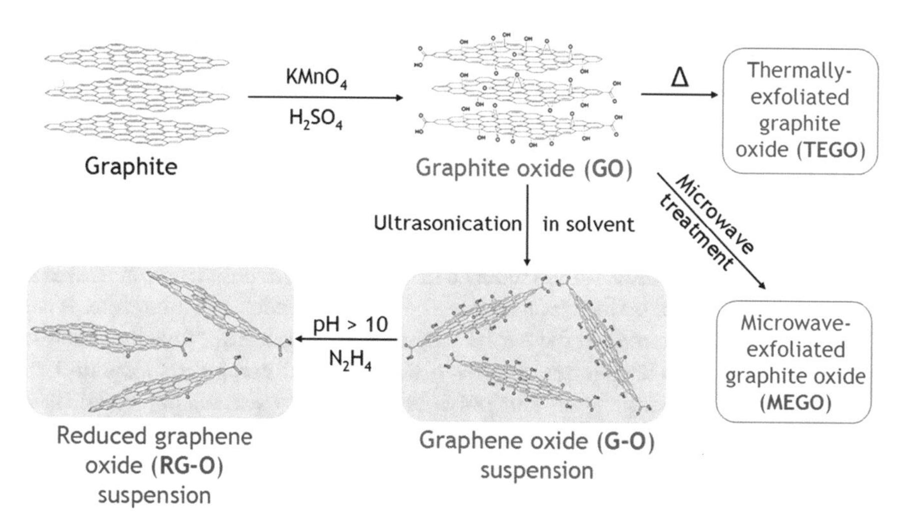

Fig. 3.9 Scheme illustrating the preparation of graphene–oxide sheets from graphite. Reprinted with permission from [18]. Copyright 2013 Elsevier Science Ltd.

exfoliation [71, 73]. The general route employed for the preparation of graphene-based materials is illustrated in Fig. 3.9.

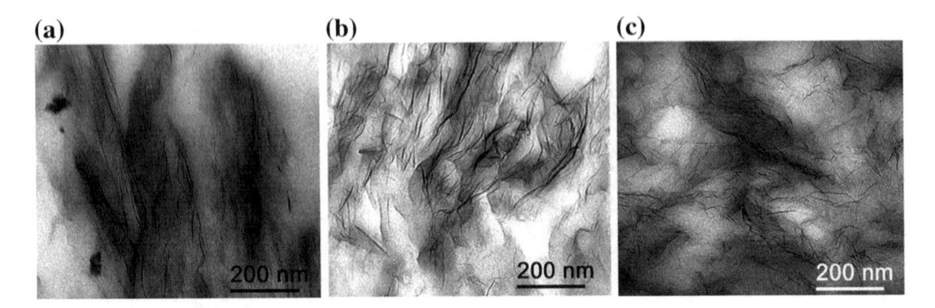

Fig. 3.10 TEM images of TPU nanocomposites loaded with thermally reduced graphene oxides (TRGs): **a** 5 wt% of TRGs, prepared via melt mixing; **b** 5 wt% of TRGs, prepared via solvent mixing; and **c** 3 wt% of TRGs, prepared via in situ polymerization. Reprinted with permission from [12]. Copyright 2010 the American Chemical Society

3.8.1 Mechanical Properties

Kim et al. [12] produced TPU/graphene oxide nanocomposites with three different methods, namely, the melt, the solution, and the in situ polymerization methods. In order to increase the exfoliation of the graphite oxide sheets, the authors first functionalized the sheets with acids (FGS). Subsequently the sheets were shocked thermally (1050 °C) to weaken the van der Waals forces between the sheets and to increase their surface area (700–1500 m^2/g), furnishing thus thermally reduced graphene oxides (TRGs). The TEM images (Fig. 3.10) of these TRGs demonstrate clearly the higher degree of dispersion of the solvent mixed and in situ polymerized nanocomposites when compared to those prepared via melt mixing.

As suggested by the TEM images, the nanocomposites prepared via solution mixing and in situ polymerization exhibited higher stiffness owing to better dispersions than the nanocomposites obtained via melt mixing. The authors suggested that in fact it is the re-aggregation and reduction in the size of the particles during the melt mixing that are responsible for the reduced dispersity and modulus. Any increases in modulus are mediated by the changes in the AR of the sheets, and, therefore, it can be expected that shortening of the ARs during melt mixing can improve the modulus slightly. By contrast, despite the improved dispersion of graphene sheets in TPU nanocomposites prepared by in situ polymerization, they exhibit poorer stiffness when compared with that of the nanocomposites prepared by solvent mixing as a result of the inhibited inter-chain hydrogen bonding in TPU. Potts et al. [74] subjected already co-coagulated latex of NR/reduced graphene oxides using both a two-roll mill mixing and solution mixing to observe the effect of curing agents on these differently processed nanocomposites. Interestingly, the results revealed that the milled samples have better dispersion and slightly higher ARs than the solution-mixed samples. The different treatments affected the mechanical properties significantly. Specifically, while the solution-treated samples displayed strong modulus dependence on RGO loadings at low elongations, the milled samples exhibited strain hardening at high

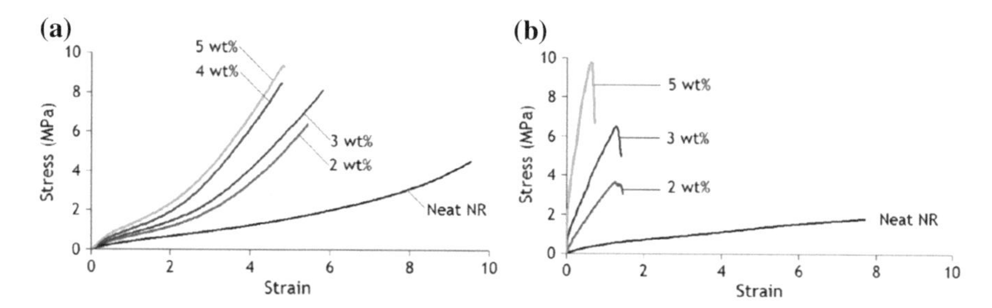

Fig. 3.11 Stress–strain curves of **a** roll-milled and **b** solution-treated NR/RGO nanocomposites. Reprinted with permission from [74]. Copyright 2012 the American Chemical Society

elongations (Fig. 3.11). The authors also showed that the more uniform dispersion of RGOs with higher ARs is in fact responsible for the mechanical reinforcement of the milled nanocomposites owing to the fact that RGOs are aligned at higher elongations, resulting in strain-induced crystallization. By contrast, the reinforcement in the solution-treated samples was suggested to be derived from the presence of a network of RGO platelets in between the latex particles.

Later, these authors examined the effect of incorporating thermally exfoliated graphite oxide into NR latex to produce a material known as TEGO/NR and subjected it to "pre-mixing" and milling [18]. Subsequently, the properties of the "pre-mixed" and milled samples were compared. Microscopy observations revealed a more uniform dispersion of GO in the case of "pre-mixed" latex TEGO/NR nanocomposites than the milled ones, with almost identical average ARs for both. The enhanced dispersion resulted in a higher tensile modulus and strength of the latex ("pre-mixed") TEGO/NR nanocomposites. The authors reported that the reinforcement obtained as a result of the incorporation of the nanofiller could not be caused solely by the higher AR since both systems exhibited virtually identical ARs. Subsequent analysis showed that the average thickness and lateral dimensions of the latex "pre-mixed" TEGO/NR nanocomposites are smaller than those of the milled nanocomposites, providing increased surface area for "rubber binding"—i.e., area where rubber chains can interact with the surface of the platelets. In a fashion similar to that described for clay based nanocomposites, the modifications of sheets can generally boost their dispersion and enhance the formation of bonds between the two sides of the sheets with larger rubber chains [75]. Table 3.4 reports the tensile properties of selected graphene-based rubber nanocomposites.

3.8.2 *Electrical Properties*

One of the greatest aspects of graphene-based nanomaterials is their excellent electrical conductivity, which creates outstanding opportunity for electronic applications. The high AR of graphene-based nanomaterials imparts them with the capacity to

Table 3.4 Tensile properties of selected graphene-based rubber nanocomposites

Rubber	Graphite/Graphene oxide	Φ wt% or phr	Preparation method	Tensile strength (MPa)	Tensile modulus (MPa)	Elongation at break (%)	References
TPU	FGS	0.5	Melt	–	9.1	–	[12]
TPU	FGS	1.0	Melt	–	11.5	–	
TPU	FGS	1.5	Melt	–	14.5	–	
TPU	FGS	2.0	Melt	–	15.7	–	
TPU	FGS	3.0	Melt	–	20.8	–	
TPU	FGS	0.0	Solution	–	6.6	–	
TPU	FGS	0.5	Solution	–	13.3	–	
TPU	FGS	1.0	Solution	–	15.5	–	
TPU	FGS	3.0	Solution	–	53.2	–	
TPU	FGS	0.0	In situ	–	7.2	–	
TPU	FGS	0.5	In situ	–	7.0	–	
TPU	FGS	0.9	In situ	–	11.7	–	
TPU	FGS	2.7	In situ	–	23.1	–	
BR	MG	1	Solution	–	0.5	–	[75]
BR	MG	3	Solution	–	0.88	–	
BR	MG	5	Solution	–	1.5	–	
BR	MG	10	Solution	–	3.4	–	
NR	RGO	0	Milled	3.1	–	860	[74] [a]
NR	RGO	2	Milled	5.2	–	550	
NR	RGO	3	Milled	6.8	–	540	
NR	RGO	4	Milled	6.4	–	373	
NR	RGO	5	Milled	8.1	–	433	
NR	RGO	0	Solution	1.5	–	624	
NR	RGO	2	Solution	2.7	–	136	
NR	RGO	3	Solution	5.3	–	102	
NR	RGO	5	Solution	7.8	–	56	
NR	TEGO	5	Milled	6.05	1.28[b]	8.96	[18]
NR	L-TEGO	5	Latex	10.90	5.19[b]	5.03	
SBR	GE	0	Latex	1.63	1.02[b]	5.46	[76]
SBR	GE	1	Latex	5.92	2.23[b]	7.00	
SBR	GE	3	Latex	13.05	4.15[b]	9.07	
SBR	GE	5	Latex	16.20	6.78[b]	7.05	
SBR	GE	7	Latex	17.81	10.51[b]	5.47	

[a]Values are roughly extracted from the curves reported in [74]
[b]Tensile modulus at 300%

transform insulating materials into conductive ones at low percolations. The electrical performances of the nanocomposites are strongly dependent on their morphology, and different factors can affect the formation of the optimal conductive network. Specifically, AR, surface area, and the dispersion of the fillers are some of the determinant parameters [71, 77]. Ozbas et al. [77] revealed that functionalized graphene sheets (FGSs) with higher surface areas (FGS–650 m^2/g) show markedly lower electrical percolation threshold at 0.8 wt% functionalization than PDMS nanocomposites loaded with lower surface area FGS (FGS–400 m^2/g) at a level of 1.6 wt%. This result indicates the higher exfoliation degree of FGS–600 than that of FGS–400. Du et al. [78] pointed out the pre-requisites for obtaining an effective network of graphene nanosheets with a high electrical conductivity. The authors stressed that the solution mixing method can be advantageous over melt mixing since the nanosheets are kept flat during the preparation, whereas they could be rolled under shear force during the melt mixing, promoting thus agglomeration. Moreover, the formation of an efficient conductive network was shown to be favored by the unidirectional alignment of parallel sheets and their plane–plane contacts within the polymer matrix [78]. The authors also stated that nanosheets are more prone to aggregation when compared to MWNTs as a result of their larger surface area and plane–plane contacts, bearing in mind that only this particular plane-to-plane contact is beneficial to electrical conductivity rather than other possible contacts namely, edge-to-edge and edge-to-plane. In addition, the nanosheets can roll up and wrinkle into tubular structures with lower ARs than those of MWNTs, thus leading to lower conductivities than those of nanocomposites loaded with MWNTs [78]. These facts highlight the difficulties associated with attaining effective conductive networks with graphene sheets. Further, despite the significant benefits associated with highly dispersed systems, it has been shown that a good dispersion is not indispensable to the formation of an effective conductive path. In fact, certain degree of agglomeration and poor dispersion can even be beneficial [79]. Studies have also reported that nanocomposites with segregated morphologies of graphene nanosheets can yield nanocomposites with reasonable electrical conductivities at very low percolation thresholds [78, 80, 81]. Potts et al. [74] reported relatively higher electrical conductivities for solution-treated NR/RGO nanocomposites than for milled samples despite the better dispersion of the latter material. This outcome corroborates further the importance of segregated ("web-like") structures for the formation of efficient conductive pathways. Figure 3.12 shows examples of NR/reduced graphene oxide nanocomposites with segregated and not-segregated morphologies.

In the support of their findings, the authors stated that in the case of highly dispersed nanocomposites, platelets are densely coated with rubber chains, thereby preventing the plate–plate contacts required for conduction. Further, they reported that in the case of "pre-mixed" latex and milled NR/TEGO nanocomposite samples with homogenous structures, the nanocomposites prepared by "pre-mixing" exhibited higher conductivities as a result of their higher degree of exfoliation [18]. Kim et al. [12] found that solution-mixed and in situ polymerized PU/thermally reduced graphene oxide nanocomposites show relatively lower resistivity than samples obtained by melt-intercalation owing to their better dispersion state. Interest-

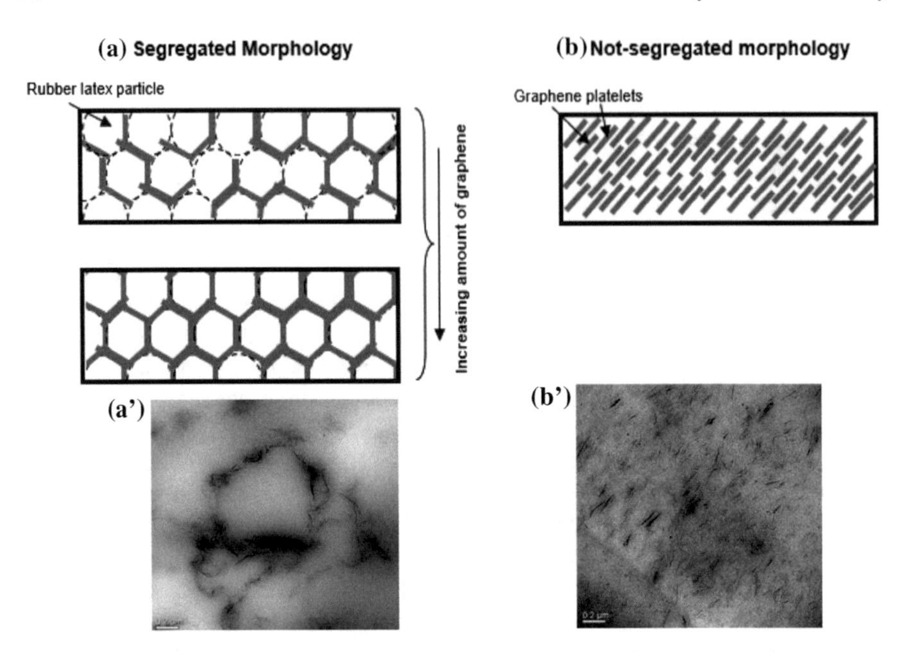

Fig. 3.12 **a**, **b** Schematic illustrations and **a'**, **b'** TEM images of natural rubber/reduced graphene oxide (NR/RGO) nanocomposites with 3.38 vol.% of RGO. **a**, **a'** Segregated morphology obtained via hot pressing and **b**, **b'** not-segregated morphology obtained via two-roll mixing. Scale bars represent a distance of 0.2 μm. Reprinted with permission from [82]. Copyright 2014 the American Chemical Society

ingly, solution mixed nanocomposites exhibited lower resistivity than in situ polymerized nanocomposites despite the better dispersion of the latter samples (Fig. 3.13). All of the above-mentioned arguments demonstrate the importance of morphology in determining the electrical properties of nanocomposites. In particular, it is apparent that different methods can be applied to obtain the desired morphology and thus also a more conductive product. It must be noted that dispersion alone may not be sufficient for the creation of an ultralow percolated system. In fact, the geometry of such network in which some type of plate–plate contacts is promoted can also impart enhanced efficiency.

3.8.3 *Barrier Properties*

Graphene based materials can be used to impart rubbers with enhanced barrier properties owing to their large ARs and surface areas. The mechanism via which the permeability is reduced is identical to that mentioned earlier in the context of clay- or CNT-based rubber nanocomposites. Naturally, the final morphology and arrangement of the graphene-based materials can influence the efficiency of the barrier

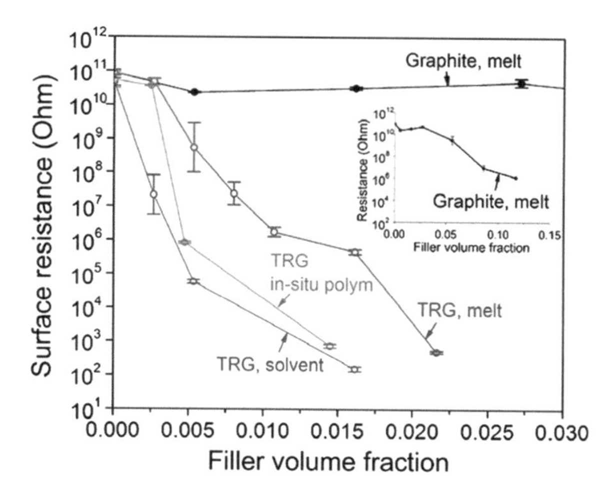

Fig. 3.13 Surface resistivity of thermally reduced graphene oxide incorporated PU nanocomposites prepared via different methods. Reprinted with permission [12]. Copyright 2010 the American Chemical Society

properties. As previously mentioned, Kim et al. [12] produced TPU/graphene oxide nanocomposites by means of conventional melt mixing, solvent mixing, and in situ polymerization. Similar to the other properties already discussed, the barrier properties of the solvent mixed nanocomposites were better than those of the in situ polymerized nanocomposites despite their better dispersion, potentially as a result of hydrogen bonding between urethane groups in TPU being impeded during the in situ polymerization. The authors stated that a high degree of dispersion is not always beneficial in terms of the final material properties. Instead, the morphology and arrangement of the particle network can be even more important. In Sect. 3.8.2, we mentioned that segregated morphology can produce finer interconnections in a network of graphene oxide sheets, and consequently higher conductivities when compared to not-segregated morphologies. The same concept can also be applied here for barrier properties. The relative permeability P_R of the segregated and not-segregated morphologies can be predicted using (3.7) and (3.8), [82–84] respectively.

$$P_R = \frac{P_{nanocomposite}}{P_{matrix}} = 1 - 3F/\delta \tag{3.7}$$

$$P_R = \frac{P_{nanocomposite}}{P_{matrix}} = \frac{1-\phi}{1+(0.5(AR))(2/3)\phi(s+0.5)} \tag{3.8}$$

Equation (3.7) for segregated morphology is based on the assumption that a nanocomposite is composed of impermeable reduced graphene oxide (RGO) coated-latex spheres embedded in a permeable matrix, and F represents the volume fraction of these coated spheres and δ is a characteristic term describing the quality of the shell coating the rubber core. In (3.8), s is a parameter describing the orientation of the particles in the nanocomposites [82]. Considering the two morphologies, Scherillo et al. [82] reported that NR/RGO nanocomposites with segregated morphology have lower permeability than nanocomposites with non-segregated morphology, indicating the effectiveness of the segregation morphology in optimizing the nanocomposite properties. A relative permeability of 0.5 was attained at 0.01-volume fraction of RGOs for the segregated nanocomposites, whereas the volume fraction had to be increased to 0.03 in order to obtain the same relative permeability for nanocomposites with the not-segregated morphology.

3.9 Applications

The most important outcome of nanoparticle incorporation into the rubber matrix is the reinforcement, which enables the resulting product to be used in a wider span of applications. The studies reported to date have shown that nanoparticles can benefit the rubber manufacturing industry in several different ways. Namely, the introduction of nanoparticles can reduce the cost of production by reducing the amount of rubber used, as well as by saving energy during the curing process by accelerating vulcanization. In addition, it can facilitate the manufacturing of products with reduced weights owing to the low nanoparticle loadings. Further, nanoparticles can improve the mechanical properties of rubbers such as abrasion, tear strength, and wear resistance, which are critical to the tire and belt industries. One of the biggest industries utilizing rubber products is the automotive industry. Figure 3.14 illustrates the many rubber parts that are required in the automotive manufacturing, including exterior and interior parts. In addition to the parts illustrated, it is important to consider the different hoses that are used for different purposes—from carrying fuel to breaks—and demonstrate the ability of rubber materials to withstand different harsh conditions and temperatures [6]. The reinforcement provided by the incorporation of nanoparticles is not only mechanical in nature. Instead, nanoparticles were also found to improve the thermal stability of the rubber products. Sealants are another product that benefits from the incorporation of nanoparticles. For instance, MWNT-reinforced rubber nanocomposites were previously utilized as O-rings for oil excavation applications in deep wells—i.e., conditions where high pressure and temperature must be taken into account [28].

Apart from the mechanical and thermal reinforcements, carbon-based particles such as nanotubes, graphite, graphene oxides, and others can find applications in electronic devices such as sensors, membranes, and electromagnetic interference (EMI) shielding owing to their excellent electrical conductivities at very low loadings. Last but not least, the improvements in the barrier properties of rubber nanocomposites facilitate their use in coating applications, in which hindering gas permeability is of critical importance [86].

The review of the work reported in the literature makes it apparent that particle inclusion into rubber materials is useful for more than simply lowering the costs of the finished product, and facilitates the creation of products with outstanding properties that can be tailored depending on the desired application. Nevertheless, in manufacturing, the price-to-performance ratio needs to be reasonable. For example, if electrical properties are not of concern, clay nanoparticles are more suitable as

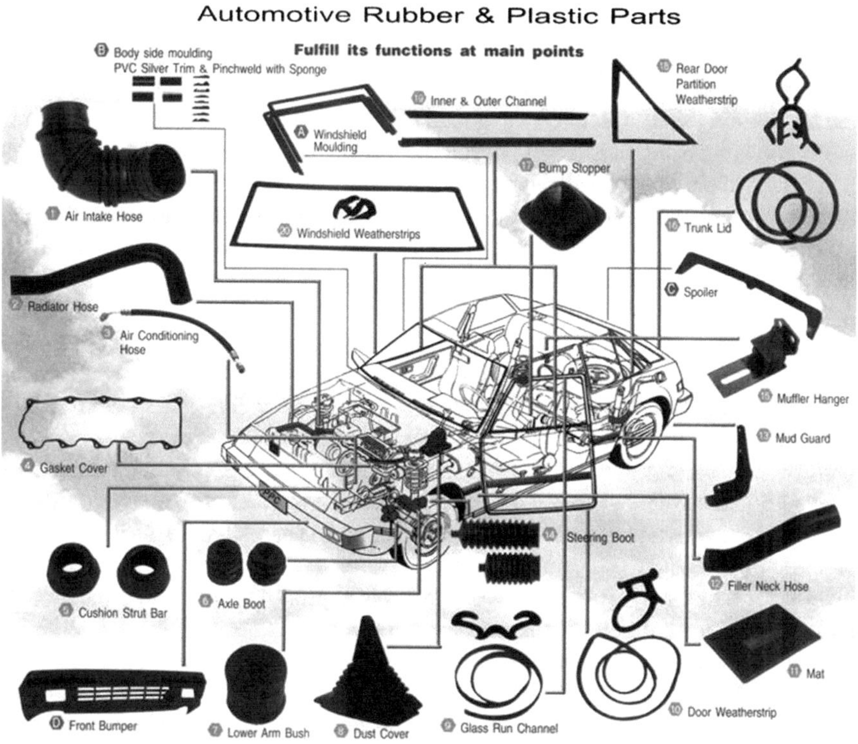

Fig. 3.14 Image showing the different parts of a typical automobile made of rubber [85]

they are available at a cheaper price and their modification is more facile and cost-effective when compared to other conductive nanoparticles. Furthermore, enhanced mechanical and barrier properties can be obtained owing to the high ARs of clay nanoparticles, which are comparable to those of carbon-based particles.

3.10 Conclusions and Future Trends

In this chapter, the different manufacturing routes to rubber nanocomposites were discussed with respect to their final morphology and properties. In addition, the mechanisms of reinforcement associated with the different types of filler particles were reviewed. As a result of the importance of the aspect ratio in determining the magnitude of the resultant reinforcement, only particles with larger aspect ratios, namely, clays, nanotubes, and graphene-based particles, were considered in this chapter instead of the more conventional carbon black particles. The introduction of nanoparticles can facilitate the curing process and serve to reinforce rubber materials. It was shown that although the degree of dispersion is typically the main factor determining the mechanical properties, the morphology of the particle arrangement within the rubber matrix also impacts the network formation and structure of the tortourous path required for conductivity and barrier properties. The key to constructing and maintaining such structures was found to lie in the process of fabrication. Specifically, solvent-mixed and in situ polymerized nanocomposites show comparatively better dispersion, and thus also better overall properties than those prepared by melt mixing. However, the better industrial viability of melt mixing and the solvent-related issues associated with solution-mixed and in situ polymerized nanocomposites also must be taken into account. Recently, the application of rubber nanocomposites in electronics has received significant attention. In this case, the main challenge is to attain percolated network structures at very low loadings. One way of attaining such results is the blending strategy, which involves the blending of two immiscible polymeric materials to make phase-separated structures. Blending is an economically efficient process that can produce materials with interesting properties by targeted tuning of morphologies. Studies have been shown that localizing conductive particles at the interface of blends tends to reduce the percolation threshold tremendously and enhance the properties as a result of the formation the so-called "double percolated structure" [87–89], which is similar to the segregated morphology discussed earlier, although in this case at lower loadings. This approach to increasing the product performance by localizing the particles at the interface is not simple since different parameters such as viscosity ratio, surface energies, size and shape of the particles, and processing protocols can all affect the results. Nevertheless, blending has demonstrated significant capacity to yield nanocomposites with ultralow percolation thresholds [87]. However, the choice of a second phase and processing conditions needs to be optimized in order to obtain the most efficient properties.

Acknowledgements The authors would like to thank the Department of Science and Technology and the Council for Scientific and Industrial Research, South Africa, for financial support.

References

1. Morton M (2013) Rubber technology. Springer Science & Business Media
2. Thomas S, Stephen R (2010) Rubber nanocomposites: preparation, properties and applications. Wiley
3. Selvin Thomas P, Abdullateef AA, Al-Harthi MA, Atieh MA, De SK, Rahaman M et al (2012) Electrical properties of natural rubber nanocomposites: effect of 1-octadecanol functionalization of carbon nanotubes. J Mater Sci 47:3344–3349
4. Bhattacharya M, Bhowmick AK (2010) Synergy in carbon black-filled natural rubber nanocomposites. Part i: mechanical, dynamic mechanical properties, and morphology. J Mater Sci 45:6126–6138
5. Bhattacharya M, Maiti M, Bhowmick AK (2008) Influence of different nanofillers and their dispersion methods on the properties of natural rubber nanocomposites. Rubber Chem Technol 81:782–808
6. Galimberti M (2011) Rubber-clay nanocomposites: science, technology, and applications. Wiley
7. Thomas S, Zaikov G (2009) Recent advances in polymer nanocomposites. CRC Press

8. Kim J-t Oh, T-s Lee D-h (2003) Morphology and rheological properties of nanocomposites based on nitrile rubber and organophilic layered silicates. Polym Int 52:1203–1208
9. Wu J, Huang G, Li H, Wu S, Liu Y, Zheng J (2013) Enhanced mechanical and gas barrier properties of rubber nanocomposites with surface functionalized graphene oxide at low content. Polymer 54:1930–1937
10. Jiang M-J, Dang Z-M, Yao S-H, Bai J (2008) Effects of surface modification of carbon nanotubes on the microstructure and electrical properties of carbon nanotubes/rubber nanocomposites. Chem Phys Lett 457:352–356
11. Gu Z, Song G, Liu W, Wang B, Li J (2009) Preparation and properties of organo-montmorillonite/cis-1,4-polybutadiene rubber nanocomposites by solution intercalation. Appl Clay Sci 45:50–53
12. Kim H, Miura Y, Macosko CW (2010) Graphene/polyurethane nanocomposites for improved gas barrier and electrical conductivity. Chem Mater 22:3441–3450
13. Mahmoud WE (2011) Morphology and physical properties of poly(ethylene oxide) loaded graphene nanocomposites prepared by two different techniques. Euro Polym J 47:1534–1540
14. Kalaitzidou K, Fukushima H, Drzal LT (2007) A new compounding method for exfoliated graphite–polypropylene nanocomposites with enhanced flexural properties and lower percolation threshold. Compos Sci Technol 67:2045–2051
15. Liao M, Zhang W, Shan W, Zhang Y (2006) Structure and properties of polybutadiene/montmorillonite nanocomposites prepared by in situ polymerization. J Appl Polym Sci 99:3615–3621
16. Koning C, Hermant MC, Grossiord N (2012) Polymer carbon nanotube composites: the polymer latex concept. CRC Press
17. Peng Z, Feng C, Luo Y, Li Y, Kong LX (2010) Self-assembled natural rubber/multi-walled carbon nanotube composites using latex compounding techniques. Carbon 48:4497–4503
18. Potts JR, Shankar O, Murali S, Du L, Ruoff RS (2013) Latex and two-roll mill processing of thermally-exfoliated graphite oxide/natural rubber nanocomposites. Compos Sci Technol 74:166–172
19. Mohamed A, Anas AK, Abu Bakar S, Aziz AA, Sagisaka M, Brown P et al (2014) Preparation of multiwall carbon nanotubes (MWCNTs) stabilised by highly branched hydrocarbon surfactants and dispersed in natural rubber latex nanocomposites. Colloid Polym Sci 292:3013–3023
20. Mohamed A, Anas AK, Bakar SA, Ardyani T, Zin WMW, Ibrahim S et al (2015) Enhanced dispersion of multiwall carbon nanotubes in natural rubber latex nanocomposites by surfactants bearing phenyl groups. J Colloid Interface Sci 455:179–187
21. Furuya M, Shimono N, Yamazaki K, Domura R, Okamoto M (2017) Evaluation on cytotoxicity of natural rubber latex nanoparticles and application in bone tissue engineering. E-J Soft Mater 12:1–10
22. George N, Bipinbal PK, Bhadran B, Mathiazhagan A, Joseph R (2017) Segregated network formation of multiwalled carbon nanotubes in natural rubber through surfactant assisted latex compounding: a novel technique for multifunctional properties. Polymer 112:264–277
23. Jose T, Moni G, Shalini S, Raju AJ, George JJ, George SC (2017) Multifunctional multi-walled carbon nanotube reinforced natural rubber nanocomposites. Indust Crops Prod 105:63–73
24. Goodyear C (1853) Gum-elastic and its varieties: with a detailed account of its applications and uses, and of the discovery of vulcanization. Published for the author
25. Medalia AI (1986) Electrical conduction in carbon black composites. Rubber Chem Technol 59:432–454
26. Karasek L, Sumita M (1996) Characterization of dispersion state of filler and polymer-filler interactions in rubber-carbon black composites. J Mater Sci 31:281–289
27. Liu Q, Zhang Y, Xu H (2008) Properties of vulcanized rubber nanocomposites filled with nanokaolin and precipitated silica. Appl Clay Sci 42:232–237
28. Endo M, Noguchi T, Ito M, Takeuchi K, Hayashi T, Kim YA et al (2008) Extreme-performance rubber nanocomposites for probing and excavating deep oil resources using multi-walled carbon nanotubes. Adv Funct Mater 18:3403–3409

29. Carli LN, Roncato CR, Zanchet A, Mauler RS, Giovanela M, Brandalise RN et al (2011) Characterization of natural rubber nanocomposites filled with organoclay as a substitute for silica obtained by the conventional two-roll mill method. Appl Clay Sci 52:56–61
30. Sun Y, Luo Y, Jia D (2008) Preparation and properties of natural rubber nanocomposites with solid-state organomodified montmorillonite. J Appl Polym Sci 107:2786–2792
31. Kim M-S, Kim D-W, Ray Chowdhury S, Kim G-H (2006) Melt-compounded butadiene rubber nanocomposites with improved mechanical properties and abrasion resistance. J Appl Polym Sci 102:2062–2066
32. López-Manchado MA, Herrero B, Arroyo M (2004) Organoclay–natural rubber nanocomposites synthesized by mechanical and solution mixing methods. Polym Inter 53:1766–1772
33. Sui G, Zhong WH, Yang XP, Yu YH (2008) Curing kinetics and mechanical behavior of natural rubber reinforced with pretreated carbon nanotubes. Mater Sci Eng, A 485:524–531
34. Sui G, Zhong W, Yang X, Zhao S (2007) Processing and material characteristics of a carbon-nanotube-reinforced natural rubber. Macromol Mater Eng 292:1020–1026
35. Liu X, Kuang W, Guo B (2015) Preparation of rubber/graphene oxide composites with in situ interfacial design. Polymer 56:553–562
36. Wu J, Xing W, Huang G, Li H, Tang M, Wu S et al (2013) Vulcanization kinetics of graphene/natural rubber nanocomposites. Polymer 54:3314–3323
37. Bueche F (1961) Mullins effect and rubber–filler interaction. J Appl Polym Sci 5:271–281
38. Govindjee S, Simo JC (1992) Mullins' effect and the strain amplitude dependence of the storage modulus. Inter J Solids Struct 29:1737–1751
39. Harwood J, Mullins L, Payne A (1965) Stress softening in natural rubber vulcanizates. Part II: stress softening effects in pure gum and filler loaded rubbers. J Appl Polym Sci 9:3011–3021
40. Harwood J, Payne A (1966) Stress softening in natural rubber vulcanizates III: carbon black filled vulcanizates. Rubber Chem Technol 39:1544–1552
41. Mullins L, Tobin N (1965) Stress softening in rubber vulcanizates. Part I: use of a strain amplification factor to describe the elastic behavior of filler-reinforced vulcanized rubber. J Appl Polym Sci 9:2993–3009
42. Payne AR (1962) The dynamic properties of carbon black-loaded natural rubber vulcanizates. J Appl Polym Sci Part I 6:57–63
43. Payne AR (1962) The dynamic properties of carbon black loaded natural rubber vulcanizates. J Appl Polym Sci Part II 6:368–372
44. Payne A, Watson W (1963) Carbon black structure in rubber. Rubber Chem Technol 36:147–155
45. Mullins L (1969) Softening of rubber by deformation. Rubber Chem Technol 42:339–362
46. Diani J, Fayolle B, Gilormini P (2009) A review on the mullins effect. Euro Polym J 45:601–612
47. Harwood J, Payne A (1966) Stress softening in natural rubber vulcanizates. Part III: carbon black-filled vulcanizates. J Appl Polym Sci 10:315–324
48. Fröhlich J, Niedermeier W, Luginsland HD (2005) The effect of filler–filler and filler–elastomer interaction on rubber reinforcement. Compos Part A Appl Sci Manufact 36:449–460
49. Luginsland H-D, Fröhlich J, Wehmeier A (2002) Influence of different silanes on the reinforcement of silica-filled rubber compounds. Rubber Chem Technol 75:563–579
50. Sadhu S, Bhowmick AK (2004) Preparation and properties of styrene–butadiene rubber based nanocomposites: the influence of the structural and processing parameters. J Appl Polym Sci 92:698–709
51. Kotal M, Bhowmick AK (2015) Polymer nanocomposites from modified clays: recent advances and challenges. Prog Polym Sci 51:127–187
52. Wu Y-P, Ma Y, Wang Y-Q, Zhang L-Q (2004) Effects of characteristics of rubber, mixing and vulcanization on the structure and properties of rubber/clay nanocomposites by melt blending. Macromol Mater Eng 289:890–894
53. Zhu L, Wool RP (2006) Nanoclay reinforced bio-based elastomers: synthesis and characterization. Polymer 47:8106–8115
54. Wang Y, Zhang L, Tang C, Yu D (2000) Preparation and characterization of rubber–clay nanocomposites. J Appl Polym Sci 78:1879–1883

55. Zhang L, Wang Y, Wang Y, Sui Y, Yu D (2000) Morphology and mechanical properties of clay/styrene-butadiene rubber nanocomposites. J Appl Polym Sci 78:1873–1878
56. Pramanik M, Srivastava SK, Samantaray BK, Bhowmick AK (2003) Rubber–clay nanocomposite by solution blending. J Appl Polym Sci 87:2216–2220
57. Malkappa K, Rao BN, Jana T (2016) Functionalized polybutadiene diol based hydrophobic, water dispersible polyurethane nanocomposites: role of organo-clay structure. Polymer 99:404–416
58. Lape NK, Nuxoll EE, Cussler E (2004) Polydisperse flakes in barrier films. J Membr Sci 236:29–37
59. Nielsen LE (1967) Models for the permeability of filled polymer systems. J Macromol Sci Part A Chem 1:929–942
60. Wang Y, Zhang H, Wu Y, Yang J, Zhang L (2005) Preparation and properties of natural rubber/rectorite nanocomposites. Euro Polym J. 41:2776–2783
61. Wu Y-P, Jia Q-X, Yu D-S, Zhang L-Q (2003) Structure and properties of nitrile rubber (NBR)–clay nanocomposites by co-coagulating NBR latex and clay aqueous suspension. J Appl Polym Sci 89:3855–3858
62. Jiang M-J, Dang Z-M, Xu H-P (2007) Giant dielectric constant and resistance-pressure sensitivity in carbon nanotubes/rubber nanocomposites with low percolation threshold. Appl Phys Lett 90:042914
63. Jiang M-J, Dang Z-M, Xu H-P (2006) Significant temperature and pressure sensitivities of electrical properties in chemically modified multiwall carbon nanotube/methylvinyl silicone rubber nanocomposites. Appl Phys Lett 89:182902
64. Jiang M-J, Dang Z-M, Xu H-P, Yao S-H, Bai J (2007) Effect of aspect ratio of multiwall carbon nanotubes on resistance-pressure sensitivity of rubber nanocomposites. Appl Phys Lett 91:072907
65. Wang H (2009) Dispersing carbon nanotubes using surfactants. Curr Opin Colloid Interface Sci 14:364–371
66. Halpin J (1969) Stiffness and expansion estimates for oriented short fiber composites. J Compos Mater 3:732–734
67. Bhattacharyya S, Sinturel C, Bahloul O, Saboungi M-L, Thomas S, Salvetat J-P (2008) Improving reinforcement of natural rubber by networking of activated carbon nanotubes. Carbon 46:1037–1045
68. Deng F, Ito M, Noguchi T, Wang L, Ueki H, K-i Niihara et al (2011) Elucidation of the reinforcing mechanism in carbon nanotube/rubber nanocomposites. ACS Nano 5:3858–3866
69. Dang Z-M, Jiang M-J, Xie D, Yao S-H, Zhang L-Q, Bai J (2008) Supersensitive linear piezoresistive property in carbon nanotubes/ silicone rubber nanocomposites. J Appl Phys 104:024114
70. Jo JO, Saha P, Kim NG, Chang Ho C, Kim JK (2015) Development of nanocomposite with epoxidized natural rubber and functionalized multiwalled carbon nanotubes for enhanced thermal conductivity and gas barrier property. Mater Des 83:777–785
71. Potts JR, Dreyer DR, Bielawski CW, Ruoff RS (2011) Graphene-based polymer nanocomposites. Polymer 52:5–25
72. McAllister MJ, Li J-L, Adamson DH, Schniepp HC, Abdala AA, Liu J et al (2007) Single sheet functionalized graphene by oxidation and thermal expansion of graphite. Chem Mater 19:4396–4404
73. Song SH, Jeong HK, Kang YG (2010) Preparation and characterization of exfoliated graphite and its styrene butadiene rubber nanocomposites. J Indus Eng Chem 16:1059–1065
74. Potts JR, Shankar O, Du L, Ruoff RS (2012) Processing–morphology–property relationships and composite theory analysis of reduced graphene oxide/natural rubber nanocomposites. Macromolecules 45:6045–6055
75. Lian H, Li S, Liu K, Xu L, Wang K, Guo W (2011) Study on modified graphene/butyl rubber nanocomposites. I: preparation and characterization. Polym Eng Sci 51:2254–22560
76. Xing W, Tang M, Wu J, Huang G, Li H, Lei Z et al (2014) Multifunctional properties of graphene/rubber nanocomposites fabricated by a modified latex compounding method. Compos Sci Technol 99:67–74

77. Ozbas B, O'Neill CD, Register RA, Aksay IA, Prud'homme RK, Adamson DH (2012) Multi-functional elastomer nanocomposites with functionalized graphene single sheets. J Polym Sci, Part B: Polym Phys 50:910–916
78. Du J, Zhao L, Zeng Y, Zhang L, Li F, Liu P et al (2011) Comparison of electrical properties between multi-walled carbon nanotube and graphene nanosheet/high density polyethylene composites with a segregated network structure. Carbon 49:1094–1100
79. Schaefer DW, Justice RS (2007) How Nano Are Nanocomposites? Macromolecules 40:8501–8517
80. Pang H, Chen T, Zhang G, Zeng B, Li Z-M (2010) An electrically conducting polymer/graphene composite with a very low percolation threshold. Mater Lett 64:2226–2229
81. Li M, Gao C, Hu H, Zhao Z (2013) Electrical conductivity of thermally reduced graphene oxide/polymer composites with a segregated structure. Carbon 65:371–373
82. Scherillo G, Lavorgna M, Buonocore GG, Zhan YH, Xia HS, Mensitieri G et al (2014) Tailoring assembly of reduced graphene oxide nanosheets to control gas barrier properties of natural rubber nanocomposites. ACS Appl Mater Interfaces 6:2230–2234
83. Bharadwaj RK (2001) Modeling the barrier properties of polymer-layered silicate nanocomposites. Macromolecules 34:9189–99192
84. Moosavi A, Sarkomaa P, Polashenski W Jr (2003) The effective conductivity of composite materials with cubic arrays of multi-coated spheres. Appl Phys A 77:441–448
85. Worldstyling (2018) Automotive rubber parts & plastic parts supply by worldstyling
86. Takahashi S, Goldberg H, Feeney C, Karim D, Farrell M, O'leary K et al (2006) Gas barrier properties of butyl rubber/vermiculite nanocomposite coatings. Polymer 47:3083–3093
87. Thankappan Nair S, Vijayan PP, Xavier P, Bose S, George SC, Thomas S (2015) Selective localisation of multi walled carbon nanotubes in polypropylene/natural rubber blends to reduce the percolation threshold. Compos Sci Technol 116:9–17
88. Wiwattananukul R, Fan B, Yamaguchi M (2017) Improvement of rigidity for rubber-toughened polypropylene via localization of carbon nanotubes. Compos Sci Technol 141:106–112
89. Sumita M, Sakata K, Hayakawa Y, Asai S, Miyasaka K, Tanemura M (1992) Double percolation effect on the electrical conductivity of conductive particles filled polymer blends. Colloid Polym Sci 270:134–139

Chapter 4
Processing Thermoset-Based Nanocomposites

Vincent Ojijo and Suprakas Sinha Ray

Abstract High-performance thermoset nanocomposites are advanced materials with applications in various industries, including aerospace, electronics, and automotive. Recent research and development have been focused on both two-phase systems consisting of a thermoset matrix and nanoscale filler and multiscale composites consisting of a matrix, nanoscale filler, and microscale continuous fiber fabric. This chapter discusses the various techniques used for fabrication of both the two-phase and multiscale composites, with an emphasis on epoxy-based systems. We also focus on only three nanoparticles: clays, carbon nanotubes (CNTs), and carbon nanofiber.

4.1 Overview of Thermosets

Unlike thermoplastics, thermosetting polymers undergo chemical reactions during heating (known as curing) and become infusible and insoluble. The properties of thermosets due to the network structure developed during curing that cross-link the resin molecules. Generally, thermosetting polymers can be derived from low molecular weight resins to yield *thermoset resins* or high molecular weight precursors to yield *elastomers* or *rubbers*. In this chapter, we limit our discussion to thermoset resins. Like thermoplastics, thermoset resins are generally grouped into three categories: (1) general purpose thermosets; (2) engineering thermosets, and (3) high-temperature specialized thermosets. Some of the most important thermoset resins, their precursors, and applications are summarized in Table 4.1. Within these categories, specific thermoset resins include phenols, epoxy, polyurethane, unsaturated polyesters, urea,

V. Ojijo (✉) · S. Sinha Ray (✉)
DST-CSIR National Centre for Nanostructured Materials, Council for Scientific and Industrial Research, Pretoria 0001, South Africa
e-mail: vojijo@csir.co.za; rsuprakas@csir.co.za

S. Sinha Ray
Department of Applied Chemistry, University of Johannesburg, Doornfontein 2028, Johannesburg, South Africa
e-mail: ssinharay@uj.ac.za

S. Sinha Ray (ed.), *Processing of Polymer-based Nanocomposites*, Springer Series in Materials Science 278, https://doi.org/10.1007/978-3-319-97792-8_4

Table 4.1 Types of thermoset resins and their applications

Category	Resin*	Precursor	Application
General purpose thermosets	Phenolic	Phenol and Formaldehyde and anhydrides (Resole and Novolac Types)	Coatings, adhesives, laminate mouldings, mouldings (e.g. billiard balls)
	Unsaturated polyesters	Dicarboxylic acids and diols and reactive diluents like styrene	Boat hulls, countertops, pipes, surface coatings
	Urea	Urea and formaldehyde	Moulding and extrusion compounds, foundry binders
	Melamine	Melamine and formaldehydes	Coatings, moulding compounds, laminations
Engineering thermosets	Epoxy	Epichlorohydrin and bisphenols	Engineering surface coatings, fibre composites for automotive, construction and aerospace industries
	Polyurethane (PU)	Diisocynate and polyols	Abrasion resistance and water proof coatings, adhesives
High temperature specialty thermosets	Polyimides	Aromatic diamines with aromatic dianhydride derivatives and unsaturated anhydrides	Composites for high temperature applications, e.g. in aerospace
	Bis-maleimides	Diamine and maleic anhydride	Circuit boards, aero-space composite structures, space re-entry vehicles, rockets

*The list is in-exhaustive and is only for illustrative purposes

melamine, polyimide, silicone, furan, cyanate ester, and bis-maleimide. One of the most common thermoset resins is diglycidyl ether bisphenol A (DGEBA), which is synthesized from the reaction between epichlorohydrin (ECD) and bisphenol A (BPA), as depicted in Fig. 4.1. Detailed discussion of the various thermoset types can be found in other texts [1, 2].

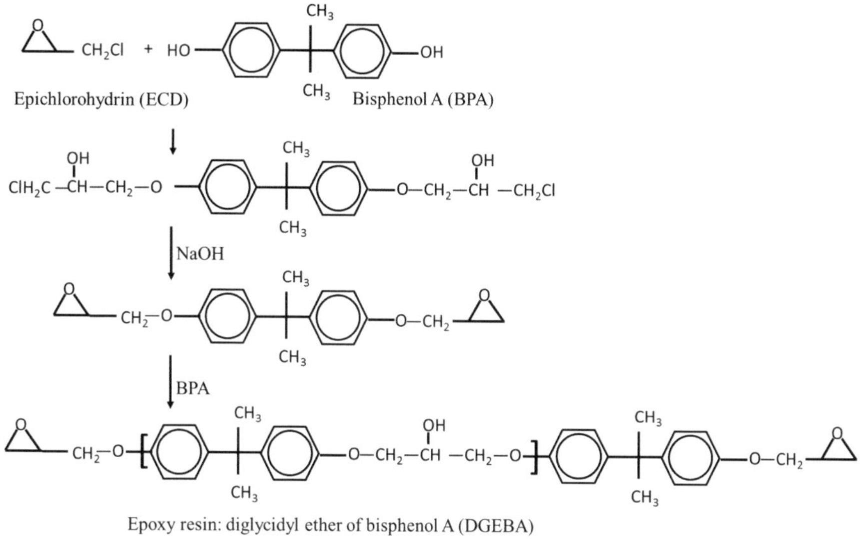

Epichlorohydrin (ECD) Bisphenol A (BPA)

NaOH

BPA

Epoxy resin: diglycidyl ether of bisphenol A (DGEBA)

Fig. 4.1 Reaction scheme for DGEBA epoxy resin

4.1.1 Network Structure in Thermosets

To form a three-dimensional (3D) cross-linked structure, the functionality (f), number of reactive moieties per mole of reactant, of the molecules has to be more than two. The reaction between precursor molecules with functionality of 2 always yields a product or intermediate molecule with a functionality of 2. The reaction yields a linear structure, as depicted in Fig. 4.2. On the other hand, reaction of molecules with a functionality of more than two results in branching, leading to the formation of a network structure, as shown in the schematic in Fig. 4.2. During polymerization, the point at which the network structure first occurs with unit probability is referred to as the *gelation* point. The gelation process can be evaluated from the viscosity curves and is determined as the point at which the viscosity increases sharply due to network formation in the resins. As the curing proceeds, the viscosity increases towards infinity as the network density increases.

4.1.2 Crosslink Density

The final properties of thermosets depend on the crosslink density, which is defined as the number of effective crosslinks per unit volume of the thermoset material, or the molecular weight between the two crosslink points (M_c). In straight-chain polymers, the molecular weight significantly affects its properties, whereas the crosslink density controls the properties of thermosets. As the crosslink density increases, so do

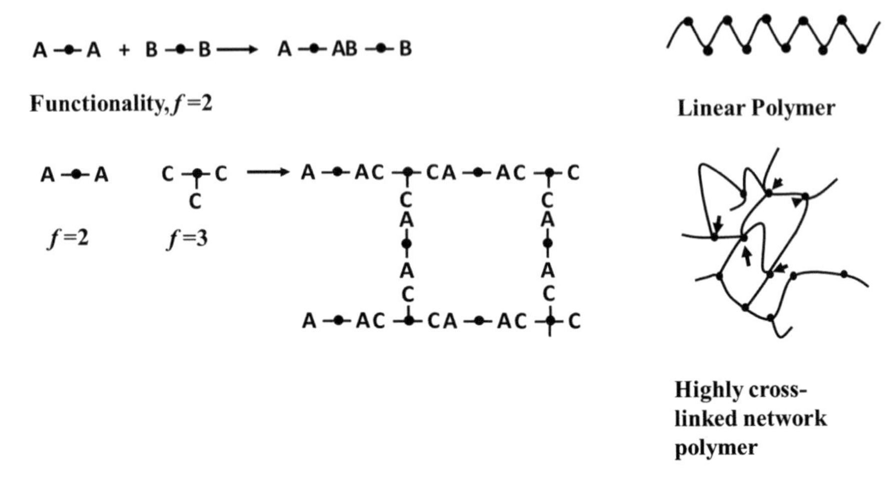

Fig. 4.2 Schematic representation of linear polymerization and network formation for molecules with functionality of >2

the modulus, mechanical strength, hardness, glass transition temperature, heat resistance, and internal stress. However, the impact strength, ductility, low temperature flex, coefficient of thermal expansion, peel strength, and thermal shock resistance decrease with increasing crosslink density. The crosslink density is influenced by many variables, including: functionality; chain length between crosslinks; the number of sites actually participating in the reaction (controlled by the process profiles); and chain mobility. The most common technique for quantifying the crosslink density is via swelling studies, as proposed by Flory-Rehner [3] theory. By evaluating the degree of swelling, the extent of crosslink can be ascertained as the former is inversely dependent on the crosslink density. Other techniques, such as differential scanning calorimetry (DSC), infrared (IR) or near-IR spectroscopy, and rheology involve measurement of the extent to which the reactive sites of the resin are consumed. Alternative methods use correlations between the crosslink density and e.g., the glass transition temperature, dynamic mechanical properties (e.g. storage modulus), refractive index, and heat distortion temperature.

To effectively apply thermoset resins or their composites, understanding their cure kinetics is necessary. Analysis of the cure kinetics requires a definition of the extent of cure, or conversion (α). This can be determined by measuring the disappearance of the reactive functional groups, via thermal and spectroscopic techniques, including IR spectroscopy, nuclear magnetic resonance (NMR) spectroscopy, and DSC methods. In the case of IR spectroscopy, α is calculated as follows:

$$\alpha = 1 - \frac{A^t}{A^o} \tag{4.1}$$

where, A^t and A^o are the normalized IR absorbance of the reactive functional group at time *t* and at the beginning of the reaction, respectively. Normalization is performed by taking the absorbance ratio of the functional group to the absorbance of a group that does not take part in the reaction.

The conversation may also be obtained from DSC using the following equation:

$$\frac{d}{dt} = \frac{1}{H_0}\frac{dH}{dt} \tag{4.2}$$

where H_0 is the total exothermic enthalpy released for a complete cure, while H is the heat realized from the start of polymerization to a time *t*. The extent of cure can be used to evaluate the kinetics of curing, as described using different models. A general kinetic model is shown in the following equation:

$$\frac{d\alpha}{dt} = K_1(1 - \alpha)^n \tag{4.3}$$

where, K_1 is the rate constant and n is the order of the reaction.

In addition to the temperature, pressure plays an important role in molding thermosets. Without application of pressure, voids can occur in thermoset parts due to curing shrinkage. Therefore, in a number of processing profiles, temperature and pressure are controlled in order to obtain parts with minimal internal stresses and are free of voids.

4.2 Thermoset Nanocomposites

Thermoset nanocomposites are a class of nanostructured materials where the constituent particles are of nanometer dimensions. These materials have demonstrated significant enhancement in the e.g., mechanical, electrical, and thermal properties at low nanoparticle (NP) loading. Compared to microcomposites, nanocomposites can exhibit a well-balanced enhancement of the properties, as opposed to the property trade-offs that characterize the microcomposites. Hence, they find applications in automotive, aerospace, and marine fields, as well as replacing metals due to their high specific strength. Some commonly used NPs include clay [4–10], silica [11–14], CNTs [15–21], graphene [22–25], and polyhedral oligomeric silsesquioxane (POSS) [26–31]. This chapter focuses on nanocomposites based on CNTs, carbon nanofibers (CNF), and clays. Like thermoplastic nanocomposites, the performance of the thermoset nanocomposites depends on the extent of dispersion and distribution of the particles in the resin matrix. Therefore, during the different processing procedures, the aim is to achieve good NP dispersion, and in the case of e.g., CNTs and graphene, percolation of the particles is equally important.

4.3 Processing Techniques for Thermoset Nanocomposites

In situ polymerization is predominantly used to process thermoset nanocomposites. The NPs are mixed in the resin before curing to form the desired part. The mixing of precursors/resins and the NPs is achieved via ultrasound stirring, ball milling, mechanical stirring, ultrasonication, and high shear mixing. Subsequent forming techniques include reactive injection molding, compression molding, casting, and 3D printing. Some of these processing techniques are briefly discussed here.

4.3.1 Reactive Injection Molding

Reaction injection molding (RIM) is a rapid production technique used to produce thermoset parts with complex features directly from monomers or oligomers. Developed by Bayer AG in the 1960s for polyurethane (PU) part production, it is considered an economical technique to fabricate relatively large and complex parts. The reaction in this process is essentially step growth polymerization with no formation of byproducts during formation of PU nanocomposites. The schematic in Fig. 4.3 shows a commercial PU RIM process as presented by Covestro AG (formerly Bayer AG).

In the case of PU, the process uses two low-viscosity liquid components, i.e., polyol and isocyanate, where one may have NPs dispersed in it. These two components are known as a *RIM system*, and depending on their formulations, the molded part can be either be flexible or rigid, as well as solid or porous (foamed). The two components are pumped into a mix-head where mixing of these reactant occurs by high-speed impingement, from where they flow into the mold. Then, curing occurs before the sample is de-molded.

4.3.2 Resin Transfer Molding

Resin transfer molding (RTM) is a low-pressure process involving injection of liquid resin and other constituents into a closed mold containing continuous fibers or a preform. It can be employed for a wide range of resins systems, including epoxies, polyesters, phenolics, and vinyl esters. Detailed analysis of the process can be found in other texts [33–35]. Vacuum-assisted RTM (VARTM) is a variant of the RTM process where vacuum is applied to evacuate the mold, as depicted in Fig. 4.4. This technique is also known as vacuum-assisted resin infusion molding (VARIM).

A recent trend has been to use RTM to manufacture hierarchical or multiscale composites, containing both continuous fibers (such as aramid, glass, or carbon) and discontinuous NPs, such as CNTs [38–44], carbon nanofibers [45–47], graphene [48, 49], clays [36, 37, 50–52], cellulose nanofibers [53, 54], and alumina nanoparticles [55]. These composites can enhance targeted performance, e.g., electrical conductiv-

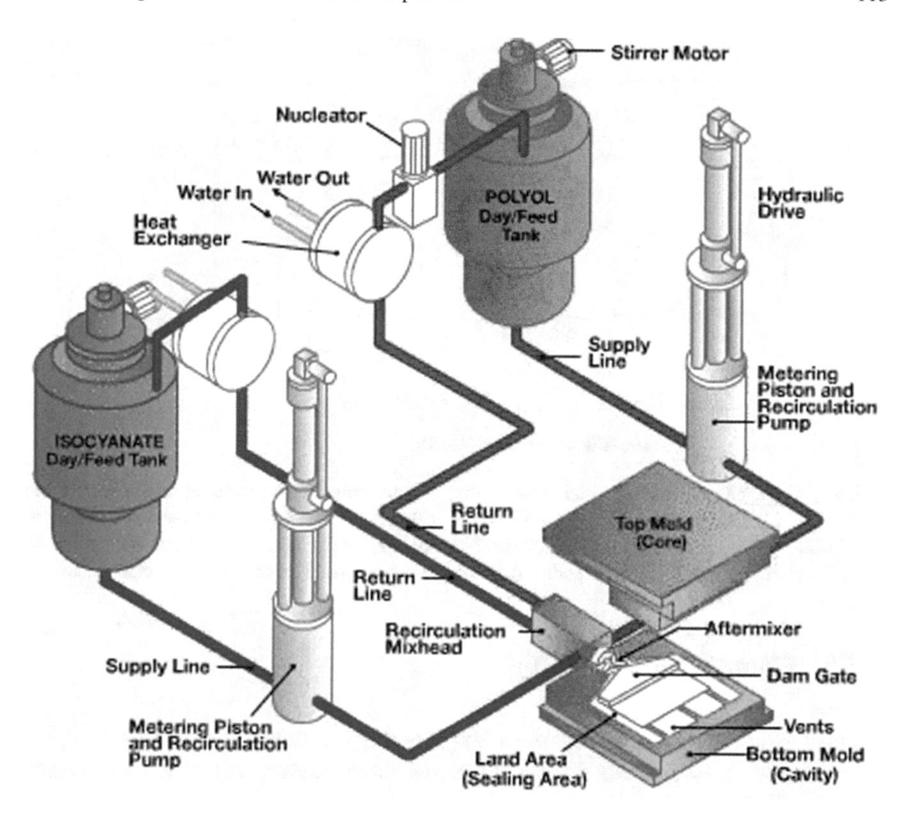

Fig. 4.3 Schematic showing polyurethane RIM process. Reproduced with permission from [32]. Copyright 2017, Covestro AG, Leverkusen

ity, inter-laminar mechanical properties, and fire retardation compared to pure polymer. Further discussion of multiscale composite processing is presented in Sect. 8.5.

For a successful defect-free RTM process, certain parameters need to be understood and controlled, including the injection pressures, viscosity of the resin, wettability of the fibers, temperature, cure kinetics, and part design [33, 56]. Controlling the key parameters ensures defects such as voids and agglomerated NPs are avoided. Flow of the resin during injection and wetting of the reinforcement fibers are key factors determining the resultant structure and subsequent properties of the composite. The flow becomes an even more critical issue when NPs are dispersed in the resin as the particles affect the flow behavior. Depth filtration of these particles can occur if the spaces between the continuous fibers are smaller than the size of the dispersed discontinuous NPs. This may lead to agglomeration of the NPs and structural defects at such points.

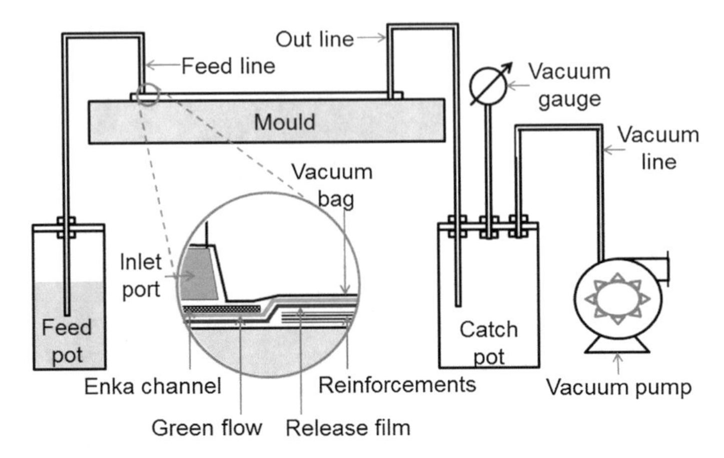

Fig. 4.4 VARTM process to fabricate the thermosetting composites processed at 99.5% vacuum level following by the curing of 10 h. The feed pot contained a dispersion of organo-modified montmorillonite (OMMT) in unsaturated polyester, while the reinforcement was a continuous biaxial glass fiber. Reproduced with permission from [36, 37]. Copyright 2016, Elsevier Science Ltd.

4.3.3 Compression Molding

Compression molding is widely used for preparing thermoset composites. A typical compression mold consists of two halves, an upper male part and a lower female part. The lower part has a cavity, while the upper part has a projection that fits into the cavity, leaving a space between them. The space between the male and female parts of the mold gives the shape of the desired composite part. A slight excess of material is fed into the cavity, and then the upper part is places, resulting in the excess material being pushed out of the mold. After curing, this excess is trimmed off. To promote curing reactions, the temperature and pressure are controlled to ensure that the formed part is of appropriate quality.

4.3.4 Casting

In the casting process, the resin, curing agent, and modifiers (e.g., CNTs) are premixed and degassed before being cast either in a die or rotational mold. For solid components, die casting is performed, whereas for hollow components, rotational molding is required. The die is preheated to the desired temperature before the resin mixture is added. At a suitable temperature, curing occurs, while the part shrinks, thus enabling easy removal after cooling. For rotational casting, the premixed and degassed resin system is poured into a rotational mold, which rotates in the x and y directions, spreading the resin mixture uniformly. Optimization of the temperature ensures correct curing. After cooling, the part is de-molded.

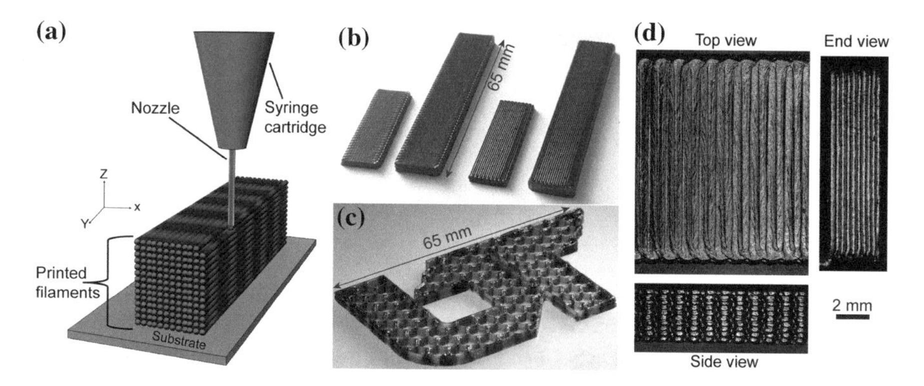

Fig. 4.5 **a** Schematic of a direct wire-3D printing and examples of 3D printed articles: **b** transversally and longitudinally printed DMA and Flexural test specimen bars; **c** University of Tennessee printed logo with honeycomb infill and **d** optical micrographs of a transverse flexure specimen printed using the 12.5 wt% nanoclay ink formulation. Reproduced with permission from [57]. Copyright 2018, Elsevier Science Ltd.

4.3.5 3D Printing

3D printing of thermoset-based nanocomposites includes additive processes of forming bottom-up 3D objects, layer-by-layer, directly from computer-aided design files. 3D printing does not involve subtractive processes like milling. Figure 4.5a shows a schematic of a 3D printing process known as *direct-wire 3D printing,* where resin held in a cartridge is ejected through a nozzle onto a substrate in a layer-by-layer manner to form the 3D object. In one instance, Hmeidat et al. [57] successfully formulated epoxy/nanoclay ink containing up to 12.5 wt% nanoclay for use in direct-wire 3D printing. The printed parts (see Fig. 4.5b) exhibited enhanced mechanical properties, suggesting that the epoxy/clay formulations could be developed further for 3D printing feedstocks.

4.4 Clay-Based Thermoset Nanocomposites

Although thermoset materials are extensively applied for structural components in various industries, some exhibit brittleness and internal stresses, depending on the quality of the curing step. Clays have been incorporated into thermoset resins, including epoxy [9, 58–60], phenolic resins [61–66], PU [67, 68], unsaturated polyester [69–71], and vinyl ester [72, 73], to enhance functionalities such as flame retardation [36]. Like thermoplastic systems, filler dispersion remains a challenge in thermoset-based clay nanocomposites. Hence, much research has been undertaken to improve dispersion of the clays, preferably to an exfoliated state. In thermoset nanocomposites, in situ polymerization is widely used; hence, pre-dispersion of the clays in either a solvent or one of the components of the resin system is vital. It is also important to understand the effect of the clays on the curing kinetics, the structure of the composites, and the final properties.

4.4.1 Surface Functionalization

Pristine clay is hydrophilic and may not be compatible with some resin systems. Therefore, before fabrication of a nanocomposite, suitable surface modification of the clay to match the chemistry of the resin is required. Commonly used surfactant modifiers include cations, such as ammonium or phosphonium ions. The mechanism of modification is cation exchange with loosely held cations, such as Na^+ and Ca^+, within the gallery spacing. The effect of this is the expansion of the gallery spacing, which enables diffusion of resin oligomers during intercalation. Silanization of the clays is another method of surface modification to enhance interactions with the organic oligomers of the resin systems [74, 75].

4.4.2 Techniques for Clay Dispersion

A critical step in fabricating clay–thermoset nanocomposites is the dispersion of clay in the resin prior to curing. A number of factors influence the extent of dispersion of the clays in the resin systems, including the clay type and modification, resin type, curing agent, temperature [75], time [76], and mixing technique [77]. Several methods are available for dispersion, including mechanical stirring, high-speed mixing, ultrasonication, ball milling, and electric field dispersion methods [78]. Some of these methods are discussed here.

In *mechanical mixing*, clay and resin system mixtures are mechanically stirred until equilibrium dispersion is achieved. Pure montmorillonite (MMT) (CNa^+) and MMT modified by various surfactants (C30B: methylbis(2-hydroxyethyl)-hydrogenated tallow quaternary ammonium; C10A: benzyldimethyl-hydrogenated tallow quaternary ammonium; CID19: 3-hexadecyl-1-methylimidazoliumbromide) were dispersed in a phenolic using high-speed mixing (5000 rpm for 5 min) [65]. Typically, a mixture of novolac resin and curing agent was dissolved in methanol and then the desired quantity of clay was dispersed in the solution. The authors observed that phenolics do not readily intercalate into either the pure or organically modified MMT; high-speed shearing resulted in significant increase in the gallery spacing of only pure MMT. The extent of dispersion defined by the tactoid sizes and the gallery spacing of the clay in the final composite followed the order, $CNa^+ \approx CID19 > C30B > C10A$. Poorer dispersion was observed for increasing clay loading. This trend indicated that the further hydrophilic modifiers, preferably containing phenolic functions, should be investigated.

Epoxy resins are able to intercalate into clays better than phenolics. Ngo et al. [79] studied the effect of temperature, time, and speed of pre-mixing on the dispersion of organically modified clay in epoxy. X-ray diffraction (XRD) and viscosity measurements were used to characterize the dispersion quality. In a typical experiment, the clay (C30B), was mixed with the epoxy (EPON™ 828) and mechanically stirred in a homogenizer at different speeds, temperatures, and mixing times. The authors observed that these factors did not affect the extent of dispersion, with the d_{001} spacing expanding from 1.85 nm for C30B to 3.72–3.81 nm for nanocomposites undergoing different pre-treatments.

Ultra-sonication involves the use of high-energy vibrational waves in a localized region which can potentially separate nanoclay platelets in a liquid medium. The vibrations result in localized heating that may initiate polymerization. Therefore, extra mechanical stirring and a cooling jacket is recommended for ultrasonication process. Some critical parameters that require optimization include: the power, frequency, duration of mixing, and temperature [80, 81]. In a typical procedure, clays are mixed with a resin system, and then sonicated before other components are added. Zhou et al. [82] obtained a homogenous mixture of epoxy resin and clay after high-intensity sonication of the dispersion. The clays were firstly mixed with diglycidyl ether of bisphenol A via sonic cavitation and then the hardener (cycloaliphatic amine) was added and mechanically mixed with a high-speed mixer. A high-intensity ultrasonic liquid processor was used to obtain a homogeneous molecular mixture of epoxy resin and nanoclay. Even though the extent of dispersion was not reported, the mechanical properties of the composite were enhanced at low clay loadings, and further addition of clay beyond 2 wt% did not show any incremental improvement in the mechanical properties. This could be due to inefficient dispersion at higher clay loadings. Other researchers also proposed that it is not possible to obtain exfoliated platelets via sonication; rather, intercalated nanoclay platelets form stacked clusters [83]. In an unsaturated polyester system, mechanical mixing and further sonication of C30B clay mixed with pre-promoted casting polyester resin resulted in a well-dispersed and stable suspension [84]. Crosslinking was initiated by adding a catalyst, but the authors [84] noticed a slower crosslinking reaction at clay loadings of >2.5 wt%.

High-shear mixing methods include the use of a three-roll mill [85–87], where the clay/resin system mixture is fed between feed and center rollers, and then collected from an apron roller. The shearing of the platelets can potentially occur as the material passes between the rollers. Yasmin et al. [85] used a three-roll mill to disperse clay NPs in an epoxy matrix. The authors achieved good dispersion of the clays, even at loadings as high as 10 wt%. The process involved a three-component epoxy system where the Diglycidyl Ether of Bisphenol A (DGEBA) resin (GY6010), anhydride hardener (Aradur 917), and accelerator (DY070) were mixed in proportions of 100:90:1. First, the DGEBA resin was placed on the roller and the motion started. Thereafter, clay was slowly added, ensuring maximum contact with the roller. Room-temperature compounding was performed for 3 h at a rolling speed of 500 rpm. The milled product was mixed with the hardener and placed on a hot plate for 1 h at 60°, before adding the accelerator. The obtained nanocomposites showed good dispersion

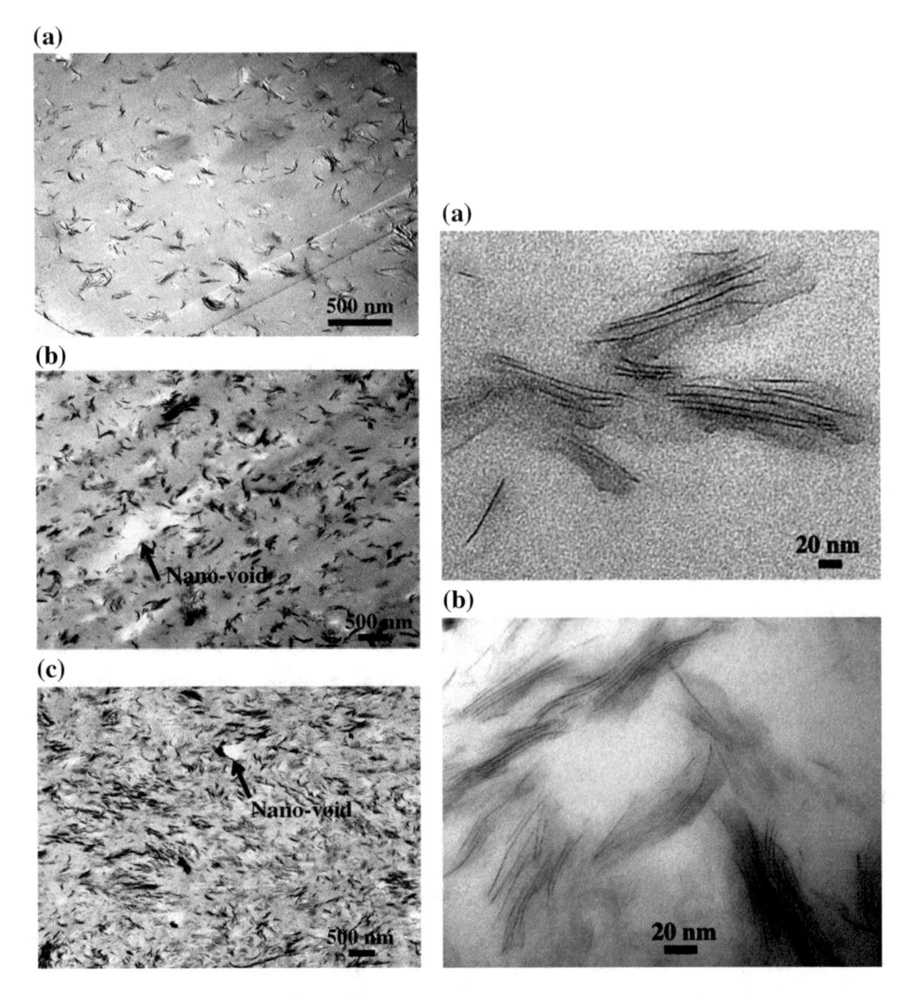

Fig. 4.6 (Left) Low magnification TEM images of clay nanocomposites: **a** 1 wt%, **b** 5 wt% and **c** 10 wt%; (Right) high magnification TEM images of: **a** 5 wt% and **b** 10 wt%. Reproduced with permission from [85]. Copyright 2013, Elsevier Science Ltd.

(exfoliation/intercalation) as shown in the TEM images in Fig. 4.6. The XRD patterns did not show any peaks for the nanocomposites, indicating amorphous structures as a result of the reduced stack thicknesses, higher basal spacing, and random orientation of the platelets.

Ball milling can be used to reduce the size of clay agglomerates in resins and is considered cost effective, versatile, and scalable. Ball milling has been used to improve the dispersion of halloysite in an epoxy matrix [88] and enhance further exfoliation of MMT [89]. Lu et al. [89] employed ball milling to help exfoliate MMT that was previously modified with dodecyl benzyl dimethyl ammonium chloride (DBDA) and *meta*-xylylenediamine (MXDA). The process involves premixing the modified clays with resin (DGEBA), then adding a solvent (ketone in this case), and pulverizing the dispersion for a given time. Ball milling exerts external shearing forces on the modified MMT in solution; hence, large agglomerates are broken down, also facilitating exfoliation of the MMT in the epoxy. Compared to stirring, an additional ball milling step results in nanocomposites with better clay dispersion (exfoliation) and better flexural and impact properties.

High-pressure mixing applies shear forces, as well as the impingement against obstacles such as vessel walls, to help disintegrate clay agglomerates. In a typical process, Nanomer I.30E was dispersed in acetone to form a paste with a 10 wt% loading. The mixture was forced under high pressures (15,000 psi) into a small chamber and then impacted on the walls of the chamber several times to break down the clay agglomerates [90]. The paste containing the dispersed clay was then added to the epoxy system, and then mixed first manually, then mechanically. The hardener was then added after raising the temperature to a suitable level. The d_{001} spacing increased from 2.37 to 3.22 nm due to the additional high-pressure treatment compared to direct mixing. A disadvantage of this technique is the formation of voids and limited clay loading due to the high viscosity.

4.4.3 Properties of Nanoclay-Containing Thermoset Nanocomposites

4.4.3.1 Mechanical Properties

Generally, the addition of clays enhances the mechanical performance of the nanocomposites, depending on the state microstructure and extent of cross-linking achieved. Yasmin et al. [85] were able to obtain good dispersion of clays (basal spacing of>5 nm for 1–10 wt% clay content), resulting in an increase in the elastic modulus, with a maximum of 80% improvement for an epoxy/clay nanocomposite containing 10 wt% clay. However, the tensile strength decreased for all clay concentrations, which was attributed to the occurrence of some nanoscale to microscale voids and clustering of nanoparticles. In a separate study, the same group [87] also studied the influence of the extent of clay dispersion on the mechanical performance of clay-based thermoset nanocomposites. Clays with different organic modifiers were compared, including octadecyl trimethyl ammonium (ODTMA)–Nanomer I.28E, methyl, tallow, *bis*-2-hydroxyethyl, and quaternary ammonium (MT2EtOH)-C30B. Due to the interaction between the hydroxyl group in C30B and the resin system, better dispersion was achieved [87]. The elastic moduli for C30B nanocomposites were superior to those of Nanomer 1.2 E nanocomposites at all loading levels.

Some researchers have hypothesized that organoclays can interfere with the curing reaction, leading to poorer mechanical properties. Bharadwaj et al. [84] observed a reduction in the tensile modulus, and the storage and loss moduli, with the addition of clay to an unsaturated polyester matrix. Although the dispersion was good, especially at 2.5 wt% clay loading, the poor mechanical properties were attributed to a reduced degree of cross-linking.

In novolac phenolic resin-based composites containing clay and woven glass fiber, both pristine and modified MMT clay improved the modulus significantly. At only 2.5 wt%, pristine and modified MMT enhanced the elastic modulus by 38% and 43%, respectively [91]. Withers et al. [92] observed a similar trend in the enhanced mechanical properties of clay-reinforced epoxy/glass fiber hierarchical composites. There was an improvement of 11.7%, 10.6%, and 10.5% in the ultimate tensile strength, tensile modulus, and tensile ductility, respectively, compared to the pure polymer.

4.4.3.2 Fire Retardation

Layered silicates are known to impart improved fire retardation behavior to thermoplastics. In particular, layered double hydroxides (LDHs) are known for their ability to enhance flame retardation. Kalali et al. [93] modified LDH with a multi-component modifier consisting of hydroxypropyl-sulfobutyl-beta-cyclodextrin sodium (sCD), dodecylbenzenesulfonate (DBS), and taurine (T), and fabricated epoxy/LDH nanocomposites. In a typical procedure, the modified LDH was dispersed in epoxy resin via a two-step process using a three-roll mill followed by sonication in acetone. The resultant nanocomposites exhibited significantly improved fire retardation properties. The limiting oxygen index (LOI) values for the nanocomposite with multiple modifiers were 26.8% compared to 23% for pure epoxy. The UL 94 nanocomposite achieved a V-0 rating, where the flame was extinguished immediately. In cone calorimetry tests, the time to ignition and peak heat release rate (pHRR) were 40 s and 318 kW m^{-2}, respectively. The corresponding values for the pure epoxy were 58 s and 318 kW m^{-2}, respectively.

Various factors were investigated by Nguyen et al. [36] for their influence on the fire retardation of multiscale composites containing organically modified montmorillonite (OMMT). The authors used a statistical design of experiments (Taguchi) to evaluate: three OMMT contents (1, 3, and 5 wt%); two types of organo-modifications (silane (3–5 wt% aminopropyl triethoxysilane) and 15–35 wt% octadecylamine treated MMT, and dimethyl dialkyl (C14–C18) amine modifier); and three types of resins (epoxy, unsaturated polyester, and vinyl ester). The fourth factor investigated was the procedure: (i) mechanical mixing of nanoclays with the resins prior to infusion; (ii) sonication of the nanoclay dispersion in the resins prior to infusion; and lastly, (iii) prior dispersion of nanoclays in acetone before mechanical mixing with resins. The general finding was that the organoclays enhanced fire performance of the thermoset composite panels; specifically, the performance was dependent on the type of resin and clay modifier. Ultrasonication dispersed the clays better than mechanical stirring, resulting in improved fire performance. The unsaturated polyester composite with 5 wt% clay modified with dimethyl dialkyl amine treated by procedure B was the most desirable factor combination to give the best fire performance. The group [36] further investigated the role of OMMT in enhancing

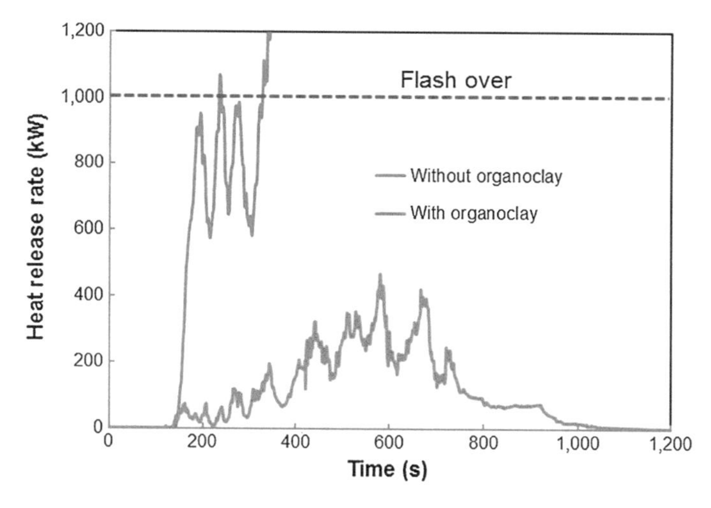

Fig. 4.7 Heat release rate as a function of time for GFRP panel without clay, and hierarchical composite containing GFRP only (turquoise color) and organoclay (orange color). Reproduced with permission from [36]. Copyright 2016, Elsevier Science Ltd.

the fire retardation in a hierarchical composite based on unsaturated polyester and continuous glass fibers, prepared using VA-RTM. Multiscale composites, comprising glass fiber reinforcement and dimethyl dialkyl (C14-C18) amine-modified OMMT, showed better fire retardation than the glass fiber reinforced polymer (GFRP) alone. Figure 4.7 shows the time evolution of the heat release rate (HRR) for the GFRP and the multiscale composite. Evidently, at only 5 wt% concentration, OMMT reduced the total HRR and prevented flash-over (defined by ISO 9750-1:2013 as the period when HRR within the room reaches 1 MW). However, the GFRP burnt more severely; flash-over was reached after 197 s of burning.

In conclusion, clay-based thermoset nanocomposites exhibit improved performances over pure resin matrices. Dispersion of the clays controls many of the properties, yet still remains a challenge to date. Emerging hybrid systems consisting of traditional continuous fibers together with clays are promising for unlocking the potential of these NPs.

4.5 Carbon Nanotube (CNT)- and Carbon Nanofiber (CNF)-Containg Thermoset Nanocomposites

CNT- and CNF-based thermoset composites are a class of advanced materials with promising properties and a wide range of practical applications in e.g., automotive and aerospace industries [94]. These NPs have high electrical conductivity, outstanding mechanical and thermal properties, large specific surface area, and very high aspect ratios. These factors make the particles attractive for composite fabrication. Due to their high specific surface area, these NPs are able to substantially modify properties of the thermoset matrices at very low concentrations. The most common resin used in these composites is epoxy [95–98], although the CNTs and CNFs have also been added to other thermoset matrices, such as phenolics [94, 99, 100], PU [101], PI [102], unsaturated polyesters [18, 103–107], and vinyl ester [19]. In this chapter more emphasis is placed on epoxies than other resins. The composites are either based on single-scale particles or multiscale systems. The latter are predominantly fabricated via RTM methods, while casting and compression molding can also be used. Some processing methods for the various CNT- and CNF-based systems are summarized in Table 4.2.

Even though CNT/thermoset nanocomposites can be fabricated without surface modification for some resins, it is sometimes necessary to functionalize the CNT and/or CNF surfaces to achieve suitable dispersion in the resins. The functionalization can be achieved via e.g., ultrasound ozonolysis [111], plasma treatment [94], acid treatment [47, 98], fluorination [98], and amine treatment [46, 109, 112]. The composites may be based on a single system of NPs, a combination of similarly scaled NPs, or a multiscale system comprising microscale continuous fibers reinforced with NPs.

4.5.1 Single-Scale System of CNT or CNF Thermoset Nanocomposites

These systems comprise a single type of NPs or a hybrid of NPs with a similar scale dispersed in a resin matrix [95–98, 112, 113]. Examples of such a system include the recent works of Li et al. [112] and Cheng et al. [95]. The former fabricated epoxy/MWCNT single system nanocomposites via compression molding. These epoxy/MWCNT nanocomposites were then used to make foamed structures. Alpha-zirconium phosphate (ZrP) was used to disentangle carboxylated MWCNT before functionalization with sulfanilamide. The process was based on techniques described by Sue's group [114, 115], as shown in the scheme in Fig. 4.8, which includes the following steps: (i) exfoliation of the MWCNT with ZrP in aqueous medium; (ii) stabilization of the dispersion with a sodium dodecyl sulfate (SDS) surfactant; (iii) removal of the ZrP through precipitation via acid and centrifugation to isolate MWCNT; and (iv) replacement of the ionic SDS surfactant with an organophilic surfactant (sulfanilamide). Finally, the functionalized MWCNT (f-MWCNT) in acetone was premixed with epoxy resin and degassed, before compression molding to fabricate an epoxy/f-MWCNT nanocomposite.

The foam structure and electrical conductivity depended on the concentration of CNTs in the composite. As shown in Fig. 4.9, the mean cell size generally decreased (and the density increased) with increasing CNT concentration. The mean cell diameter and density of pure epoxy (EP) foams were 12.3 μm and 3.35×10^8 cells/cm^3, respectively. However, at 1 wt% CNT concentration, the corresponding

Table 4.2 Preparation techniques for Carbon nanotube/nanofibre–based thermoset nanocomposite

	CNT/CNF type/modification	Technique	Comments	References
Epoxy	Hexanediamine functionalised electrospun carbon nanofiber (0.1–0.3wt%)	Casting for CNF/epoxy and VARTM for multiscale CNF/carbon fibre composites	Improved inter-laminar shear strength (ILSS), impact and flexural strength	[46]
	0.5–1wt% carbon nanofibre (CNF)	Sonication of CNF/resin then RIM	3–6 orders enhancement in electrical conductivity	[108]
	Carboxylic acid- and amin-functionalised carbon nanofibre	Electrophoretic deposition of CNF on carbon fibres then VARTM	Hierarchical composites exhibit better ILSS and compressive strength	[47]
	0.1–0.3 wt% Amino functionalised and non-functionalised CNT	Calendaring of CNT/resin before VARTM	Amino-functionalised CNT inhibit shear deformation and exhibit higher flexural modulus and strength	[109]
	0.5wt% MWCNT	Injection double VARTM (Glass fibre + MWCNT) and flow flooding chamber (FFC)	No improvement with VARTM; 21.3% improvement with FFC	[38]
Unsaturated polyester	Unmodified CNT	Pre-dispersion in solvent THF then casting	Tensile strength, modulus, impact strength, and elongation-at-break, of nanocomposite increased to 22, 20, 28, and 87%	[103]
	0.5–1.5wt% surfactant coated CNF + untreated CNF	Mechanically mixed CNF/polyester infused into glass fibre preform (VARTM)	Filtration observed > 1wt% CNF. VARTM successful for surfactant coated CNF but not the untreated. Improvement in mode-I delamination resistance	[110]

(continued)

Table 4.2 (continued)

	CNT/CNF type/modification	Technique	Comments	References
Phenolic	CNT grown on Carbon fibre substrate via CVD	Wet Mixing of phenolic/CF-CNT	20–75% increase in flexural strength; and 28–46% increase in flexural modulus	[99]
	0.5–4wt% MWCNT (unmodified)	Wet impregnation of CNT by phenolic	Better mechanical properties for network MWNTs than dispersed MWNTs composites	[100]

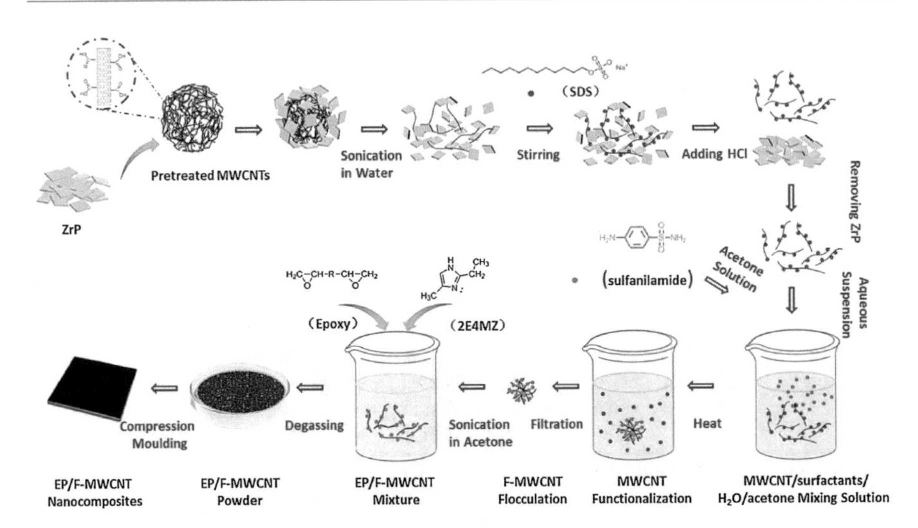

Fig. 4.8 Schematic detailing the surface functionalisation process and fabrication of nanocomposites of epoxy/f-MWCNT. Reproduced with permission from [112]. Copyright 2018, Elsevier Science Ltd.

values were 3.4 μm and 7.79×10^9 cells/cm^3, respectively. Above 2 wt% CNT content, no improvement in the cell structure was observed. The foamed structure also showed higher electrical conductivity than that of the solid nanocomposite structure. At 5 wt% CNT loading, the electrical conductivity of the solid EP/f-MWCNT nanocomposite was 5.37×10^{-7} S/cm, compared to 3.76×10^{-5} S/cm for the equivalent foam. However, these values are lower than those reported by Cheng et al. [95] for RTM-formed CNT/epoxy composites.

Cheng et al. [95] described a process for fabricating epoxy/CNT nanocomposites via RTM. Firstly, the authors synthesized highly aligned MWCNTs on a 4-inch silicon

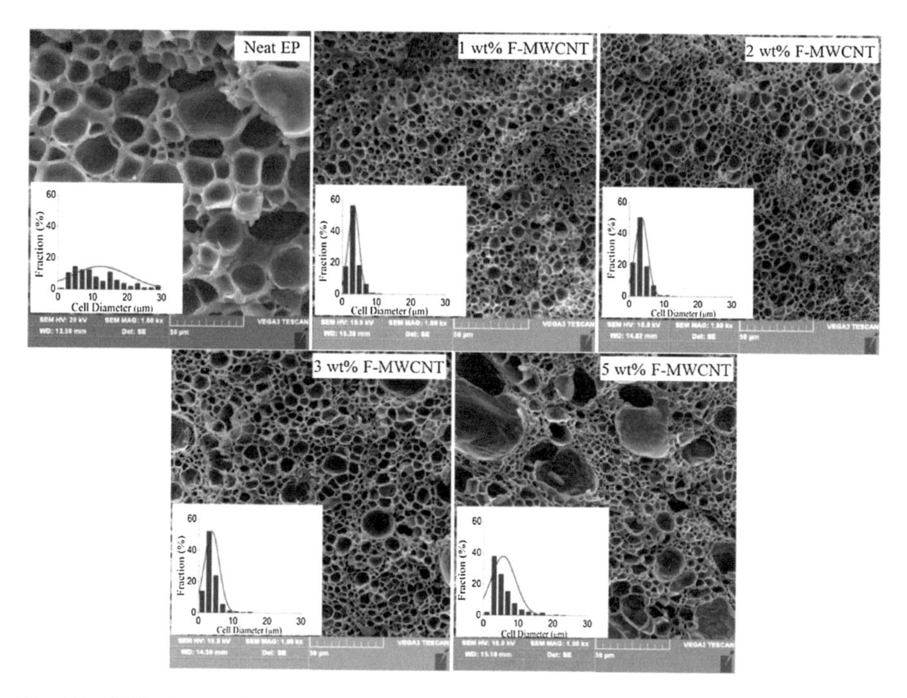

Fig. 4.9 SEM micrographs depicting the effect of concentration of the MWCNT on the cell diameter distribution of neat epoxy and EP/F-MWCNT foams. Reproduced with permission from [112]. Copyright 2018, Elsevier Science Ltd.

wafer in a low-pressure chemical vapor deposition (LPCVD) using Fe as a catalyst and acetylene as the precursor, following a method developed by the same group [116]. To form continuous and aligned MWCNT sheets, the CNTs were drawn from highly aligned arrays and were held end to end by the van der Waals forces. These MWCNT sheets were then used to make CNT preforms with different orientations, 0° and 0°/90°, similar to standard carbon fiber preforms. Figure 4.10a shows the orientation of the macroscopic MWCNT preforms, with about 2000-sheet stacks, used for the RTM process. The preforms were placed in the mold as illustrated in Fig. 4.10b, followed by infusion of the epoxy resin. The resin consisted of glycidyl ester, and resorcinol diglycidyl ethers, while hexahydrophthalic anhydride was the curing agent. The resulting composite parts contained highly aligned CNTs with a loading of 16.5 wt%, which exhibited dramatic increases in electrical and mechanical properties. The electrical conductivity along the fiber length was over 1×10^4 S/m, whereas the Young's modulus and tensile strength improved by 716% and 160%, respectively. CNTs and CNFs can also be used to further reinforce and add extra functionality, e.g. electrical conductivity to microscale composites. In the following sub-section, these hierarchical composites are discussed.

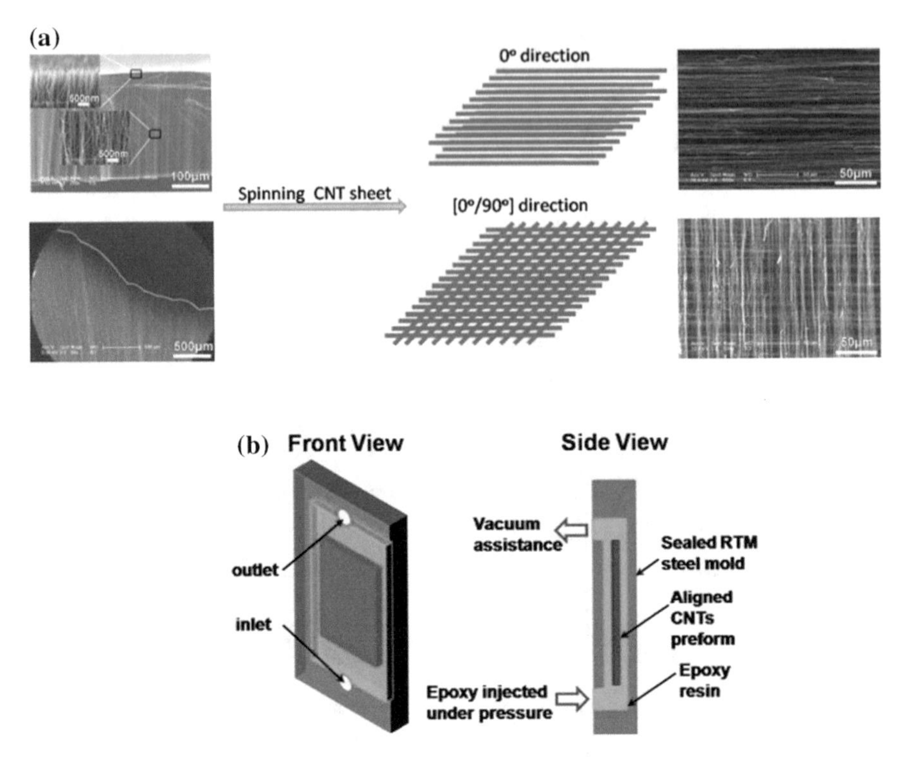

Fig. 4.10 **a** SEM images of CNT preforms with [0°] and [0°/90°] alignment of CNT sheets and **b** RTM process schematic for moulding CNT/epoxy composites. Reproduced with permission from [95]. Copyright 2010, Elsevier Science Ltd.

4.5.2 Multiscale Composites

These composites contain hierarchical constituents of microscale continuous fiber reinforcement and discontinuous NPs/nanofibers within the thermoset matrix [44, 117]. The in-plane specific tensile strength and stiffness of microscale fiber composites are excellent; however, their out-of-plane properties and in-plane compressive strength are relatively low. The incorporation of CNTs and CNFs aims to enhance the in-plane compressive strength and out-of-plane tensile strength and stiffness; the literature contains numerous examples of such improvements [44]. CNT particles also enhance the electrical properties of the composites. To fabricate good multiscale composites, the processing techniques need to consider three major challenges: (i) the need for uniform dispersion of the nanosized constituents in the matrix system; (ii) impregnation of the preform by the CNFs and CNTs; and (iii) compatibility between the microscale fibers, NPs, and matrix.

Multiscale composites consisting of CNTs and microscale fibers are fabricated mostly by VARTM/VARIM (refer to Table 4.2). Various methods for incorporating CNTs and CNFs into microscale composites have been reported, which can be grouped into four main categories: (i) infusion of CNF or CNT resin mixtures into preforms [110, 118–121]; (ii) growth of the CNT or CNF on the fiber fabric via CVD, followed by RTM [41, 122–128]; (iii) direct layering of the CNFs or CNTs between preform laminates [129–133]; and (iv) electrophoretic deposition of the nanoparticles onto the fabric [44, 47, 111, 134–140].

In the first method, the resins are premixed with CNTs or CNFs before infusion into the pre-laid continuous fabric via RTM. This method is widely used due to its simplicity, scalability, and practicality. However, the CNT loading is limited, due to its influence on the flow properties during infusion. At high loading, the viscosity increases and the flow is severely altered, with increased chances of deep-bed filtration resulting in defects [44, 45, 110, 118]. Another disadvantage of this method is the less effective dispersion of the CNTs and CNFs in the resin matrix when sonication, calendaring, or mechanical mixing is used.

In the other process variations, the continuous fibers are impregnated with CNTs via different routes, before resin transfer. This process allows higher CNT loadings, as reported by He et al. [41]; they investigated three methods of pre-impregnation of CNTs on glass fiber constructions before RTM, as depicted in Fig. 4.11: (i) one-step CVD method; (ii) two-step CVD process involving pre-wetting the fibers with a catalyst before CVD; and (iii) casting of CNT dispersion on the fibers. Epoxy resin was then infused into the composite laminates comprising ten hybrid fabric bands. As shown in Fig. 4.12, the structure of the CNTs on the fibers was dependent on the technique used. In the case of the CVD processes (one-step or two-step), the CNTs were aligned nearly vertically on the fiber surface; casting resulted in entangled CNTs on the fiber top surface, with poor dispersion. The CVD samples showed higher electrical conductivity (190 S/m for the one-step method, at 6 wt% CNT concentration), compared to the casting method with poor dispersion of CNT on only the top side (8.2 S/m for 6 wt% CNT concentration) due to the network structure of the CNTs with reduced contact resistance obtained during CVD. Lower conductivity was obtained when surface-functionalized CNTs were pre-dispersed in the resin and infused into pure glass fibers (e.g., 3.02×10^{-4} S/m, reported by Li et al. [112]). The CVD method shows several advantages, the most significant being the ability to control dispersion and alignment of the CNT and CNFs on the fiber surface. With an optimum CNT content, good dispersion leads to a percolating network structure that enhances electrical conductivity and out-of-plane properties of the hierarchical composites. However, CVD processes have some disadvantages; the extreme processing conditions can damage the fibers, resulting in degradation of the properties dependent on fiber quality. The other major obstacle to commercial adoption is the challenge of scalability due to limitations in growing large quantities of CNTs and CNFs.

The third method involves layering of the CNTs or CNFs in between laminates of the preform before resin injection and curing. Although impregnation of the preform with higher quantities of CNTs of CNFs is possible, and in some instances CNT alignment can be achieved, the main disadvantage of this method is that it is

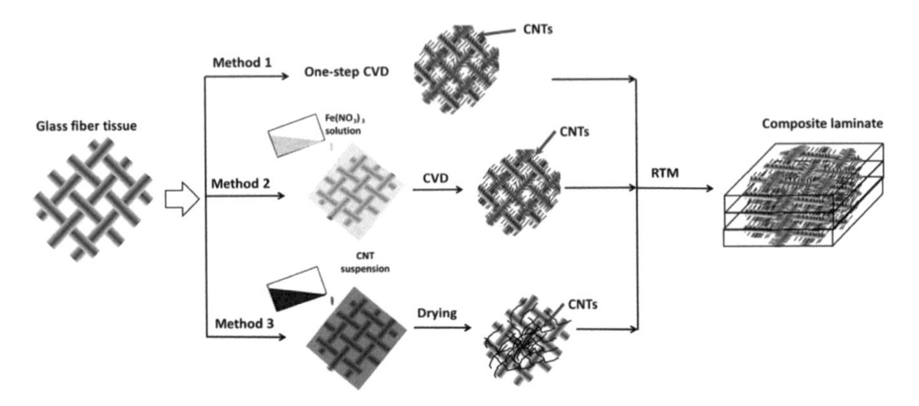

Fig. 4.11 Schematic showing the preparation GFRP composites reinforced with CNT-loaded glass fibre fabrics. Reproduced with permission from [41]. Copyright 2017, Elsevier Science Ltd.

impractical for producing large sample sizes and variations in the thicknesses of the molded part can occur.

The fourth technique, electrophoretic deposition (EPD), has the advantages of being scalable, practical, and cost effective [47]. However, there is no chemical bonding between the CNTs/CNFs and the microscale fibers, and poor control of the alignment of the former. Briefly, for EPD, the CNTs or CNFs are firstly dispersed in a medium and are deposited on an electrode when an electric current is applied; charged particles move to either the cathode or anode, depending on their charge, and get deposited on the surface. Pre-treatment of the CNTs/CNFs is vital to ensure that they are charged, which also achieves stable dispersion [138]. Surface functionalization can be performed by introducing charged groups via oxidation. Typical oxidation processes involve refluxing of the CNTs in a 3:1 (v/v) mixture of concentrated sulfuric and nitric acids at 120 °C for 30 min [138], or ultrasound ozonolysis [111]. The CNTs are then dispersed in a liquid medium in an electrophoresis tank, into which two counter electrodes are immersed. In this case, the negative electrode was a graphite plate, while the fiber acted as the positive electrode [138]. Since the surface is negatively charged after acid treatment, the CNTs move towards the positive electrode, where they get deposited on the fiber.

Multiscale composites developed via EPD have exhibited enhanced out-of-plane performances. Zhang et al. [134] reported ~30% enhancement of the inter-laminar shear strength (ILSS) and improved out-of-plane electrical conductivity for a CNT-decorated carbon fiber hierarchical composite, compared to that of carbon fiber/epoxy composites without CNTs. Similarly, Guo et al. [138] reported improved single fiber tensile strength and Weibull modulus by 16% and 41%, respectively, with the deposition of CNTs on carbon fibers. The authors also reported up to 68% improvement in interfacial shear strength, from a single fiber pull-out test for the hierarchical composites containing electrodeposited CNTs. Deposition of CNTs on the carbon fibers decreased the surface energy and enhanced wettability of the epoxy resin.

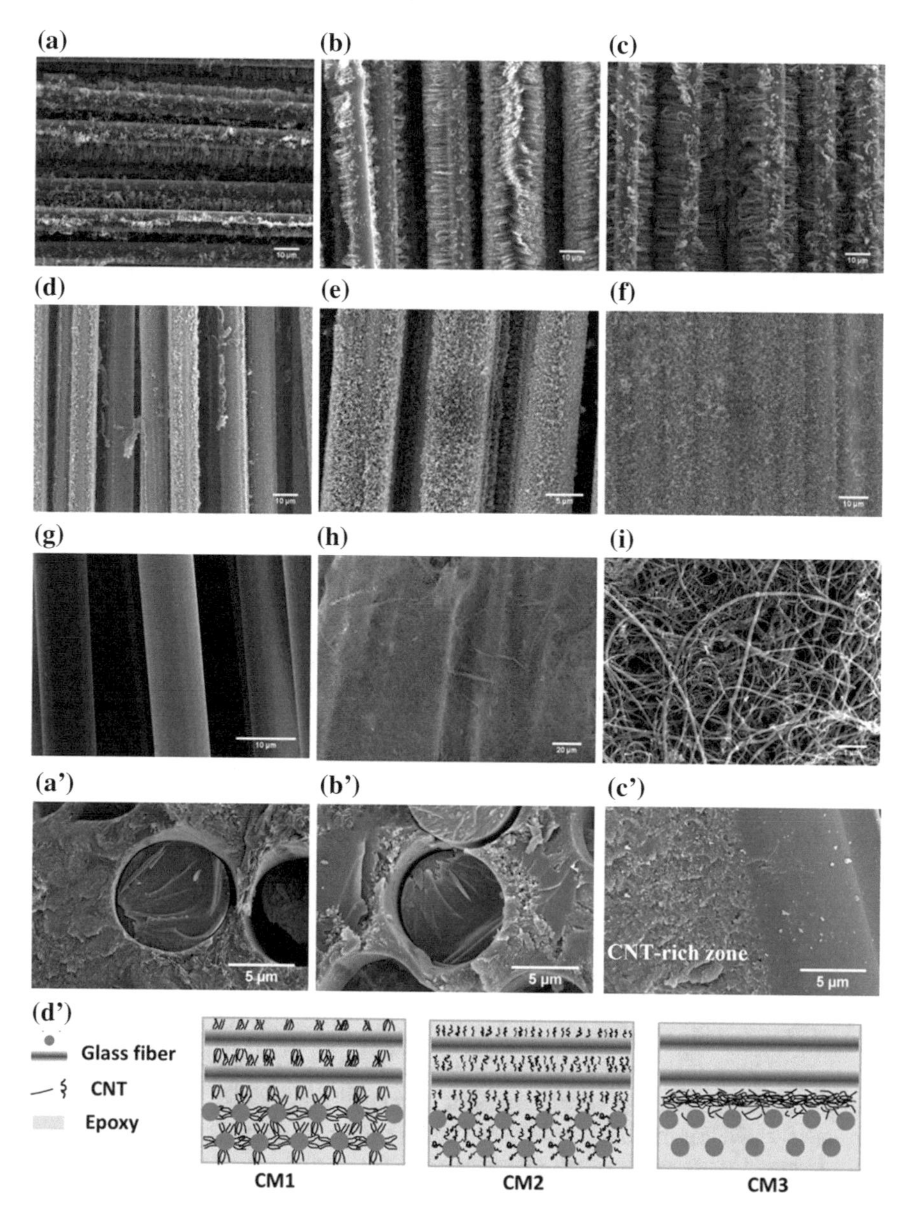

Fig. 4.12 SEM micrographs of the composites (**a**–**c**, **a**′), (**d**–**f**, **b**′) and (**g**–**I**, **c**′) corresponding to methods 1, 2 and 3, respectively, as shown in Fig. 4.11; and (**d**′) the schematic of CNT distribution in each kind of the composites. Reproduced with permission from [41]. Copyright 2017, Elsevier Science Ltd.

Recently, significant enhancement in the interfacial and anti-fatigue performance of multiscale carbon fiber/epoxy composites reinforced with oxidized CNTs was presented [139]. The authors used EPD to introduce oxidized MWCNTs to the carbon fibers, which acted as a transition layer between the fibers and the epoxy matrix. They observed 33.3%, 10.5%, and 9.5% improvement in interfacial shear strength, ILSS, and flexural strengths, respectively, for the hierarchical composite. Similarly, these materials showed a 4.5% enhancement in residual bending strength retention post fatigue tests. These findings suggest that the EPD technique offers a practical and feasible methodology for enhancing the interfacial properties of advanced multiscale composites.

In conclusion, it was observed that multiscale systems are superior for various applications compared to single-scale systems. The NPs not only impart additional properties, including electrical conductivity, but also enhance out-of-plane mechanical performance.. To enhance properties such as the electrical conductivity in multiscale composite systems, the dispersion and loading of the CNTs must be adequate enough to minimize contact resistance. Non-functionalized CNTs grown in situ via CVD processes have shown promising results and could be used to achieve high loading and good dispersion of CNTs, resulting in higher electrical conductivity compared to casting techniques. However, the CVD process may not be practical and could have detrimental effects on continuous fibers; hence, EPD is potentially a more suitable process to achieve the desired performance enhancement in multiscale composite systems.

4.6 Conclusions and Outlook

Clay-based thermoset nanocomposites are now quite a well-established field. An existing challenge is dispersion of the clays within the thermoset matrices. This is being addressed by modifying the surface chemistry, as well as developing cost effective, environmentally friendly, versatile, and commercially feasible dispersion methods (preferably exfoliation) of the platelets in the matrix. Recent focus on the hierarchical hybrid systems comprising thermoset fiber composites reinforced with clay particles promises to create superior composite materials. Layered silicates, particularly LDHs infused into traditional fiber-based composites, could substantially improve fire retardance of these advanced materials.

CNT and CNF materials have low density and outstanding physical and mechanical properties and are hence suitable for reinforcement of thermosets. They impart useful properties to the composites, including electrical conductivity, and improved thermal, thermomechanical, and mechanical properties. Much research has been undertaken into the two-phase systems containing only CNTs or CNFs within a thermoset matrix, where the focus has been to achieve good dispersion of the particles within the matrix in order to take advantage of their useful properties. On the other hand, the emerging field of multiscale composite system (comprising the traditional continuous fiber phase, thermoset phase, and an additional nanoscale phase of CNTs or CNFs) still requires further research to realize its full potential. Both two-phase systems with only the nanoscale constituents and multiscale systems are still limited by dispersion challenges. Even though this chapter focused on CNTs, CNFs, and clays, other particles, such as graphene, are increasingly being investigated, especially for multiscale systems, which is seemingly the future of advanced composites.

Acknowledgements The authors would like to thank the South African Department of Science and Technology (DST) and the Council for Scientific and Industrial Research (CSIR) for financial support.

References

1. Ratna D (2009) Handbook of thermoset resins. Smithers Shawbury
2. Dodiuk H, Goodman SH (2013) Handbook of thermoset plastics. William Andrew
3. Flory PJ (1953) Principles of polymer chemistry. Cornell University Press
4. Qi B, Zhang Q, Bannister M, Mai Y-W (2006) Investigation of the mechanical properties of DGEBA-based epoxy resin with nanoclay additives. Compos Struct 75:514–519
5. Becker O, Varley R, Simon G (2002) Morphology, thermal relaxations and mechanical properties of layered silicate nanocomposites based upon high-functionality epoxy resins. Polymer 43:4365–4373
6. Kornmann X, Lindberg H, Berglund LA (2001) Synthesis of epoxy–clay nanocomposites: influence of the nature of the clay on structure. Polymer 42:1303–1310
7. Wang K, Chen L, Wu J, Toh ML, He C, Yee AF (2005) Epoxy nanocomposites with highly exfoliated clay: mechanical properties and fracture mechanisms. Macromolecules 38:788–800
8. Azeez AA, Rhee KY, Park SJ, Hui D (2013) Epoxy clay nanocomposites–processing, properties and applications: a review. Compos B Eng 45:308–320
9. Messersmith PB, Giannelis EP (1994) Synthesis and characterization of layered silicate-epoxy nanocomposites. Chem Mater 6:1719–1725
10. Ratna D, Becker O, Krishnamurthy R, Simon G, Varley RJ (2003) Nanocomposites based on a combination of epoxy resin, hyperbranched epoxy and a layered silicate. Polymer 44:7449–7457
11. Haddadi SA, Kardar P, Abbasi F, Mahdavian M (2017) Effects of nano-silica and boron carbide on the curing kinetics of resole resin. J Therm Anal Calorim 128:1217–1226
12. Baller J, Becker N, Ziehmer M, Thomassey M, Zielinski B, Müller U et al (2009) Interactions between silica nanoparticles and an epoxy resin before and during network formation. Polymer 50:3211–3219
13. Rashti A, Yahyaei H, Firoozi S, Ramezani S, Rahiminejad A, Karimi R et al (2016) Development of novel biocompatible hybrid nanocomposites based on polyurethane-silica prepared by sol gel process. Mater Sci Eng C 69:1248–1255
14. Luo P, Xu M, Wang S, Xu Y (2017) Structural, dynamic mechanical and dielectric properties of mesoporous silica/epoxy resin nanocomposites. IEEE Trans Dielectr Electr Insul 24:1685–1697
15. Noparvar-Qarebagh A, Roghani-Mamaqani H, Salami-Kalajahi M, Kariminejad B (2017) Nanohybrids of novolac phenolic resin and carbon nanotube-containing silica network: Two different approaches for improving thermal properties of resin. J Therm Anal Calorim Int Forum Therm Stud 128:1027–1037
16. Abdalla M, Dean D, Robinson P, Nyairo E (2008) Cure behavior of epoxy/MWCNT nanocomposites: the effect of nanotube surface modification. Polymer 49:3310–3317
17. Thostenson ET, Chou T-W (2006) Processing-structure-multi-functional property relationship in carbon nanotube/epoxy composites. Carbon 44:3022–3029

18. Battisti A, Skordos AA, Partridge IK (2010) Percolation threshold of carbon nanotubes filled unsaturated polyesters. Compos Sci Technol 70:633–637
19. Gryshchuk O, Karger-Kocsis J, Thomann R, Konya Z, Kiricsi I (2006) Multiwall carbon nanotube modified vinylester and vinylester–based hybrid resins. Compos A Appl Sci Manuf 37:1252–1259
20. Li X, Gao H, Scrivens WA, Fei D, Xu X, Sutton MA et al (2004) Nanomechanical characterization of single-walled carbon nanotube reinforced epoxy composites. Nanotechnology 15:1416
21. Kim J-W, Sauti G, Siochi EJ, Smith JG, Wincheski RA, Cano RJ et al (2014) Toward high performance thermoset/carbon nanotube sheet nanocomposites via resistive heating assisted infiltration and cure. ACS Appl Mater Interfaces 6:18832–18843
22. Naebe M, Wang J, Amini A, Khayyam H, Hameed N, Li LH et al (2014) Mechanical property and structure of covalent functionalised graphene/epoxy nanocomposites. Sci Rep 4:4375
23. Tang L-C, Wan Y-J, Yan D, Pei Y-B, Zhao L, Li Y-B et al (2013) The effect of graphene dispersion on the mechanical properties of graphene/epoxy composites. Carbon 60:16–27
24. Teng C-C, Ma C-CM, Lu C-H, Yang S-Y, Lee S-H, Hsiao M-C et al (2011) Thermal conductivity and structure of non-covalent functionalized graphene/epoxy composites. Carbon 49:5107–5116
25. Monti M, Rallini M, Puglia D, Peponi L, Torre L, Kenny J (2013) Morphology and electrical properties of graphene–epoxy nanocomposites obtained by different solvent assisted processing methods. Compos A Appl Sci Manuf 46:166–172
26. Heid T, Fréchette M, David E (2015) Nanostructured epoxy/POSS composites: enhanced materials for high voltage insulation applications. IEEE Trans Dielectr Electr Insul 22:1594–1604
27. Huang X, Li Y, Liu F, Jiang P, Iizuka T, Tatsumi K et al (2014) Electrical properties of epoxy/POSS composites with homogeneous nanostructure. IEEE Trans Dielectr Electr Insul 21:1516–1528
28. Zhang Z, Gu A, Liang G, Ren P, Xie J, Wang X (2007) Thermo-oxygen degradation mechanisms of POSS/epoxy nanocomposites. Polym Degrad Stab 92:1986–1993
29. Longhi M, Zini LP, Kunst SR, Zattera AJ (2017) Influence of the type of epoxy resin and concentration of glycidylisobutyl-poss in the properties of nanocomposites. Polym Polym Compos 25:593
30. Xu Y, Chen J, Huang J, Cao J, Gérard J-F, Dai L (2017) Nanostructure of reactive polyhedral oligomeric silsesquioxane-based block copolymer as modifier in an epoxy network. High Perform Polym 29:1148–1157
31. Mishra K, Pandey G, Singh RP (2017) Enhancing the mechanical properties of an epoxy resin using polyhedral oligomeric silsesquioxane (POSS) as nano-reinforcement. Polym Testing 62:210–218
32. Covestro AG, Leverkusen (2018) Reaction injection molding. https://www.polyurethanes.covestro.com/Technologies/Processing/RIM.aspx. Acceseed 18 Apr 2018
33. Potter K (2012) Resin transfer moulding. Springer Science & Business Media
34. Rudd CD, Long AC, Kendall K, Mangin C (1997) Liquid moulding technologies: resin transfer moulding, structural reaction injection moulding and related processing techniques. Elsevier
35. Rachmadini Y, Tan VBC, Tay TE (2010) Enhancement of mechanical properties of composites through Incorporation of CNT in VARTM—a review. J Reinf Plast Compos 29:2782–2807
36. Nguyen QT, Ngo T, Tran P, Mendis P, Zobec M, Aye L (2016) Fire performance of prefabricated modular units using organoclay/glass fibre reinforced polymer composite. Constr Build Mater 129:204–215
37. Nguyen QT, Ngo TD, Tran P, Mendis P, Bhattacharyya D (2015) Influences of clay and manufacturing on fire resistance of organoclay/thermoset nanocomposites. Compos A Appl Sci Manuf 74:26–37
38. Chandrasekaran VCS, Advani SG, Santare MH (2010) Role of processing on interlaminar shear strength enhancement of epoxy/glass fiber/multi-walled carbon nanotube hybrid composites. Carbon 48:3692–3699

39. Tehrani M, Boroujeni AY, Hartman TB, Haugh TP, Case SW, Al-Haik MS (2013) Mechanical characterization and impact damage assessment of a woven carbon fiber reinforced carbon nanotube–epoxy composite. Compos Sci Technol 75:42–48
40. González-Julián J, Iglesias Y, Caballero AC, Belmonte M, Garzón L, Ocal C et al (2011) Multi-scale electrical response of silicon nitride/multi-walled carbon nanotubes composites. Compos Sci Technol 71:60–66
41. He D, Salem D, Cinquin J, Piau G-P, Bai J (2017) Impact of the spatial distribution of high content of carbon nanotubes on the electrical conductivity of glass fiber fabrics/epoxy composites fabricated by RTM technique. Compos Sci Technol 147:107–115
42. Wang B-C, Zhou X, Ma K-M (2013) Fabrication and properties of CNTs/carbon fabric hybrid multiscale composites processed via resin transfer molding technique. Compos B Eng 46:123–129
43. Chandrasekaran VCS, Advani SG, Santare MH (2011) Influence of resin properties on interlaminar shear strength of glass/epoxy/MWNT hybrid composites. Compos A Appl Sci Manuf 42:1007–1016
44. Bekyarova E, Thostenson E, Yu A, Kim H, Gao J, Tang J et al (2007) Multiscale carbon nanotube—carbon fiber reinforcement for advanced epoxy composites. Langmuir 23:3970–3974
45. Green KJ, Dean DR, Vaidya UK, Nyairo E (2009) Multiscale fiber reinforced composites based on a carbon nanofiber/epoxy nanophased polymer matrix: Synthesis, mechanical, and thermomechanical behavior. Compos A Appl Sci Manuf 40:1470–1475
46. Chen Q, Wu W, Zhao Y, Xi M, Xu T, Fong H (2014) Nano-epoxy resins containing electrospun carbon nanofibers and the resulting hybrid multi-scale composites. Compos B Eng 58:43–53
47. Rodriguez AJ, Guzman ME, Lim C-S, Minaie B (2011) Mechanical properties of carbon nanofiber/fiber-reinforced hierarchical polymer composites manufactured with multiscale-reinforcement fabrics. Carbon 49:937–948
48. Kamar NT, Hossain MM, Khomenko A, Haq M, Drzal LT, Loos A (2015) Interlaminar reinforcement of glass fiber/epoxy composites with graphene nanoplatelets. Compos A Appl Sci Manuf 70:82–92
49. Qin W, Vautard F, Drzal LT, Yu J (2015) Mechanical and electrical properties of carbon fiber composites with incorporation of graphene nanoplatelets at the fiber–matrix interphase. Compos B Eng 69:335–341
50. Lin L-Y, Lee J-H, Hong C-E, Yoo G-H, Advani SG (2006) Preparation and characterization of layered silicate/glass fiber/epoxy hybrid nanocomposites via vacuum-assisted resin transfer molding (VARTM). Compos Sci Technol 66:2116–2125
51. Sharma B, Mahajan S, Chhibber R, Mehta R (2012) Glass fiber reinforced polymer-clay nanocomposites: processing, structure and hygrothermal effects on mechanical properties. Proc Chem 4:39–46
52. Dean D, Obore AM, Richmond S, Nyairo E (2006) Multiscale fiber-reinforced nanocomposites: synthesis, processing and properties. Compos Sci Technol 66:2135–2142
53. Aitomäki Y, Moreno-Rodriguez S, Lundström TS, Oksman K (2016) Vacuum infusion of cellulose nanofibre network composites: Influence of porosity on permeability and impregnation. Mater Des 95:204–211
54. Barari B, Ellingham TK, Ghamhia II, Pillai KM, El-Hajjar R, Turng L-S et al (2016) Mechanical characterization of scalable cellulose nano-fiber based composites made using liquid composite molding process. Compos B Eng 84:277–284
55. Rajanish M, Nanjundaradhya NV, Sharma RS (2015) An investigation on ILSS properties of unidirectional glass fibre/alumina nanoparticles filled epoxy nanocomposite at different angles of fibre orientations. Proc Mater Sci 10:555–562
56. Lundström TS, Gebart BR (1994) Influence from process parameters on void formation in resin transfer molding. Polym Compos 15:25–33
57. Hmeidat NS, Kemp JW, Compton BG (2018) High-strength epoxy nanocomposites for 3D printing. Compos Sci Technol 160:9–20
58. Zabihi O, Ahmadi M, Nikafshar S, Chandrakumar Preyeswary K, Naebe M (2018) A technical review on epoxy-clay nanocomposites: structure, properties, and their applications in fiber reinforced composites. Compos B Eng 135:1–24

59. Azeez AA, Rhee KY, Park SJ, Hui D (2013) Epoxy clay nanocomposites—processing, properties and applications: a review. Compos B Eng 45:308–320
60. Lan T, Pinnavaia TJ (1994) Clay-reinforced epoxy nanocomposites. Chem Mater 6:2216–2219
61. Wang D-C, Chang G-W, Chen Y (2008) Preparation and thermal stability of boron-containing phenolic resin/clay nanocomposites. Polym Degrad Stab 93:125–133
62. Koo J, Stretz H, Bray A, Wootan W, Mulich S, Powell B et al (2002) Phenolic-clay nanocomposites for rocket propulsion system. Int SAMPE Symp Exhib SAMPE 1999:1085–1099
63. Wu Z, Zhou C, Qi R (2002) The preparation of phenolic resin/montmorillonite nanocomposites by suspension condensation polymerization and their morphology. Polym Compos 23:634–646
64. Jiang W, Chen SH, Chen Y (2006) Nanocomposites from phenolic resin and various organo-modified montmorillonites: preparation and thermal stability. J Appl Polym Sci 102:5336–5343
65. Zhang Z, Ye G, Toghiani H, Pittman CU (2010) Morphology and thermal stability of novolac phenolic resin/clay nanocomposites prepared via solution high-shear mixing. Macromol Mater Eng 295:923–933
66. Pappas J, Patel K, Nauman E (2005) Structure and properties of phenolic resin/nanoclay composites synthesized by in situ polymerization. J Appl Polym Sci 95:1169–1174
67. Xiong J, Liu Y, Yang X, Wang X (2004) Thermal and mechanical properties of polyurethane/montmorillonite nanocomposites based on a novel reactive modifier. Polym Degrad Stab 86:549–555
68. Chang JH, An YU (2002) Nanocomposites of polyurethane with various organoclays: thermomechanical properties, morphology, and gas permeability. J Polym Sci Part B Polym Phys 40:670–677
69. Ollier R, Rodriguez E, Alvarez V (2013) Unsaturated polyester/bentonite nanocomposites: influence of clay modification on final performance. Compos A Appl Sci Manuf 48:137–143
70. Mironi-Harpaz I, Narkis M, Siegmann A (2005) Nanocomposite systems based on unsaturated polyester and organo-clay. Polym Eng Sci 45:174–186
71. Kornmann X, Berglund LA, Sterte J, Giannelis E (1998) Nanocomposites based on montmorillonite and unsaturated polyester. Polym Eng Sci 38:1351–1358
72. Tsai T-Y, Bunekar N, Yen C-H, Lin Y-B (2016) Synthesis and characterization of vinyl ester/inorganic layered material nanocomposites. RSC Adv 6:102797–102803
73. Mohaddespour A, Ahmadi SJ, Abolghassemi H, Mahjoub SM, Atashrouz S (2018) Irradiation of poly (vinyl ester)/clay nanocomposites. J Compos Mater 52:17–25
74. Ryu SH, Reddy MJK, Shanmugharaj A (2017) Role of silane concentration on the structural characteristics and properties of epoxy-/silane-modified montmorillonite clay nanocomposites. J Elastomers Plast 49:665–683
75. Su L, Zeng X, He H, Tao Q, Komarneni S (2017) Preparation of functionalized kaolinite/epoxy resin nanocomposites with enhanced thermal properties. Appl Clay Sci 148:103–108
76. Tolle TB, Anderson DP (2004) The role of preconditioning on morphology development in layered silicate thermoset nanocomposites. J Appl Polym Sci 91:89–100
77. Hutchinson JM, Montserrat S, Román F, Cortés P, Campos L (2006) Intercalation of epoxy resin in organically modified montmorillonite. J Appl Polym Sci 102:3751–3763
78. Jae-Jun P, Jae-Young L (2010) A new dispersion method for the preparation of polymer/organoclay nanocomposite in the electric fields. IEEE Trans Dielectr Electr Insul:17
79. Ngo TD, Ton-That MT, Hoa SV, Cole KC (2009) Effect of temperature, duration and speed of pre-mixing on the dispersion of clay/epoxy nanocomposites. Compos Sci Technol 69:1831–1840
80. Zhao L, Li J, Guo S, Du Q (2006) Ultrasonic oscillations induced morphology and property development of polypropylene/montmorillonite nanocomposites. Polymer 47:2460–2469
81. Shokrieh MM, Kefayati AR, Chitsazzadeh M (2012) Fabrication and mechanical properties of clay/epoxy nanocomposite and its polymer concrete. Mater Des 40:443–452

82. Zhou Y, Pervin F, Biswas MA, Rangari VK, Jeelani S (2006) Fabrication and characterization of montmorillonite clay-filled SC-15 epoxy. Mater Lett 60:869–873
83. C-k Lam, K-t Lau, H-y Cheung, H-y Ling (2005) Effect of ultrasound sonication in nanoclay clusters of nanoclay/epoxy composites. Mater Lett 59:1369–1372
84. Bharadwaj RK, Mehrabi AR, Hamilton C, Trujillo C, Murga M, Fan R et al (2002) Structure–property relationships in cross-linked polyester–clay nanocomposites. Polymer 43:3699–3705
85. Yasmin A, Abot JL, Daniel IM (2003) Processing of clay/epoxy nanocomposites by shear mixing. Scripta Mater 49:81–86
86. Yasmin A, Abot JL, Daniel IM (2002) Processing of clay/epoxy nanocomposites with a three-roll mill machine. MRS Online Proceedings Library Archive:740
87. Yasmin A, Luo J, Abot J, Daniel I (2006) Mechanical and thermal behavior of clay/epoxy nanocomposites. Compos Sci Technol 66:2415–2422
88. Deng S, Zhang J, Ye L (2009) Halloysite–epoxy nanocomposites with improved particle dispersion through ball mill homogenisation and chemical treatments. Compos Sci Technol 69:2497–2505
89. Lu HJ, Liang GZ, Ma XY, Zhang BY, Chen XB (2004) Epoxy/clay nanocomposites: further exfoliation of newly modified clay induced by shearing force of ball milling. Polym Int 53:1545–1553
90. Liu W, Hoa SV, Pugh M (2005) Organoclay-modified high performance epoxy nanocomposites. Compos Sci Technol 65:307–316
91. Eesaee M, Shojaei A (2014) Effect of nanoclays on the mechanical properties and durability of novolac phenolic resin/woven glass fiber composite at various chemical environments. Compos A Appl Sci Manuf 63:149–158
92. Withers GJ, Yu Y, Khabashesku VN, Cercone L, Hadjiev VG, Souza JM et al (2015) Improved mechanical properties of an epoxy glass–fiber composite reinforced with surface organomodified nanoclays. Compos B Eng 72:175–182
93. Kalali EN, Wang X, Wang D-Y (2015) Functionalized layered double hydroxide-based epoxy nanocomposites with improved flame retardancy and mechanical properties. J Mater Chem A 3:6819–6826
94. Park J-M, Wang Z-J, Kwon D-J, Gu G-Y, Lee W-I, Park J-K et al (2012) Optimum dispersion conditions and interfacial modification of carbon fiber and CNT–phenolic composites by atmospheric pressure plasma treatment. Compos B Eng 43:2272–2278
95. Cheng QF, Wang JP, Wen JJ, Liu CH, Jiang KL, Li QQ et al (2010) Carbon nanotube/epoxy composites fabricated by resin transfer molding. Carbon 48:260–266
96. Ma J, Larsen RM (2014) Effect of concentration and surface modification of single walled carbon nanotubes on mechanical properties of epoxy composites. Fibers Polym 15:2169–2174
97. Ganguli S, Bhuyan M, Allie L, Aglan H (2005) Effect of multi-walled carbon nanotube reinforcement on the fracture behavior of a tetrafunctional epoxy. J Mater Sci 40:3593–3595
98. Abdalla M, Dean D, Adibempe D, Nyairo E, Robinson P, Thompson G (2007) The effect of interfacial chemistry on molecular mobility and morphology of multiwalled carbon nanotubes epoxy nanocomposite. Polymer 48:5662–5670
99. Mathur R, Chatterjee S, Singh B (2008) Growth of carbon nanotubes on carbon fibre substrates to produce hybrid/phenolic composites with improved mechanical properties. Compos Sci Technol 68:1608–1615
100. Yeh M-K, Tai N-H, Liu J-H (2006) Mechanical behavior of phenolic-based composites reinforced with multi-walled carbon nanotubes. Carbon 44:1–9
101. McClory C, McNally T, Brennan GP, Erskine J (2007) Thermosetting polyurethane multi-walled carbon nanotube composites. J Appl Polym Sci 105:1003–1011
102. Schlea MR, Meree CE, Gerhardt RA, Mintz EA, Shofner ML (2012) Network behavior of thermosetting polyimide/multiwalled carbon nanotube composites. Polymer 53:1020–1027
103. Beg M, Alam AM, Yunus R, Mina M (2015) Improvement of interaction between pre-dispersed multi-walled carbon nanotubes and unsaturated polyester resin. J Nanopart Res 17:53

104. Ureña-Benavides EE, Kayatin MJ, Davis VA (2013) Dispersion and rheology of multiwalled carbon nanotubes in unsaturated polyester resin. Macromolecules 46:1642–1650
105. Natsuki T, Ni QQ, Wu SH (2008) Temperature dependence of electrical resistivity in carbon nanofiber/unsaturated polyester nanocomposites. Polym Eng Sci 48:1345–1350
106. Wu Z, Meng L, Liu L, Jiang Z, Xing L, Jiang D et al (2014) Chemically grafting carbon nanotubes onto carbon fibers by poly (acryloyl chloride) for enhancing interfacial strength in carbon fiber/unsaturated polyester composites. Fibers Polym 15:659–663
107. Makki MS, Abdelaal MY, Bellucci S, Abdel Salam M (2014) Multi-walled carbon nanotubes/unsaturated polyester composites: Mechanical and thermal properties study. Fullerene Nanotubes Carbon Nanostruct 22:820–833
108. Bal S (2010) Experimental study of mechanical and electrical properties of carbon nanofiber/epoxy composites. Mater Des (1980–2015) 31:2406–2413
109. Sánchez M, Campo M, Jiménez-Suárez A, Ureña A (2013) Effect of the carbon nanotube functionalization on flexural properties of multiscale carbon fiber/epoxy composites manufactured by VARIM. Compos B Eng 45:1613–1619
110. Sadeghian R, Gangireddy S, Minaie B, Hsiao K-T (2006) Manufacturing carbon nanofibers toughened polyester/glass fiber composites using vacuum assisted resin transfer molding for enhancing the mode-I delamination resistance. Compos A Appl Sci Manuf 37:1787–1795
111. An Q, Rider AN, Thostenson ET (2012) Electrophoretic deposition of carbon nanotubes onto carbon-fiber fabric for production of carbon/epoxy composites with improved mechanical properties. Carbon 50:4130–4143
112. Li J, Zhang G, Zhang H, Fan X, Zhou L, Shang Z et al (2018) Electrical conductivity and electromagnetic interference shielding of epoxy nanocomposite foams containing functionalized multi-wall carbon nanotubes. Appl Surf Sci 428:7–16
113. Ghosh PK, Kumar K, Chaudhary N (2015) Influence of ultrasonic dual mixing on thermal and tensile properties of MWCNTs-epoxy composite. Compos B Eng 77:139–144
114. Sun D, Chu C-C, Sue H-J (2010) Simple approach for preparation of epoxy hybrid nanocomposites based on carbon nanotubes and a model clay. Chem Mater 22:3773–3778
115. Zhang X, Sue H-J, Nishimura R (2013) Acid-mediated isolation of individually dispersed SWCNTs from electrostatically tethered nanoplatelet dispersants. Carbon 56:374–382
116. Liu K, Sun Y, Chen L, Feng C, Feng X, Jiang K et al (2008) Controlled growth of super-aligned carbon nanotube arrays for spinning continuous unidirectional sheets with tunable physical properties. Nano Lett 8:700–705
117. Qian H, Greenhalgh ES, Shaffer MS, Bismarck A (2010) Carbon nanotube-based hierarchical composites: a review. J Mater Chem 20:4751–4762
118. Qiu J, Zhang C, Wang B, Liang R (2007) Carbon nanotube integrated multifunctional multiscale composites. Nanotechnology 18:275708
119. Morales G, Barrena M, De Salazar JG, Merino C, Rodríguez D (2010) Conductive CNF-reinforced hybrid composites by injection moulding. Compos Struct 92:1416–1422
120. Gojny FH, Wichmann MH, Fiedler B, Bauhofer W, Schulte K (2005) Influence of nano-modification on the mechanical and electrical properties of conventional fibre-reinforced composites. Compos A Appl Sci Manuf 36:1525–1535
121. Reia da Costa EF, Skordos AA, Partridge IK, Rezai A (2012) RTM processing and electrical performance of carbon nanotube modified epoxy/fibre composites. Compos Part A Appl Sci Manuf 43:593–602
122. Thostenson E, Li W, Wang D, Ren Z, Chou T (2002) Carbon nanotube/carbon fiber hybrid multiscale composites. J Appl Phys 91:6034–6037
123. Sager R, Klein P, Lagoudas D, Zhang Q, Liu J, Dai L et al (2009) Effect of carbon nanotubes on the interfacial shear strength of T650 carbon fiber in an epoxy matrix. Compos Sci Technol 69:898–904
124. Qian H, Bismarck A, Greenhalgh ES, Shaffer MS (2010) Carbon nanotube grafted carbon fibres: a study of wetting and fibre fragmentation. Compos A Appl Sci Manuf 41:1107–1114
125. Kepple K, Sanborn G, Lacasse P, Gruenberg K, Ready W (2008) Improved fracture toughness of carbon fiber composite functionalized with multi walled carbon nanotubes. Carbon 46:2026–2033

126. Gong Q-J, Li H-J, Wang X, Fu Q-G (2007) Wang Z-w, Li K-Z. In situ catalytic growth of carbon nanotubes on the surface of carbon cloth. Compos Sci Technol 67:2986–2989
127. Duan H, Liang J, Xia Z (2010) Synthetic hierarchical nanostructures: growth of carbon nanofibers on microfibers by chemical vapor deposition. Mater Sci Eng B 166:190–195
128. Tzeng S-S, Hung K-H, Ko T-H (2006) Growth of carbon nanofibers on activated carbon fiber fabrics. Carbon 44:859–865
129. Abot J, Song Y, Schulz M, Shanov V (2008) Novel carbon nanotube array-reinforced laminated composite materials with higher interlaminar elastic properties. Compos Sci Technol 68:2755–2760
130. Arai M, Noro Y, Sugimoto KI, Endo M (2008) Mode I and mode II interlaminar fracture toughness of CFRP laminates toughened by carbon nanofiber interlayer. Compos Sci Technol 68:516–525
131. Garcia EJ, Wardle BL, Hart AJ (2008) Joining prepreg composite interfaces with aligned carbon nanotubes. Compos A Appl Sci Manuf 39:1065–1070
132. Wicks SS, de Villoria RG, Wardle BL (2010) Interlaminar and intralaminar reinforcement of composite laminates with aligned carbon nanotubes. Compos Sci Technol 70:20–28
133. Li Y, Hori N, Arai M, Hu N, Liu Y, Fukunaga H (2009) Improvement of interlaminar mechanical properties of CFRP laminates using VGCF. Compos A Appl Sci Manuf 40:2004–2012
134. Zhang J, Zhuang R, Liu J, Mäder E, Heinrich G, Gao S (2010) Functional interphases with multi-walled carbon nanotubes in glass fibre/epoxy composites. Carbon 48:2273–2281
135. An Q, Rider AN, Thostenson ET (2013) Hierarchical composite structures prepared by electrophoretic deposition of carbon nanotubes onto glass fibers. ACS Appl Mater Interfaces 5:2022–2032
136. Schaefer JD, Rodriguez AJ, Guzman ME, Lim C-S, Minaie B (2011) Effects of electrophoretically deposited carbon nanofibers on the interface of single carbon fibers embedded in epoxy matrix. Carbon 49:2750–2759
137. Guo J, Lu C (2012) Continuous preparation of multiscale reinforcement by electrophoretic deposition of carbon nanotubes onto carbon fiber tows. Carbon 50:3101–3103
138. Guo J, Lu C, An F (2012) Effect of electrophoretically deposited carbon nanotubes on the interface of carbon fiber reinforced epoxy composite. J Mater Sci 47:2831–2836
139. Sui X, Shi J, Yao H, Xu Z, Chen L, Li X et al (2017) Interfacial and fatigue-resistant synergetic enhancement of carbon fiber/epoxy hierarchical composites via an electrophoresis deposited carbon nanotube-toughened transition layer. Compos A Appl Sci Manuf 92:134–144
140. Rodriguez AJ, Guzman ME, Lim C-S, Minaie B (2010) Synthesis of multiscale reinforcement fabric by electrophoretic deposition of amine-functionalized carbon nanofibers onto carbon fiber layers. Carbon 48:3256–3259

Chapter 5
Processing of Sustainable Polymer Nanocomposites

Orebotse Joseph Botlhoko and Suprakas Sinha Ray

Abstract This chapter provides a brief overview on the processing methods and structure-property relationships of polylactide, polyhydroxybutyrate, and starch nanocomposites as well as the challenges faced in their development, in order to further the state-of-the-art of green nanochemistry. The concept of biodegradable polymer nanocomposites and the role of nanofillers are discussed in detail. Further, the performances and potential industrial applications of these sustainable polymer nanocomposites are also discussed in brief.

5.1 Introduction

Developments in nanoscience and nanotechnology have resulted in sharp reductions in the size of the filler particle sizes used as reinforcements in polymer composites. On the other hand, recent changes in environmental policies have led renewed research interest in the production and applicability of biodegradable polymers [1]. Hence, nowadays, sustainable polymers and their nanocomposites are among the most widely investigated materials from both an academic and an industrial perspective. The development of sustainable polymeric materials is basically based on the development of biodegradable polymers, which can be classified based on their synthesis process, as shown in Fig. 5.1. Briefly, biodegradable polymers can be classified into main categories: agro-polymers and biodegradable polyesters. Agro-polymers can be subdivided into polysaccharides, proteins, and lignin (i.e., starch

O. J. Botlhoko · S. Sinha Ray (✉)
DST-CSIR National Centre for Nanostructured Materials, Council for Scientific and Industrial Research, Pretoria 0001, South Africa
e-mail: rsuprakas@csir.co.za

O. J. Botlhoko
e-mail: obotlhoko@csir.co.za

S. Sinha Ray
Department of Applied Chemistry, University of Johannesburg, Doornfontein 2028, Johannesburg, South Africa
e-mail: ssinharay@uj.ac.za

S. Sinha Ray (ed.), *Processing of Polymer-based Nanocomposites*, Springer Series in Materials Science 278, https://doi.org/10.1007/978-3-319-97792-8_5

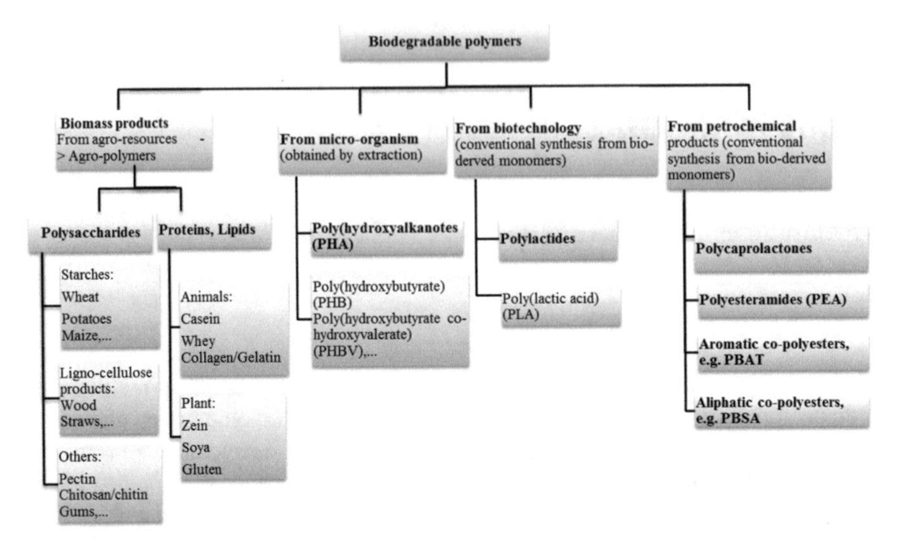

Fig. 5.1 Classification of main biodegradable polymers. Redesigned on the basis of information available in [2]

and cellulose, among others), whereas biodegradable polyesters include polyhydroxyalkanoates (PHAs), polylactides, and polycaprolactones (PCL) (i.e., polyhydroxybutyrate (PHB)), and polylactic acid (PLA) [2].

The main advantage of using biodegradable polymers is their low environmental impact at the time of disposal since they generally degrade under standard environmental conditions. However, despite this important advantage over conventional nondegradable polymers, the poor processability and material properties of biodegradable polymers limit their applicability. The incorporation of either micro- or nanofillers into the polymer matrix has been investigated widely as a method of overcoming these issues. However, nanofillers are more effective reinforcements than microfillers, owing to their surface area per unit volume being higher than that of microfillers. Therefore, processing strategies for polymer nanocomposites as well as the optimization of the nanofiller content have become subject of significant interest, with the aim of improving the structure and properties of biodegradable polymer nanocomposites (i.e., mechanical, thermal, and rheological characteristics and conductivity) for a wide range of applications. These nanofillers include clay, titanium oxide, carbon nanotubes, and graphene and its derivatives. However, most of these nanofillers agglomerate during processing. Hence, the development of improved processing methods that yield sustainable polymer nanocomposites filled with nanofillers would be highly desirable.

Given this background, this chapter presents a critical review of the processing methods currently available for producing sustainable polymer nanocomposites and, in particular, PLA-, PHB-, and starch-based nanocomposites. In addition, the relationships between their structure and properties as well as their performances, the challenges faced in their production, and future trends in their development are discussed at length. These biodegradable polymers were chosen not only based on their biodegradability, biocompatibility, and sustainability but because of their desirable mechanical properties when they are filled with nanofillers. More attention should be paid to nanofillers of clay, graphene, and their derivatives.

In brief, PLA exhibits a Young's modulus (3–4 GPa) and tensile strength (50–70 MPa) comparable to those of most industrially available polymers, namely, polyethylene (PE), polypropylene (PP), polystyrene (PS), and polyethylene terephthalate (PET) [3]. On the other hand, PHB is attractive because of its short chain length and good barrier properties. In addition, PHB exhibits a high melting point (170–180 °C) as well as physical and mechanical properties (tensile strength of 43 MPa) comparable to those of PP and a Young's modulus of about 3 GPa, comparable to that of PS [4–8]. On the other hand, starch is also attractive given its low cost, high abundant, and renewability [9, 10]. However, all these components (PLA, PHB, and starch) are brittle in nature.

Despite the advantages of both PLA and PHB, these biodegradable polymers are costlier than conventional nondegradable ones. For instance, commercial grade PHB is approximately 10 times more expensive than conventional polymers. In addition, PLA shows poor gas barrier properties. Further, PLA has poor crystallinity whereas PHB crystallizes rapidly; these behaviors limit their widescale applicability in the industry [10–13]. Therefore, the addition of clays or graphene to PLA, PHB, and starch is a potentially suitable way of developing sustainable polymer nanocomposites. However, the optimization of the processing technique in order to produce ideal biodegradable polymer nanocomposites remains a challenge. This explains why there nanofillers, especially those mentioned above, are being explored widely for use as reinforcements. The various key paths suggested by many other researchers should also be explored.

5.2 Processing Methods for Polylactide, Polyhydroxybutyrate, and Starch Nanocomposites

5.2.1 General Overview of Processing Methods for Polymer Nanocomposites

Several processing methods are available for fabricating biodegradable polymer nanocomposites (PNCs) that exhibit enhanced properties and are suitable for use in various industrial applications. Often, the two challenges involved are choosing the appropriate method for de-aggregating or disperse and homogeneously distributing the nanofillers in the polymeric system without degrading the mechanical and other important properties of the virgin polymer and ensuring that the desired properties are transferred to the polymer. Generally, the degree to which these properties are transferred depends on the dispersion level of the nanoparticles within the polymer

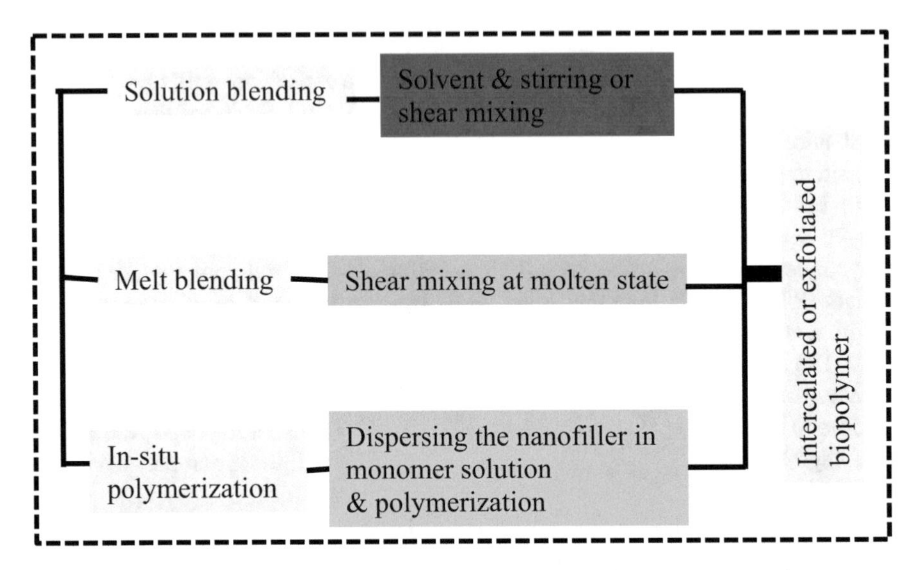

Fig. 5.2 Three primary methods for fabricating biodegradable polymer nanocomposites

matrix—the greater the degree of dispersion, the more significant the level of reinforcement. From this viewpoint, general processing methods such as solution blending, melt blending, and in situ polymerization will be briefly discussed (Fig. 5.2). Of these, solution processing and melt processing have received more attention [14, 15].

5.2.1.1 Solution Blending

In the solution blending method, a polymer is dissolved separately in the appropriate solvent, whereas the nanofillers are dispersed in the appropriate solvent prior to mixing. Next, the nanofiller suspension is mixed with the polymer solution by vigorous agitation or shear mixing to initiate the intercalation of the polymer chains and the exfoliation of the nanofillers. Ultrasonication is normally performed to improve the dispersion of the nanofillers [16]. Finally, the nanofillers/polymer solution is poured into molds and allowed to homogenize. Subsequently, the solvent is allowed to evaporate naturally or is evaporated through heating. The remaining intercalated structure is the polymer nanocomposite. This method is particularly suited in the case of graphene-filled nanocomposites, as it allows the nanofiller particles, especially functionalized graphene particles, to be dispersed with the polymer matrix more homogeneously than does the melt blending method. However, the use of non-environmentally friendly solvents and the difficulties arising from the presence of the residual solvent limit its industrial applicability. On the other hand, slower solvent evaporation normally induces particle reaggregation [16].

5.2.1.2 Melt Blending Method

In the melt blending method, the nanofiller particles are either dry mixed with the polymer or are introduced within the process at specific time intervals or at once when the polymer is in the molten state. In this method, intercalated polymer nanocomposites are produced by heating a mixture of a polymer and the nanofillers above the melting point of the polymer but below its decomposition temperature, either statically or under a shear force. The mixing equipment generally used includes an extruder, an internal mixer, and a two-roll mill [17]. The melt blending method has several advantages over solution blending and in situ polymerization. For example, this method does not involve the use of a non-environmentally friendly or any other solvent. Further, it can be used with biopolymers that cannot be subjected to in situ polymerization. Hence, this method is suitable for industrial use [16, 18]. However, this method is not appropriate for materials that are prone to thermal degradation [16].

5.2.1.3 In Situ Polymerization Method

In the in situ polymerization method, the particles of the reinforcement material are normally dispersed in the monomer or in its solution. This step is followed by the polymerization process, which is initiated by heat, radiation, or an appropriate initiator [18]. The in situ polymerization method exhibits several advantages over the solution and melt blending methods, such as a higher degree of reinforcement material dispersion and better compatibility. However, the in situ polymerization method finds limited use as it is very time consuming [16].

5.2.2 Processing Strategies of PLA Nanocomposites

PLA is aliphatic polyester that is generally produced through the polymerization of lactic acid. Prior to the polymerization of PLA, corn, sugarcane or other microbial carbohydrates are fermented into lactic acid. The polymerization process is performed through polycondensation or ring-opening polymerization (ROP). ROP has attracted the most attention as it allows for the production of high-molecular-weight PLA. PLA can be processed using the conventional methods used for processing PP. In addition, PLA can be produced using the existing manufacturing equipment initially designed and used for the preparation of conventional plastics at the industrial level. This makes PLA production relatively cost efficient. However, PLA's low heat resistance, brittleness, and slow crystallization ability limit its usability. In addition, it is also possible to injection mold PLA; however, there some shortfalls in this case too, such as poor material flowability and the rapid thermal degradation, owing to which injection molding is not used widely in the case of PLA. The thermal degradation of

PLA is generally due to the hydrolysis, depolymerization, oxidative random-chain scission, and inter- and intramolecular transesterification [19].

Recently, Picard et al. [20] investigated the development of PLA/clay nanocomposites. PLA nanocomposites containing 3.8 and 7.9 wt% clay were prepared using the melt blending method. The samples were denoted as PLANC4 and PLANC8, respectively. The results of XRD and TEM analyses indicated that the samples had complex morphologies, with the diffraction peaks centered at 5 and 7° showing comparable intensities, this phenomenon signifies the presence of clay particles in the PLA matrix (Fig. 5.3). More importantly, PLANC4 exhibited a high-intensity diffraction peak centered at 2.5°; this peak was attributable to the large number of tactoid structures. On the other hand, this peak was not visible in the case of PLACN8. This can be attributed to the coexistence of an exfoliated/intercalated structure in this sample owing to the disorganization of the clay tactoids. These results suggest that the distribution of clay (as well as other nanofillers) has a significant influence on whether a micro- or nanostructure is formed as well as on the properties of the structure formed (see Sect. 5.3.1.1 below). In addition, the authors reported that an annealing process improved the structure and properties of the formed nanocomposite. They indicated that the samples annealed at 90, 100, and 110 °C exhibited two melting peaks, whereas those annealed at 120 °C had a single melting peak. More importantly, the samples annealed at 120 °C exhibited the highest crystallinity values. Therefore, optimized processing is essential for improving the performance and properties of polymer nanocomposites.

Similarly, Shameli et al. [21] investigated the development of PLA/clay nanocomposites using both solution and melt blending. XRD and TEM analysis results indicated the coexistence of an intercalated/exfoliated structure, wherein organoclay wall well dispersed within the PLA matrix. More importantly, the mechanical characteristics of the organoclay-filled PLA nanocomposites processed by the melt blending method were superior to those of the organoclay-filled PLA nanocomposites processed by the typical solution blending method. These results highlight the suitability of the melt blending processing method for the fabrication of sustainable polymer nanocomposites on the industrial scale. In addition, the 5 wt% organoclay-filled PLA nanocomposite samples fabricated using the two methods showed the optimal elongation at break value as well as an increased Young's modulus.

5.2.3 *Processing Strategies for PHB Nanocomposites*

Melt blending and solution processing are the methods for used widely for producing PHB nanocomposites. However, when processing PHB by the melt blending method, a few drawbacks must be kept in mind, such as its thermal degradation and viscosity. PHB has a high melting point (approximately 170–180 °C), which is very close to its degradation temperature (approximately 200–270 °C). Hence, it is thermally unstable during melt processing and its viscosity and molar mass decrease. These decreases are in accordance with the zero-shear rate viscosity, as given by (5.1). This

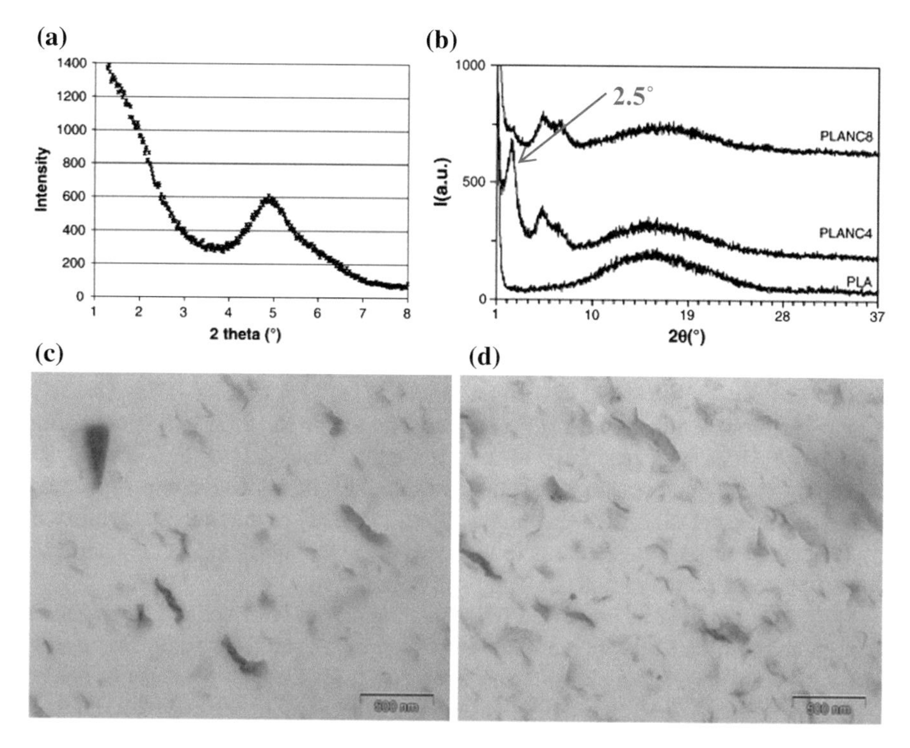

Fig. 5.3 XRD patterns of **a** organically modified clay, **b** PLA, PLANC4, and PLANC8, and TEM micrographs of **c** PLANC4 and **d** PLANC8. Reproduced with permission from [20]. Copyright 2011, Elsevier Science Ltd.

equation indicate that the zero-shear viscosity of a polymer is related to its molecular weight [4, 22, 23].

$$\eta_0 = K\ Mw^{\alpha} \tag{5.1}$$

where η_0 is the zero-shear viscosity, K is a material constant, and α is the exponent. Therefore, common thermal processing methods, namely, extrusion, compression molding, and injection molding, should be performed within a reasonable timeframe in order to prevent significant thermal degradation of the polymer. This statement is supported by the time sweep-complex viscosity results reported by D'Urso et al. [23], who showed that the viscosity of virgin PLA is relatively higher than that of extruded PLA. Thus, the narrow thermal processing window for PHB prevents it from being used widely in industrial applications. However, in order to broaden the thermal processing window of PHB, it is generally copolymerized with hydroxyvalerate (HV), resulting in a highly thermally stable poly(hydroxybutyrate-co-hydroxybutyrate) (PHBV), which has a lower processing temperature [24]. Despite the disadvantages of PHB, it is a perfectly isotactic biopolymer without any chain branching, in contrast to PE and hence flows easily during melt processing.

Maiti et al. [25] investigated the development of PHB/clay nanocomposites by the melt extrusion method (Fig. 5.4). Both XRD and TEM analysis results showed that the nanocomposites had a well-ordered intercalated structure. However, the degree of intercalation depended on the amounts of silicate and modifier present. TEM images showed that stacks of silicate layers were homogenously dispersed within the PHB matrix. More importantly, the distribution of the stacks of the silicate layers was in keeping the characteristic clay-related XRD peak observed at approximately 2.5°. This structural development resulted in enhanced thermal stability and improved mechanical characteristics. In addition, Lim et al. [26] prepared PHB/clay nanocomposites by the solution blending method (solvent casting). Clay (namely, organo-montmorillonite (OMMT)) was incorporated into the PHB matrix at loading rates of 3–9 wt%, with neat PHB used as a control. The XRD and FTIR analyses results suggested the formation of well-intercalated PHB/clay nanocomposites, with the *d*-spacing increasing with the clay content. In addition, the thermal stability of the PHB/clay nanocomposites was influenced by the incorporation of clay particles as well as the clay loading rate and degree of dispersion.

Thus, it is possible to produce PHB nanocomposites with enhanced properties through solution and melt processing. However, care should be taken in the case of melt processing. In the above-described study [25], the clay particles protected PHB from rapid thermal degradation during processing and hence improved its thermal stability. Fillers belonging to the graphene family can also be used in this manner. For example, graphene oxide has a lower thermal stability but has been shown to be suitable for producing polymer nanocomposites with enhanced thermal stability [27].

5.2.4 Processing Strategies for Starch Nanocomposites

Starch can also be processed using the conventional processing methods such as solution and melt blending (the melting processes include internal mixing, extrusion, injection molding, and compression molding) [28]. However, the production of thermoplastic starch through melt extrusion is complex as comparison to the case for other polymers (Fig. 5.5). The production of thermoplastic starch through melt processing involves high energy inputs (>10^2 kWh/t) and large amounts of water [28]. Native starch is not very processable, exhibits poor dimensional stability, and reasonably high mechanical properties. Therefore, it can be processed using a broad range of techniques, including being mixed in dry starch form with a filler prior to processing [29, 30]. Native starch can be strategically injection molded to produce thermoplastic starch at its glass transition temperature (approximately 60–80 °C) for a water weight fraction of 0.12–0.14 [29]. The water content during the production of thermoplastic starch is a critical parameter, since the water lowers the melting temperature and plasticizes the starch during processing. Hence, in order to ensure

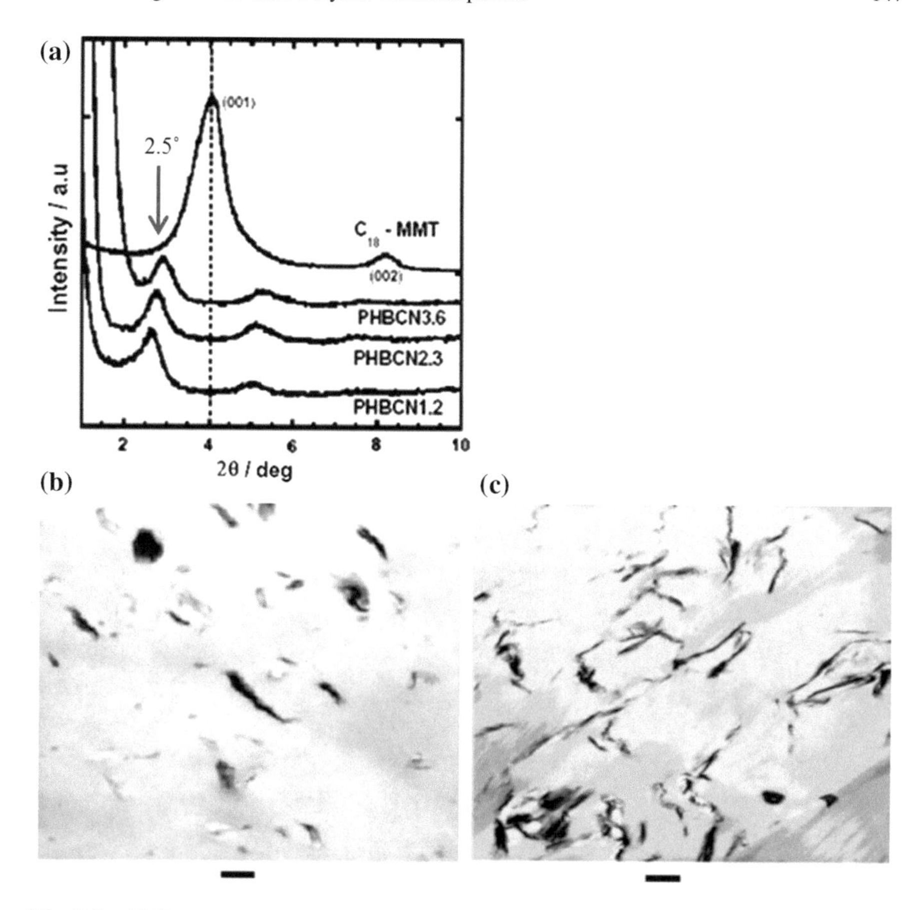

Fig. 5.4 **a** XRD patterns of organically modified clay (C18 MMT) and various PHB nanocomposites and TEM images of **b** PHBCN2.3 (2.3 wt% MMT), **c** PHBCN2 (2 wt% MAE). Reprinted with permission from [25]. Copyright 2007, The American Chemical Society

a high degree of plasticization of the starch granules, the water evaporation must be limited.

Muller et al. [31] reported the processing of starch/clay nanocomposites using melt extrusion. Both unmodified and modified nanoclay-filled starch composites exhibited a more crystalline structure as compared to that of the reference sample. Interestingly, the XRD analysis results indicated that the unmodified nanoclay-filled starch composites had an intercalated structure. Further, the unmodified nanoclay-filled starch composites exhibited a uniform surface morphology, with no insoluble particles being present, owing to the inherent hydrophilic nature of unmodified nanoclay (Fig. 5.6). On the other hand, modified nanoclay-filled starch composites exhibited a uniform surface morphology, with the surface containing many insoluble particles; this was due to the lower compatibility between the hydrophilic starch matrix and the hydrophobic nanoclay particles.

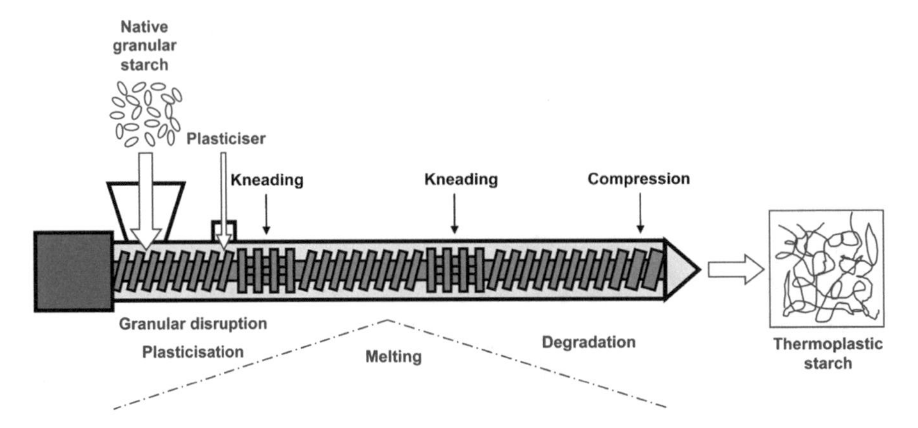

Fig. 5.5 Schematic diagram of starch processing. Reprinted with permission from [28]. Copyright 2012, Elsevier Science Ltd.

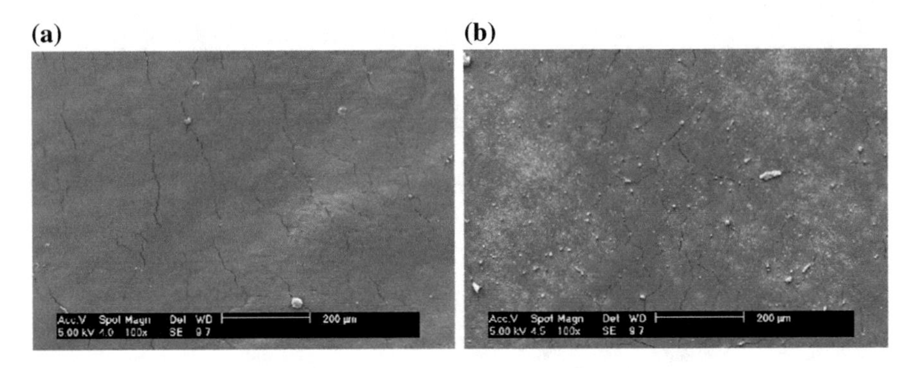

Fig. 5.6 SEM images of starch composites: **a** unmodified nanoclay and **b** organically modified nanoclay. Reprinted with permission from [31]. Copyright 2012, Elsevier Science Ltd.

Moreover, Chung et al. [32] prepared starch/clay nanocomposites using solution blending. The clay particles were incorporated into a starch matrix at loading rates of 1–7 wt%. No XRD peaks characteristic of clay were observed in any of the nanocomposites. This indicated that exfoliation had occurred because of the homogeneous dispersion of clay and simultaneous presence of small tactoids. Interestingly, the 5 wt% clay-filled starch nanocomposite exhibited enhanced modulus (65%) and tensile strength (30%) as compared to those of the unfilled starch materials, with there being no changes in the elongation at break. In addition, the degree of clay dispersion decreased with an increase in the clay content, leading to a deterioration of the mechanical properties.

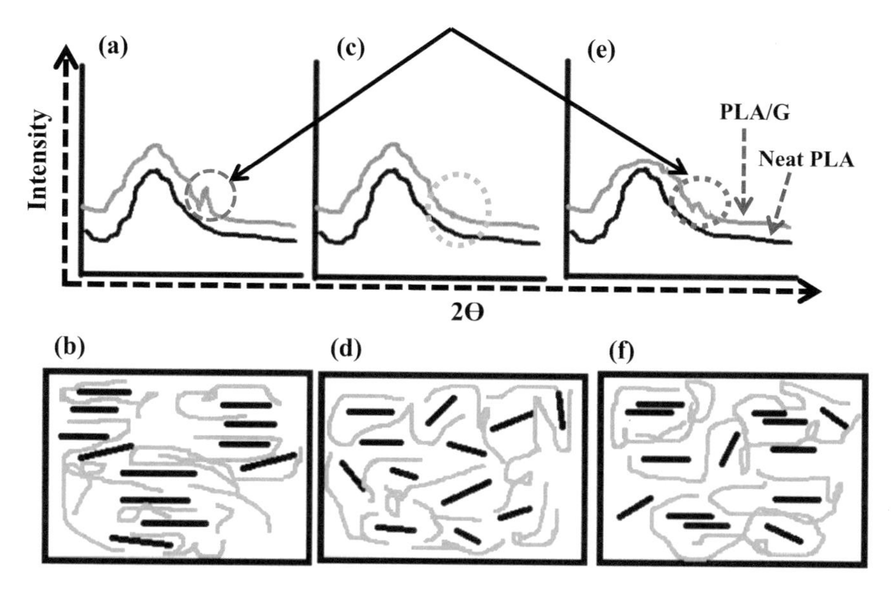

Fig. 5.7 Schematic XRD spectrum based on dispersion of nanofillers of different sizes within polymer matrix (PLA/graphene-based nanocomposite) and various structural morphologies possible: **a** and **b** intercalated, **c** and **d** exfoliated, and **e** and **f** coexistence of intercalated and exfoliated nanofiller sheets

5.3 Relationships Between Structure and Properties of Polylactide, Polyhydroxybutyrate, and Starch Nanocomposites

5.3.1 *General Overview of Relationships Between Structure and Properties of Polymer Nanocomposites*

The incorporation of nanofillers into the polymer matrix allows one to control the structure of the resulting polymer nanocomposites. Therefore, various types of nanoparticles have been used efficiently as reinforcement materials for improving the mechanical (e.g., brittleness) properties of different biodegradable polymers (i.e., PLA, PHB, and starch) in order to widen their applicability. Polymer/sheet filler nanocomposites can exhibit a range of structures, including separated, intercalated, and exfoliated structures as well as structures that are both intercalated and exfoliated (Fig. 5.7). The best tools for characterizing the morphologies of these nanocomposites are XRD and TEM analyses. The schematic diagram in Fig. 5.7 shows the differences between the various possible polymer nanocomposites structures (the PLA/graphene-based system is used as an example).

A wide diffraction peak is usually observed in the case of a neat PLA matrix. However, the wideness of the diffraction peak is reflective of the amorphousness/crystallinity of the sample, which depends on the contents of the D- and L-enantiomers within the PLA matrix. It is known that the presence of a new diffraction peak indicated the inclusion of a secondary component within the polymer matrix. However, the positions and intensities of the characteristic diffraction peaks can be used to elucidate the structure of the polymer nanocomposite system being investigated. When polymer chains are intercalated in the layers of the filler (i.e., clay or graphene oxide) the characteristic XRD peak of the filler generally undergoes a shift or a decrease in intensity. On the other hand, in the case of complete exfoliation, no diffraction peak is observed in the system. This can be usually be attributed to the absence of re-aggregation of the particles. In this case, the distance between the sheets is generally larger than that detectable using wide-angle X-ray scattering. The existence of the filler in the exfoliated state results in a stronger reinforcement effect than that in the case of an intercalated or separated structure [33, 34]. On the other hand, when a intercalated/exfoliated structure is present, the degree of re-agglomeration is low, and the degree of exfoliation is generally higher.

The overall structure of polymer nanocomposites is generally determined by the size of the filler particles and their degree of dispersion within the polymer matrix as well as the compositions of the component, particle-polymer compatibility, processing conditions (see Sect. 5.2), and particle localization [35, 36]. Knowing the structure-property relationships is essential for the development of high-performance sustainable polymer nanocomposites. Therefore, the following questions must be asked: which structure results in better properties? How does the extent of incorporation of the nanofiller within the polymer matrix change the thermal, mechanical, rheological, and other properties of the neat polymer? How can the incorporation of the nanofiller help overcome the various fundamental processing challenges in order to ensure that the reinforcement effect for the polymer nanocomposites is optimized?

5.3.1.1 Structure-Property Relationship for PLA Nanocomposites

PLA is a hydrophobic polymer with $-CH_3$ side groups (Fig. 5.8). Thus, it is soluble in a wide range of solvents (i.e., chloroform, dichloromethane, tetrahydrofuran, methylene chloride, and acetonitrile). The properties of PLA strongly depend on the number of D- and L- enantiomers present, as this determines the final crystallinity of PLA, including its grade. In brief, the various grades of PLA include poly(DL-lactide) (DL-PLA), which contains both D- and L-stereoisomers, poly(D-lactide) (D-PLA), and poly(L-lactide) (L-PLA). The production of high-molecular-weight PLA samples of all three grades is possible though the most industrially attractive polymerization method, namely, ring-opening polymerization. However, low-molecular-weight PLA can be obtained through polycondensation [37, 38]. Water must be removed completely to ensure polymerization [39]. Because of the structure and properties of PLA can be tailored based on its molecular weight, there is a lot of leeway with respect to the incorporation of nanofiller reinforcement materials. The degree of the reinforcement effect depends on the processing method and conditions used as well

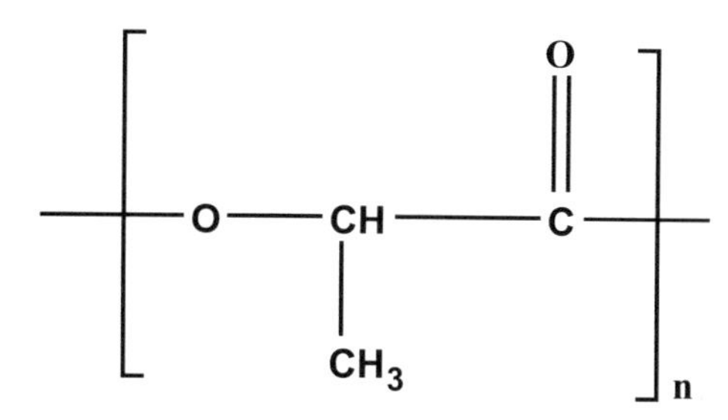

Fig. 5.8 Molecular structure of polylactide

as the type of the filler used and its content. That is, micro- and nanofillers affect the structure of the polymer differently.

Picard et al. [20] evaluated the gas transport properties (helium and oxygen) of PLA and PLA/clay nanocomposites (PLANC4 and PLANC8) with and without annealing at 120 °C. They reported that clay reduced the permeability of the polymer for both gases, with the decrease in the gas permeability being higher for a higher clay content. However, in addition to the effect of clay and its content, the morphology and structure may be responsible for this phenomenon (Fig. 5.3 above). Moreover, the stabilized morphology and coexistence of an exfoliated/intercalated structure have the ability to significantly decrease the gas permeability due to the regular tortuous path that is created, whereas the tactoid structure has poor gas permeability. TEM imaging that, at high contents, clay was well distributed within the nanocomposite, with no XRD peak being observed at approximately 2.5°. This may be due to there being an equilibrium between the shear force and the clay particles content. In addition, the oxygen permeability decreased further with annealing. This was due to the clay and the crystalline phase. These results indicate that the properties of nanocomposites are not only dependent on the filler content but also on their structure and morphological characteristics as well as the processing parameters used.

5.3.1.2 Structure-Property Relationship for PHB Nanocomposites

PHB is a perfectly isotactic biodegradable and biocompatible polymer that belongs to the family of polyhydroxyalkanoate (PHAs) and thus does not contain any branching chains, in contrast to PE (Fig. 5.9). Interestingly, PHB is generally produced through biosynthesis, which involves the use of glucose or agricultural waste products as the as the raw material. It can also be produced through cyclic bacterial fermentation. Further, it can be produced in large quantities from PHAs. On the other hand, it exhibits high crystallinity, brittleness, and high thermal stability. These factors greatly limit its processing window and applicability. Hence, there is need to modify its structure in order to improve its properties [8, 40]. Its crystallinity, glass temperature, and microstructure can be optimized to reduce its brittleness. This can be done by the incorporation of nanofillers within the polymer matrix, such that the nanofillers are exfoliated. However, the nanofiller must be added in the optimized amount.

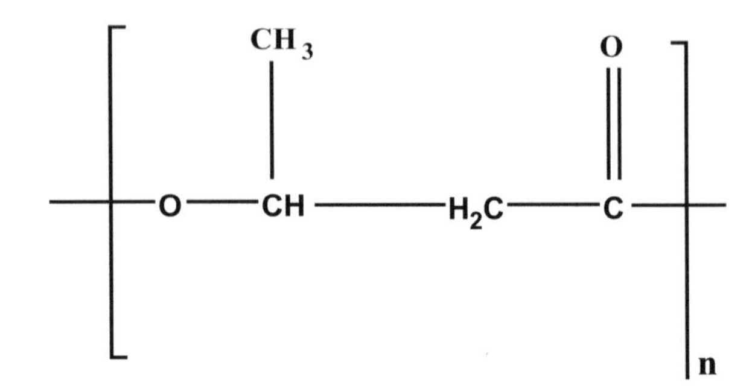

Fig. 5.9 Molecular structure of polyhydroxybutyrate

Akin et al. [41] were able to fabricate PHB/clay nanocomposites though solution blending while using clay in contents of 1–5 wt%. XRD analysis revealed that the PHB/clay nanocomposites exhibited an exfoliated structure at low clay loading rates while an intercalated structure was seen at high clay loading rates. Improvements of 152 and 73% in the tensile strength and strain at break were observed in the 3 wt% clay-filled PHB nanocomposite. This indicates that the mechanical properties of the nanocomposites depended on their structure, which was determined by the level of dispersion of the filler, which, in turn, was based on the amount of filler added. In addition, the nanocomposites exhibited increased thermal stability, good optical properties, and were suitable for use as water vapor barriers, owing to the restricted mobility of the polymer chains. Therefore, these nanocomposites were considered suitable for food and medical packaging.

5.3.1.3 Structure-Property Relationship for Starch Nanocomposites

Starch is a hydrophilic low-density biodegradable polymer that is regenerated from carbon dioxide and water in plants through the photosynthesis pathway. It is well known that starch is composed of D-glucose residues arranged either in linear α-(1 → 4) linkages, as in the case of r amylose or in a combination of linear α-(1 → 4) and ~5% α-(1 → 6) branch linkages, as in the case of amylopectin [29, 42, 43] (Fig. 5.10). The hydroxyl groups on the starch chains can potentially participate in the formation of hydrogen bonds, ethers, and esters [29]. The use of a filler with a greater number of polar groups may increase the starch–filler interactions, resulting in a stable morphology, which, in turn, can result in significant improvements in the mechanical properties.

Ma et al. [44] compared graphene oxide-filled starch composites and reduced graphene oxide-filled starch composites. It is known that graphene oxide contains a greater number of oxygenated functional groups than does reduced graphene oxide, owing to which the former is more compatible with the polymer matrix and disperses more effectively in solvents such as water. Ma and coworkers reported a more pronounced improvement in the mechanical and barrier properties in the case of the graphene oxide-filled starch composites as compared to those of the reduced

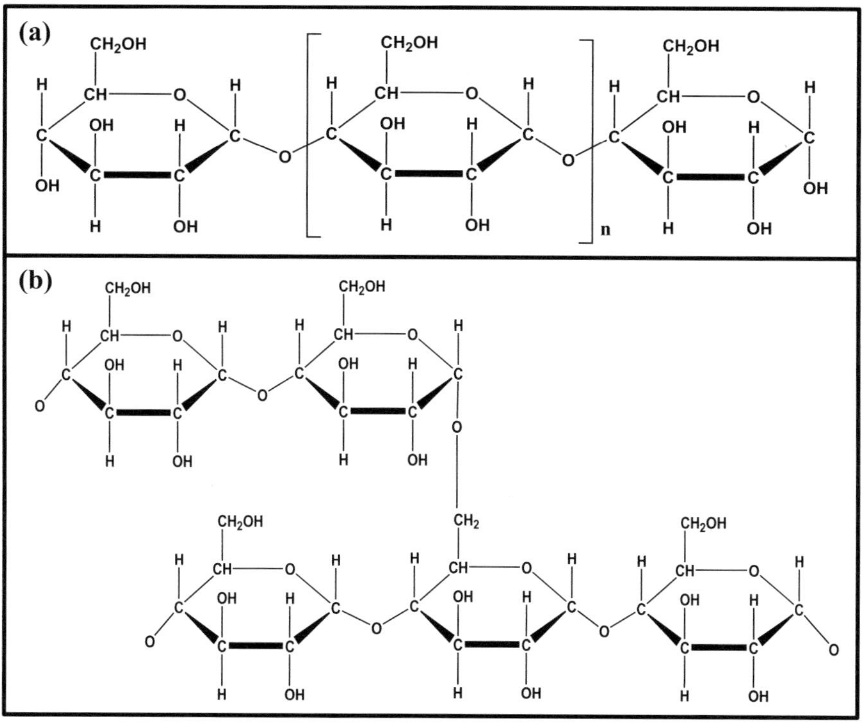

Fig. 5.10 Chemical structures of starch molecules: **a** amylose and **b** amylopectin

graphene oxide-filled starch composites. They suggested that this was due to the presence of a greater number of oxygen functionalities on the GO surface in the former case and higher dispersibility of graphene oxide particles. In addition, the increase in the tensile strength of the starch composite containing 2 wt% graphene oxide was comparable to that in the case of the composite containing 6 wt% reduced graphene oxide (66% and 67%, respectively). This indicates that efficient nanofiller-polymer interactions as well as the efficient dispersion/distribution of the nanofillers are the key factors determining the properties of the nanocomposites. Thus, this section highlights the importance of the presence of a sufficient number of polar groups with respect to the compatibility of the nanofillers and the polymer and the loading characteristics of the nanofillers as well as their effect on the properties of end product (i.e., graphene oxide-filled starch composites).

5.4 Performances and Applications of Polylactide, Polyhydroxybutyrate, and Starch Nanocomposites

5.4.1 Performance and Applications of Polymer Nanocomposites

Biodegradable polymers such as PLA and PHB primarily find use in biomedical, agricultural, and packaging products. Since these polymers are generally fabricated from renewable resources, they can be used as fertilizers for crops or as other agricultural products after disposal. PLA takes approximately one to two years to degrade. However, this depends on the size and shape of the product, the hydrolysis temperature, the catalyst concentration, and the stereoisomers involved [19]. In contrast, PHB has a higher biodegradation rate and takes only three months to degrade [36].

5.4.2 PLA Nanocomposites

PLA has significant potential for use in consumer products, particularly for packaging, owing to its high transparency, low toxicity, and completely eco-friendly nature. Visible light can pass through PLA more readily than through PET and PS; this is particularly true for light with wavelengths of 540–560 nm [19]. While PLA exhibits good mechanical properties (tensile strength and modulus), its inherent brittleness, poor crystallization, susceptibility to undergo hydrolysis, and poor barrier properties limit its usability. Thus, the large-scale production and commercial applications of PLA flexible films or injected parts with high impact strength are limited. On the other hand, its high hydrolysis rate and low thermal stability limit the use of PLA in hot food packaging while its poor gas barrier properties with respect to O_2, CO_2, and H_2O make to prone to reacting with food [19]. Furthermore, its low glass transition temperature makes materials based on it (i.e., plastic cups) unsuitable for holding hot liquids. Thus, it is clear that PLA must be reinforced with various nanoparticles and its fabrication process must be optimized.

Gao et al. [3] prepared PLA/graphite nanoplatelets-based nanocomposites using nanoplatelets of different particle sizes (small and large) by melt blending (graphite nanoplatelets-small (GNP-S) and graphite nanoplatelets-large (GNP-L)). The results of XRD and SEM analysis indicated that, the GNP-S particles were dispersed for contents of 5–10 wt%, with slight agglomeration being observed at 15 wt%. On the other hand, the GNP-L particles started to agglomerate even at a loading rate of 7 wt%. This was in keeping with the high-intensity characteristic peak of the graphene-based particles observed at $2\theta = 26°$ in the case of the GNP-L-filled PLA nanocomposites. These results indicate that it is possible to control the structure-property relationship by varying the particle size of the filler material.

5.4.3 PHB Nanocomposites

Unlike PLA, PHB shows excellent barrier properties, which include low permeability with respect to O_2, H_2O, and CO_2. It also exhibits high crystallinity, which ranges from 50 to 70% [4, 45]. Thus, it is the best material for packaging applications and is suitable for manufacturing high-performance industrial products such as degradable bottles, laminated foils, fishnets, potted flowers, hygiene, food packaging, cell and tablet packaging, one-way use cups, agricultural foils, fibers in textile, and microcapsules [8]. On the other hand, the inherent biodegradability and biocompatibility of PHB allow it to be used as a surgical implant material, as well as for the seam threads for the healing of wounds and blood vessels.

Follain et al. [46] reported the fabrication of PHA films by the solution and compression method. They also measured the thermal properties, water vapor sorption characteristics, and permeability to water and gas molecules. The films prepared by solution casting exhibited high water permeability and allowed the transfer of gas molecules owing to the formation of additional free space between the polymer chains during processing. Because of this, the compressed films exhibited improved properties. Further, the film properties depended not only on the processing method used but also on the crystallinity of the end product, with the lowest water permeability being recorded for the films with the highest crystallinity. They concluded that the PHA-based polymers are highly suited for forming biocomposites as the water mass gain of the films was found to be less than 2%. The films demonstrated good moisture resistance and high gas barrier characteristics. Therefore, PHA-based polymers are promising materials for use in sustainable packaging.

Given the extensive literature available on the use of nanofillers such as clay and graphene-based materials, it can be concluded that nanofillers will be effective in improving the barrier properties of PHA-based polymers even further and will also enhance their mechanical properties to the desired industrial levels when dispersed homogenously.

5.4.4 Starch Nanocomposites

Starch granules was initially used to increase the surface area available for attacking microorganisms in other polymer system [29]. However, because of its low cost, high viscosity, availability (second most abundant biomass material in nature), and possible hydrogen bonding site, it is recognized as a potential candidate for matrix biodegradable polymer in the presence of nanofillers for various high-performance industrial applications. Substitution of hydroxyl groups with other groups is a possible and effective method to improve the properties of a starch-based system [28, 29, 43]. Regarding this, an attempt was made by Avella et al. [47] to investigate the potential food applications of starch/clay nanocomposite films. The biodegradable samples were prepared through melt blending. Their results showed the formation of a good intercalated starch/clay structure, which led to increased modulus and tensile strength. More importantly, the samples demonstrated a low overall migration limit; therefore, starch/clay nanocomposite samples were regarded as potential candidate materials for food packaging.

This indicates that homogeneously dispersed montmorillonite clay nanoparticles influence the type of structure and mechanical properties of starch/clay nanocomposites because of their surface chemistry and reinforcement effects. More importantly, melt processing method, the most industrially favorable method, was utilized to successfully produce sustainable polymer nanocomposite materials for food packaging (Table 5.1).

5.5 Challenges and Opportunities with Polylactide, Polyhydroxybutyrate, and Starch Nanocomposites

Both PLA and PHB suffer from some common drawbacks such as brittleness (PLA has < 10% elongation at break) and poor thermal stability [26, 58, 59]. Extensive applications of PHB are limited because of its high cost. Thermal decomposition temperature of PHB is slightly higher than its melting temperature. As a result, its viscosity and molar mass decrease, which limit its melt processing window [26]. However, PHB can be easily processed. In contrast, the poor gas barrier properties of PLA limit its usage in packaging applications [59]. Moreover, PLA has a slow degradation rate, which leads to a long in vivo lifetime and prolongs the life span of PLA plastic commodities at disposal. It is also hydrophobic, which results in low cell affinity. Another challenge with the PLA is a lack of reactive side chain groups, which makes its modification very difficult [37].

Starch is also known to suffer from certain drawbacks such as poor mechanical properties and hydrophilic nature [14]. Techniques such as copolymerization, blending, and addition of fillers have been examined to overcome these limitations. However, the reinforcement effect of the filler strongly depends on the size, shape, and density of the undertaken filler. Moreover, the localization of the filler in the polymer system affects the interfacial interaction between the polymer matrices and the filler, filler loading concentration, and degree of filler dispersion and distribution in relation to polymer matrices [60–62]. Our researcher group has recently prepared a thermally shocked graphene oxide as a nanofiller with a higher surface area (523.2 m^2/g) and minimum number of graphene sheets (~4 layers) and demonstrated that it can fine-tune and stabilize the surface morphology of a PLA blend system with a significant increase in elongation at break and little effect on both the tensile strength and modulus. The resulting enhanced mechanical performance can be attributed to stable morphology, which is greatly associated with the nanofiller type, particle size or number of graphene layers, and surface area [63]. The attained stable morphology and enhanced mechanical performance are associated with the presence of a small amount of oxygen-containing functional groups such as carboxyl and

Table 5.1 Summary of processing strategy and structure-property relations based on sustainable polymer nanocomposites

Polymer nanocomposites	Processing method	Structure	Properties	References
PLA/clay	In situ intercalative polymerization	Exfoliated	Exfoliated PLA/clay nanocomposites yielded high thermal stability due fine dispersion of individual silicate layers into the polymer matrix	[48]
PLA/clay	Melt and solution (separate)	Coexistence of intercalated and exfoliated in both methods	Good dispersion was attained in both methods. However, mechanical properties of nanocomposites prepared by melt blending methods were more enhanced than those prepared by solution method due to the greatest organclay dispersion	[21]
PLA/clay (organically modified montmorillonite (OMMT))	Solution	Intercalation	TEM images revealed anisotropic intercalated structure with aggregated clay particles. Moreover, the UV-Vis measurements for neat PLA, nanocomposites recorded. Neat PLA exhibited good transmittance, whereas nanocomposites exhibited poor transmittance (red-shift). However, after Au-coating nanocomposites demonstrated much improvement in the reflective properties, hence PLA-based films are recognised as potential candidates for fabrication of biosensors	[49]
PLA/graphene (Two types of GNP-M and GNP-C at 0.25 wt%)	Melt	–	XRD analysis indicated the presence of the graphitic characteristic peak for all composites before and after hydrolytic degradation for six months in phosphate-buffered saline. The incorporation of either GNP-M or GNP-C nanofillers into PLA matrices improved the mechanical properties of neat PLA. After six months degradation, a molecular weight decrease of about 85% was noted for samples. Neat PLA exhibited a 10-fold decrease in toughness, whereas the GNP-M- and GNP-C-filled PLA composites exhibited 3.3 and 1.7-fold decrease in toughness. No toxic products were noted after the degradation period	[50]

(continued)

Table 5.1 (continued)

Polymer nanocomposites	Processing method	Structure	Properties	References
PLA/graphene	Melt reprocessing (reprocessed up five times by means of a single screw extruder)	Exfoliation particularly after reprocessing	The incorporation of GNP into PLA matrices decreased the degradation rate as a function of the melt reprocessing cycles relative to neat PLA; however, significant molecular Weight, viscosity decrease was noted. Interestingly, the melt reprocessing improved the particle dispersion, thus particle aggregation reduction. In this rational, the reprocessing samples (5 cycles) demonstrated the exfoliated structure this behavior was then attributed to the existence of low amount of ordered layer structure of GNP particles, hence stable morphology was attained	[51]
PLA/graphene	Solution	–	The incorporation of graphene into PLA matrices enhanced the thermal stability of PLA as well as its spherulite morphology. This improvement behavior is associated with the well-dispersed structure for the PLA/graphene nanocomposites	[52]
PHB/clays: comparison of montmorillonites (MT), Na-M (MT) and 30B-M (organically modified MT)	Melt	Intercalation/exfoliation	XRD and TEM results revealed that the 30B-M exhibited higher degree of intercalated/exfoliation structure than NA-M, hence, high moduli of the PHB-30B-M nanocomposites were noted. In addition, increased crystallization temperature and a decreased spherullites size were also noted	[36]
PHB/clay	Melt	Intercalated	The increase of thermal stability and mechanical properties of the nanocomposites was noted, the increase depends the degree of intercalation	[25]
PHB/clay (Clay loading 1, 2, 3, and 5 wt%)	Solution	Exfoliation at ≤2 wt% of clay and intercalation at ≥3 wt% of clay	An optimum increase of about 152 and 73% in tensile strength and strain at break was attained for 3wt% clay–filled PHB nanocomposite	[41]
PHB/clay	Solution	intercalation	The incorporation of clay increased the thermal stability	[26]

(continued)

Table 5.1 (continued)

Polymer nanocomposites	Processing method	Structure	Properties	References
PHB/clay (unmodified and chemically modified clays)	Solution	–	Modified clay-filled PHB nanocomposites exhibited better clay particle dispersion than unmodified clay-filled PHB nanocomposites. However, the XRD results indicated that the PHB crystalline lattice had little effects on the presence of both clays	[53]
PHB/graphite oxide	Solution	Exfoliation	Incorporation of graphite oxide into the PHB matrix significantly increased the dynamic shear modulus of the neat PHB; this behavior was attributed to the micromechanical action of the graphite oxide nanoparticles and the attained exfoliated structure. The increasing behavior increases with the increasing content of the graphite oxide nanoparticles (1–5 wt%)	[54]
PHB/graphene oxide	Solution	Exfoliation	The XRD and SEM results revealed the well exfoliated and uniformly dispersed modified graphene oxide due to the presence of hydrogen bonds. As results the enhanced mechanical and thermal stability of the nanocomposites was attained. In addition, the electrical conductivity was increased by 14 orders relative to that of neat PHB, particularly for modified graphene oxide	[55]
Starch/nanoclay	Melt extrusion	Intercalated composite	Modified water vapor permeability (WVP) and mechanical properties were noted, however, the behavior depends on the filler dispersion the compatibility. (e.g. 0.05 g of unmodified nanoclay/starch exhibited 2.22, 76, and 50, whereas 0.05 g of organically modified nanoclay/starch exhibited 1.51, 74, and 20; tensile strength (MPa), elongation at break (%), and young's modulus (MPa), respectively)	[31]
Starch/clay	Melt extrusion	Coexistence of intercalated and exfoliated structures	Gelatinized starch film displayed the highest degree of exfoliation due to the optimum level of plasticizer and nanoclay, leading to starch/clay nanocomposites with superior mechanical properties	[56]

(continued)

Table 5.1 (continued)

Polymer nanocomposites	Processing method	Structure	Properties	References
Starch/clay	Solution	Exfoliation	Good clay dispersion was noted. Nanocomposites containing 1 and 5 wt% clay (montmorillonite) content exhibited significate tensile strength and young's modulus increase, due to the good dispersion of clay, the enhanced interaction behavior between starch molecules and clays. Whereas, an increase of clay up to 7 wt% decreased the mechanical properties of the starch nanocomposite, relative to unfilled starch	[32]
Starch/graphene oxide and reduced graphene oxide	Solution	–	The presence of more oxygen functionalities on the GO surface structure enhanced the interaction between the GO and starch matrices more efficiently than RGO. In this rational, GO-filled starch composites exhibited high degree of tensile strength, elongation at break and moisture barrier properties than RGO-filled starch composites. However, RGO-filled starch composites exhibited better thermal stability than GO-filled starch composites	[44]
Starch/graphene oxide	Solution	Exfoliated	The results indicated good compatibility of GO and starch due to the possible hydrogen bonding between GO functionalities and starch. The starch biocomposite containing 2 wt% GO exhibited tensile strength increased from 4.56 to 13.79 MPa, and the young's modulus increased from 0.11 to 1.05 GPa, however, the elongation at break decreased from 36.06 to 12.11% was noted. In addition, the thermal stability of the starch was enhanced in the presence of GO particles	[57]

hydroxyl groups from the graphene-based material, which enhanced the interfacial interaction between the polymer matrices and the filler. On the contrary, $\leq$5 wt% of graphene-based particles is usually used to prepare polymer nanocomposite materials; the addition of a small amount of graphene helps maintain the eco-friendliness of biodegradable polymers. Furthermore, mono-graphene sheets can be used to produce lightweight and high-performance biopolymer nanocomposites with a wide range of novel applications [14]. In this regard, similar volume fraction of the nanofiller has a much higher filler–polymer interaction than microfillers because of their superior surface area to volume ratio (nanofiller volume fraction against microfiller volume fraction) [1].

The main challenges in the development of high-performance biodegradable polymer nanocomposites include achieving a high level of nanofiller dispersion and distribution in polymer matrices, discovering suitable plasticizers, and controlling the interfacial strength between the polymer and the nanofiller (i.e., starch and clay) [32]. However, all these factors are based on the structure and type of nanofillers (i.e., nanoparticles, nanotubes, nanolayers, or nanosheets) associated with the processing conditions [56].

Biopolymer nanocomposites such as PLA nanocomposites, PHB nanocomposites, and starch nanocomposites have various potential applications in the biomedical field and in the fabrication of environmentally biodegradable plastics that could effectively address waste management issues. So far, there is no report of the presence of such biopolymers or their monomers and their toxicity in the human body, and this makes them all the more important for broad medical applications. These biopolymers are not dangerous to many animals. Moreover, PLA is recognized as one of the safest polymers by the United States Food and Drug Administration. Therefore, it is safe for most of the packaging and medical applications [37]. Interestingly, from the energy usage point of view, PLA requires 25–55% less energy for production compared to that of petroleum-based polymers, which can be further reduced, depending on the progress made in current and future research [37]. In addition, recycling PLA products to produce new PLA plastics allows large energy savings [39]. This will invariably increase the PLA production and industrialization, will help in achieving a clean environment, and better health, which will exponentially increase PLA usage. This makes PLA nanocomposites the leading candidates of biodegradable polymer nanocomposites.

Graphene oxide has a higher content of oxygen-containing functional groups compared to that of most of the carbon-based nanofillers and a relatively higher specific surface area than that of the oxidized multiwalled carbon nanotubes (MWCNT–COOH). Hence, graphene oxide is expected to be more promising than MWCNT–COOH for polymer nanocomposite reinforcement [64]. However, this is not the case because of poor mechanical strength and extremely high surface area of graphene sheets compared to that of graphene oxide; in this regard, graphene is considered superior to polymer nanocomposites. Graphene as a single nanosheet is also better than both clay and graphite because of its lightweight and better mechanical properties. However, unreactive sides of single nanosheets of pristine graphene cause poor dispersion and agglomeration of graphene nanoparticles in polymer matrices; therefore, functionalizing graphene is considered the best way to widen the applications of graphene [64]. However, the high cost, limited availability, and difficult synthesis protocols limit the application of graphene. Some limitations of graphene synthesis protocols include poor large-scale production and time-consuming process. This makes it very important to perform extensive research to achieve large-scale and effective production of graphene with reasonable industrial time and resources. Hence, biopolymer nanocomposite production through an industrially favorable method, melt blending, and extensive usage of final bio-end products should be investigated.

5.6 Conclusions and Future Trends in Biodegradable Polymer Nanocomposites

In summary, this chapter discussed potential industrial applications of biodegradable polymer nanocomposites as the most high-performance sustainable materials of future, particularly those of graphene-filled PLA and graphene-filled PHB nanocomposites. The current environmental policies emphasize on the reduction of the usage of non-biodegradable polymers; hence, the investigation and reinforcement of degradable polymer nanocomposites are attracting increasing interest in both academia and industry. Thus, the development of significantly reinforced biodegradable polymers using graphene-based materials is a major stepping stone toward achieving a greener and sustainable environment. This chapter briefly discussed the structure and processing methods for biodegradable polymer nanocomposites, particularly PLA, PHB, and starch containing either graphene or clay nanocomposites. Therefore, the relations between the components, structure, properties, and processing must be investigated to produce excellent sustainable polymer nanocomposite materials.

Acknowledgements The authors are grateful for the financial support from the Department of Science and Technology and the Council for Scientific and Industrial Research (DST-CSIR NCNSM).

References

1. Zaman I, Manshoor B, Khalid A, Araby S (2014) From clay to graphene for polymer nanocomposites—a survey. J Polym Res 21:429
2. Avérous L (2004) Biodegradable multiphase systems based on plasticized starch: a review. J Macromol Sci Part C Polym Rev 44:231–274
3. Gao Y, Picot OT, Bilotti E, Peijs T (2017) Influence of filler size on the properties of poly(lactic acid) (PLA)/graphene nanoplatelet (GNP) nanocomposites. Eur Polym J 86:117–131
4. Pachekoski WM, Agnelli JAM, Belem LP (2009) Thermal, mechanical and morphological properties of poly(hydroxybutyrate) and polypropylene blends after processing. Mater Res 12:159–164

5. Inan K, Sal FA, Rahman A, Putman RJ, Agblevor FA, Miller CD (2016) Microbubble assisted polyhydroxybutyrate production in *Escherichia coli*. BMC Res Notes 9:338
6. Khanna S, Srivastava AK (2005) Recent advances in microbial polyhydroxyalkanoates. Process Biochem 40:607–619
7. Wang B, Sharma-Shivappa RR, Olson JW, Khan SA (2012) Upstream process optimization of polyhydroxybutyrate (PHB) by *Alcaligenes latus* using two-stage batch and fed-batch fermentation strategies. Bioprocess Biosyst Eng 35:1591–1602
8. Arrieta MP, Samper MD, Aldas M, López J (2017) On the use of PLA-PHB blends for sustainable food packaging applications. Materials 10:1008
9. Sreedevi S, Unni KN, Sajith S, Priji P, Josh MS, Benjamin S (2014) Bioplastics: advances in polyhydroxybutyrate research. Advances in polymer science. Springer, Berlin, pp 1–30
10. Luckachan GE, Pillai CKS (2011) Biodegradable polymers—a review on recent trends and emerging perspectives. J Polym Environ 19:637–676
11. Mudliar SN, Vaidya AN, Kumar MS, Dahikar S, Chakrabarti T (2008) Techno-economic evaluation of PHB production from activated sludge. Clean Techn Env Policy 10:255–262
12. Suryanegara L, Nakagaito AN, Yano H (2009) The effect of crystallization of PLA on the thermal and mechanical properties of microfibrillated cellulose-reinforced PLA composites. Compos Sci Technol 69:1187–1192
13. Nanthananon P, Seadan M, Pivsa-Art S, Suttiruengwong S (2015) Enhanced crystallization of poly (lactic acid) through reactive aliphatic bisamide. IOP Conf Ser Mater Sci Eng 87:012067
14. Rouf TB, Kokini JL (2016) Biodegradable biopolymer–graphene nanocomposites. J Mater Sci 51:9915–9945
15. Najafi N, Heuzey MC, Carreau PJ (2012) Polylactide (PLA)-clay nanocomposites prepared by melt compounding in the presence of a chain extender. Compos Sci Technol 72:608–615
16. Cui Y, Kundalwal SI, Kumar S (2016) Gas barrier performance of graphene/polymer nanocomposites. Carbon 98:313–333
17. Sengupta R, Bhattacharya M, Bandyopadhyay S, Bhowmick AK (2011) A review on the mechanical and electrical properties of graphite and modified graphite reinforced polymer composites. Prog Polym Sci 36:638–670
18. Ojijo V, Ray SS (2013) Progress in polymer science processing strategies in bionanocomposites. Prog Polym Sci 38:1543–1589
19. Armentano I, Bitinis N, Fortunati E, Mattioli S, Rescignano N, Verdejo R, Lopez-manchado MA, Kenny JM (2013) Multifunctional nanostructured PLA materials for packaging and tissue engineering. Prog Polym Sci 38:1720–1747
20. Picard E, Espuche E, Fulchiron R (2011) Effect of an organo-modi fi ed montmorillonite on PLA crystallization and gas barrier properties. Appl Clay Sci 53:58–65.
21. Shameli K, Zakaria Z, Hara H, Ahmad MB, Mohamad SE, Nordin MFM, Iwamoto K (2015) Poly(lactic acid)/organoclay blend nanocomposites: structural, mechanical and microstructural properties. J Nanomater Biostr 10:323–329
22. Moad G, Dagley IJ, Habsuda J, Garvey CJ, Li G, Nichols L, Simon GP, Nobile MR (2015) Aqueous hydrogen peroxide-induced degradation of polyole fi ns: a greener process for controlled-rheology polypropylene. Polym Degrad Stab 117:97–108
23. D'Urso L, Acocella MR, Guerra G, Iozzino V, Santis FD, Pantani R (2018) PLA melt stabilization by high-surface-area graphite and carbon black. Polymers 10:139
24. Sridhar V, Lee I, Chun HH, Park H (2013) Graphene reinforced biodegradable poly(3-hydroxybutyrate-co-4-hydroxybutyrate) nano-composites. Express Polym Lett 7:320–328
25. Maiti P, Batt CA, Giannelis EP (2007) New biodegradable polyhydroxybutyrate/layered silicate nanocomposites. Biomacromol 8:3393–3400
26. Lim ST, Hyun YH, Lee CH, Choi HJ (2003) Preparation and characterization of microbial biodegradable poly (3-hydroxybutyrate)/organoclay nanocomposite. J Mater Sci Lett 22:299–302
27. Botlhoko OJ, Ramontja J, Ray SS (2017) Thermal, mechanical, and rheological properties of graphite- and graphene oxide-filled biodegradable polylactide/poly(ε-caprolactone) blend composites. J Appl Polym Sci 134:45373

28. Xie F, Halley PJ, Avérous L (2012) Rheology to understand and optimize processibility, structures and properties of starch polymeric materials. Prog Polym Sci 37:595–623
29. Lu DR, Xiao CM, Xu SR (2009) Starch-based completely biodegradable polymer materials. Express Polym Lett 3:366–375
30. Bari SS, Chatterjee A, Mishra S (2016) Biodegradable polymer nanocomposites: an overview. Polym Rev 56:287–328
31. Müller CMO, Laurindoa JB, Yamashita F (2012) Composites of thermoplastic starch and nanoclays produced by extrusion and thermopressing. Carbohydr Polym 89:504–510
32. Chung Y, Ansari S, Estevez L, Hayrapetyan S, Giannelis EP, Lai H (2010) Preparation and properties of biodegradable starch–clay nanocomposites. Carbohydr Polym 79:391–396
33. Chieng BW, Ibrahim NA, Zin W, Yunus W, Hussein MZ, Then YY, Loo YY (2014) Effects of graphene nanoplatelets and reduced graphene oxide on poly(lactic acid) and plasticized poly(lactic acid): a comparative study. Polymers 6:2232–2246
34. Paul DR, Robeson LM (2018) Polymer nanotechnology: nanocomposites. Polymer 49:3187–3204
35. Iturrondobeitia M, Ibarretxe J, Okariz A, Jimbert P, Fernandez-martinez R, Guraya T (2018) Semi-automated quantification of the microstructure of PLA/clay nanocomposites to improve the prediction of the elastic modulus. Polym Test 66:280–291
36. Botana A, Mollo M, Eisenberg P, Sanchez RMT (2010) Effect of modified montmorillonite on biodegradable PHB nanocomposites. Appl Clay Sci 47:263–270
37. Farah S, Anderson DG, Langer R (2016) Physical and mechanical properties of PLA, and their functions in widespread applications—a comprehensive review. Adv Drug Deliv Rev 107:367–392
38. Raquez J, Habibi Y, Murariu M, Dubois P (2013) Polylactide (PLA)-based nanocomposites Jean-Marie. Prog Polym Sci 38:1504–1542
39. Gironi F, Frattari S, Piemonte V (2016) PLA chemical recycling process optimization: PLA solubilization in organic solvents. J Polym Environ 24:328–333
40. Bordes P, Pollet E, Bourbigot S, Averous L (2008) Structure and properties of PHA/clay nano-biocomposites prepared by melt intercalation. Macromol Chem Phys 209:1473–1484
41. Akin O, Tihminlioglu F (2018) Effects of organo-modified clay addition and temperature on the water vapor barrier properties of polyhydroxy butyrate homo and copolymer nanocomposite films for packaging applications. J Polym Environ 26:1121–1132
42. Xie F, Luckman P, Milne J, McDonald L, Young C, Tu CY, Pasquale TD, Faveere R, Halley PJ (2014) Thermoplastic starch: current development and future trends. J Renew Mater 2:95–106
43. Corre DL, Bras J, Dufresne A (2010) Starch nanoparticles: a review. Biomacromol 11:1139–1153
44. Ma T, Chang PR, Zheng P, Ma X (2013) The composites based on plasticized starch and graphene oxide/reduced graphene oxide. Carbohydr Polym 94:63–70
45. Khosravi-darani K, Bucci DZ (2015) Application of poly(hydroxyalkanoate) in food packaging: improvements by nanotechnology. Chem Biochem Eng 29:275–285
46. Follain N, Chappey C, Dargent E, Chivrac F, Crétois R, Marais S (2014) Structure and barrier properties of biodegradable polyhydroxyalkanoate films. J Phys Chem C 118:6165–6177
47. Avella M, De Vlieger JJ, Errico ME, Fischer S, Vacca P, Volpe MG (2005) Biodegradable starch/clay nanocomposite films for food packaging applications. Food Chem 93:467–474
48. Paul MA, Alexandre M, Degée P, Calberg C, Jerome R, Dubois P (2003) Exfoliated polylactide/clay nanocomposites by in situ coordination–insertion polymerization. Macromol Rapid Commun 24:561–566
49. Cele HM, Ojijo V, Chen H, Kumar S, Land K, Joubert T, De Villiers MFR, Ray SS (2014) Effect of nanoclay on optical properties of PLA/clay composite films. Polym Test 36:24–31
50. Pinto AM, Gonçalves C, Gonçalves IC, Magalhães FD (2016) Effect of biodegradation on thermo-mechanical properties and biocompatibility of poly (lactic acid)/graphene nanoplatelets composites. Eur Polym J 85:431–444
51. Botta L, Scaffaro R, Sutera F, Mistretta MC (2018) Reprocessing of PLA/graphene nanoplatelets nanocomposites. Polymers 10:18

52. Chen Y, Yao X, Zhou X, Pan Z, Gu Q (2011) Poly (lactic acid)/graphene nanocomposites prepared via solution blending using chloroform as a mutual solvent. J Nanosci Nanotechnol 11:7813–7819
53. Puglia D, Fortunati E, Amico DAD, Manfredi LB, Cyras VP, Kenny JM (2014) Influence of organically modified clays on the properties and disintegrability in compost of solution cast poly (3-hydroxybutyrate) films. Polym Degrad Stab 99:127–135
54. Arza CR, Jannasch P, Maurer FHJ (2014) Network formation of graphene oxide in poly(3-hydroxybutyrate) nanocomposites. Eur Polym J 59:262–269
55. Bian J, Lan H, Wang G, Zhou Q, Wang ZJ, Zhou X, Lu Y, Zhao XW (2016) Morphological, mechanical and thermal properties of chemically bonded graphene oxide nanocomposites with biodegradable poly(3-hydroxybutyrate) by solution intercalation. Polym Polym Compos 24:133–141
56. Dean K, Yu L, Wu DY (2007) Preparation and characterization of melt-extruded thermoplastic starch/clay nanocomposites. Compos Sci Technol 67:413–421
57. Li R, Liu C, Ma J (2011) Studies on the properties of graphene oxide-reinforced starch biocomposites. Carbohydr Polym 84:631–637
58. Liu H, Song W, Chen F, Guo L, Zhang J (2011) Interaction of microstructure and interfacial adhesion on impact performance of polylactide (PLA) ternary blends. Macromolecules 44:1513–1522
59. Arrieta MP, Lopez J, Ferrandiz S, Peltzer MA (2013) Characterization of PLA-limonene blends for food packaging applications. Polym Test 32:760–768
60. Mofokeng TG, Ray SS, Ojijo V (2018) Structure—property relationship in PP/LDPE blend composites: the role of nanoclay localization. J Appl Polym Sci 135:46193
61. Salehiyan R, Ray SS, Bandyopadhyay J, Ojijo V (2017) The distribution of nanoclay particles at the interface and their influence on the microstructure development and rheological properties of reactively processed biodegradable blend nanocomposites. Polymers 9:350
62. Botlhoko OJ, Ramontja J, Ray SS (2018) Morphological development and enhancement of thermal, mechanical, and electronic properties of thermally exfoliated graphene oxide- filled biodegradable polylactide/poly(ε-caprolactone) blend composites. Polymer 139:188–200
63. Botlhoko OJ, Ramontja J, Ray SS (2017) Thermally shocked graphene oxide-containing biocomposite for thermal management applications. RSC Adv 7:33751–33756
64. Phiri J, Gane P, Maloney TC (2017) General overview of graphene: production, properties and application in polymer composites. Mater Sci Eng B 215:9–28

Chapter 6
Processing of Polymer Blends, Emphasizing: Melt Compounding; Influence of Nanoparticles on Blend Morphology and Rheology; Reactive Processing in Ternary Systems; Morphology–Property Relationships; Performance and Application Challenges; and Opportunities and Future Trends

Reza Salehiyan and Suprakas Sinha Ray

Abstract This chapter discusses the structure-properties of immiscible polymer blends, focusing on the effects of compatibilization. It has been discussed that the morphology of immiscible blends governs their final properties and thus end-use applications. Therefore, refining the morphologies via different routes such as reactive or physical compatibilization methods was suggested. Among the possible compatibilization methods, the use of nanoparticles has recently gained popularity as their large surface areas lend them additional reinforcing characteristics. However, it has been shown that localization of nanoparticles within blends plays a determinant role in refining the blend morphologies. In comparison, nanoparticles located at interfaces exhibit the most efficient contribution to both compatibilization and the blend properties, by acting as a physical shield against coalescence.

R. Salehiyan · S. Sinha Ray (✉)
DST-CSIR National Centre for Nanostructured Materials, Council for Scientific and Industrial Research, Pretoria 0001, South Africa
e-mail: rsuprakas@csir.co.za

R. Salehiyan
e-mail: rsalehiyan@csir.co.za

S. Sinha Ray
Department of Applied Chemistry, University of Johannesburg, Doornfontein 2028, Johannesburg, South Africa
e-mail: ssinharay@uj.ac.za

S. Sinha Ray (ed.), *Processing of Polymer-based Nanocomposites*, Springer Series in Materials Science 278, https://doi.org/10.1007/978-3-319-97792-8_6

6.1 Introduction

Polymer blends and their associated manufacturing techniques have attracted much research interest in industry and academia owing to their diverse properties. Polymer blends can be produced by mixing of two or more polymeric phases, and the final properties can be adjusted by varying different parameters before and during compounding to attain the characteristics required for the particular end-use application. Mixing two or more polymers does not necessarily comply with the thermodynamics of mixing, leading to phase-separated structure in most polymer pairs with a balance between the properties of each polymer component depending on the fraction of each phase. According to the classical second law of thermodynamic, the Gibbs free energy of a mixture (ΔG_{mix}) must be negative $\Delta G_{mix} = \Delta H_{mix} - T\Delta S_{mix}$, where ΔH_{mix} and ΔS_{mix} are the enthalpy and entropy of mixing, respectively, and T is temperature. However, most polymers are immiscible and heterogeneous, due to the low entropy of mixing, causing the blend to have two distinct glass transition temperatures [1]. However, immiscible polymer blends are more attractive for industrial usage, as multiple structures can be tailored by varying the processing conditions. Hence, it can be inferred that morphology is a key parameter affecting the properties of all multiphase polymer blends, where differing morphologies lead to particular properties. The range of parameters affecting morphology includes blend composition ratio, viscosity ratio, interfacial tension, mixing protocol, and temperature. This highlights the importance of controlling morphology and of developing appropriate technologies to efficiently tune these parameters. Recently, inorganic particles have generated much interest for controlling the morphologies of blends, as they are economically effective in multi-fold reinforcing of the thermo-mechanical properties of blends compared to those of unfilled nanocomposites [2]. The nanocomposite approach, however, needs to be carefully designed in order to achieve optimum results, since the properties of nanocomposites are directly proportional to the degree of intercalation/exfoliation in polymers as well as to their localization in the polymer blends. Hence, quite a number of studies have investigated the distributive optimization of particles and the significant influences that different distributions that can have on the resulting polymers. Accordingly, this chapter focuses on contemporary efforts and results concerning the effects of particles (mainly nano-sized) on the morphology-properties of immiscible blends.

6.2 Melt Blending

Blends can be prepared in solution or melt state; however, despite the simplicity of solution blending, it is of less interest from an industrial perspective, mostly due to associated solvent issues. Firstly, identifying a suitable solvent for polymer pairs is not always a straightforward matter, and sometimes a combination of solvents at specific ratios is required for a mixture to be completely dissolved. Furthermore, solvent residuals in the blend are inevitable: even after great care in preparation, some solvent fractions remain in the blend, which can degrade the final properties, especially when targeting biomedical applications [3]. Last but not least, from an industrial perspective, quantity always matters, hence solution casting may not be an efficient technique as it yields quite small amounts of product which could only be useful for analytical observations. With regard to the issues associated with the solution casting method for blending, melt blending seems a cost-effective technique as it is a solvent-free method and capable of mass production at a faster rate. This process can be performed as batch mixing in internal mixers or as continuous methods in twin-screw extruders and injection molding machines. Moreover, regarding the localization of nanoparticles (discussed later), the masterbatch method would be more convenient based on the melt compounding. In such a technique, nanoparticles are pre-mixed with a less-compatible polymer phase before blending with the other (more-compatible) polymer phase. During the second mixing stage (blending), kinetics and thermodynamic driving forces will cause nanoparticles to migrate towards the phase with which they are more compatible. Hence, controlling nanoparticle localization using melt blending appears quite promising and more straightforward.

6.3 Mixing Mechanism (Breakup and Coalescence)

Mixing is an important step in blending processes, as it attempts to enhance the homogeneity of the mixture, such that the final morphology and the resulting properties mirror the quality of mixing. Therefore, it is important to know the parameters dictating the degree of mixing in immiscible blends. When a polymer with lesser volume fraction is dispersed in a polymer matrix it forms droplets of minor phase in continuous phase. These droplets deform, elongate, and eventually break into smaller droplets during the mixing process. The process of droplet deformation during mixing is controlled by two fundamental stresses: one shear stress, hydrodynamic force, (τ), the other interfacial stress, cohesive forces, (α/R with α as interfacial tension and R droplet radius) where one exerts the affine motions (convection) of the dispersed phase and the other interfacially restores the stress. Therefore, there is always competition between the two forces to determine the extent of droplet deformation. The ratio of these two stresses can be defined in a single, dimensionless number in a simple shear flow, termed capillary number:

$$Ca = \frac{\tau}{\alpha/R} = \frac{\eta_m \dot{\gamma} R}{\alpha} \tag{6.1}$$

Here, η_m is the viscosity of the matrix and $\dot{\gamma}$ the shear rate. At this point, mixing mechanisms can be pointed out with respect to the applied forces, and therefore capillary number. If the capillary number exceeds a critical value ($Ca > Ca_{crit}$) it indicates domination of shear forces over interfacial stress, where droplets break into smaller drops; conversely, if ($Ca_{crit} > Ca$), then interfacial stress withstands the shear stress and droplets only deform in an ellipsoidal fashion in the direction of flow field. The so-called critical capillary number is governed by the shape of the droplet as well as the viscosity ratio of the blend, where $K = \eta_d/\eta_m$ with η_d being the viscosity of the dispersed phase. Generally, with $K<1$ (highly viscous polymer matrix) droplet breakup occurs faster than in the case of $K>1$ (highly viscous dispersed phase) in as much as breakup will not occur beyond $K \sim 4$ [4]. Consequently, the mixing of a multi-phase structured polymer can be defined in two ways with respect to the capillary numbers: (1) distributive mixing, and (2) dispersive mixing [5, 6].

i. Distributive mixing when ($Ca \gg Ca_{crit}$): At this stage droplets show affine deformation, where interfacial stress is the dominant force holding the droplets from breakup. Such mixing mode enhances the distribution of the dispersed phase in the bulk matrix, while the size of the minor phase will not change.
ii. Dispersive mixing when ($Ca \geq Ca_{crit}$): This mode of mixing promotes breakup into smaller droplet, where applied shear stress can exert perturbation at the interface. This type can also be used to break down agglomerations and hence can assist the dispersion of the minor phase.

The following empirical fitting equation can be applied to evaluate such critical breakup conditions [7].

$$\log(Ca_{cr}) = -0.506 - 0.0994\ \log(K) + 0.124 \log^2(K) - \frac{0.115}{\log(K) - \log(K_{cr})} \tag{6.2}$$

Here, $K_{cr} \sim 4$ is the critical viscosity ratio beyond which breakup will not occur under shear flows. Note that both distributive and dispersive mixing can occur simultaneously in real situations during the mixing process. In mixing two thermodynamically unfavored polymers, in the case of immiscible blends, coalescence cannot be neglected. Two individual droplets can collide during the mixing and coalesce into a larger one.

It must be noted that these two phenomena, namely coalescence and breakup, may occur concurrently during mixing when smaller droplets collide and merge to bigger one, and that this process can go on within a specific time. For an effective collision to occur, the film between two droplets needs to be drained, a process that is highly dependent on interfacial mobility. Models have been developed to assess interfacial mobility during collisions: the fully mobile interface (FMI), the partially mobile interface (PMI), and the immobile interface (IMI), which can be arranged as in (6.3–6.5) [7, 8]. These models define the critical mean radii of two approaching droplets, beyond which coalescence cannot be expected.

$$\text{FMI}: R \ln\left(\frac{R}{h_{cr}}\right) = \frac{2}{3}\frac{\alpha}{\eta_m \dot{\gamma}} \tag{6.3}$$

$$\text{PMI}: R = \left(\frac{4}{\sqrt{3}}\frac{h_{cr}}{K}\right)^{2/5}\left(\frac{\alpha}{\eta_m \dot{\gamma}}\right)^{3/5} \tag{6.4}$$

$$IMI : R = \left(\frac{32}{9}\right)^{1/4}\left(\frac{h_{cr}\alpha}{\eta_m\dot{\gamma}}\right)^{1/2} \tag{6.5}$$

Here, h_{cr} is the critical thickness of the matrix film being squeezed off the interface between the colliding droplets. A more detailed discussion on the mechanisms of breakup and coalescence during mixing with respect to film drainage is elaborated elsewhere [9, 10]. At this point, where the boundaries of droplet deformations are defined with respect to breakup and coalescence, the limiting curves can be obtained. It can be observed that neither breakup nor coalescence will occur if the droplets are larger than the critical coalescence condition but smaller than the critical breakup limit. The above discussion elucidates the importance of the interface layer in controlling morphology during the mixing process, where careful choice of a stabilizer, termed a "compatibilizer," can play a vital role in improving the final morphology.

6.4 Compatibilization

Immiscible blends usually exhibit unstable morphologies due to the low interfacial adhesion between the phases, which leads to poor physical, mechanical, and rheological properties of the blends. This interfacial compatibility can be increased by adding a third component to the blend in the form of an interfacial agent or emulsifier, often termed a compatibilizer. Such compatibilizers could be block or graft copolymers added before blending (ex situ) or can be formed during the blending via in situ reactions. The latter process has been practiced extensively, as identifying an appropriate copolymer for particular blends is not always an easy task, whereas creating copolymers during reactive blending via functionalized polymers could be a relatively cheap and easy method [3, 11].

6.4.1 *In Situ (Reactive) Compatibilization*

In an in situ method, compatibilization is achieved by initiation of chemical interactions such as ionic associations, dipole–dipole interaction, hydrogen bonding, cross-linking, etc. Some examples of reactive blending and their reaction mechanisms are summarized in Table 6.1.

All previous studies report a common finding, which is that compatibilization effect strongly favors increased toughness of the final polymer. However, this effect is dependent on factors such as the concentration of the reactive species used with respect to the ratio of the blends and processing conditions (e.g., mixing time, temperature, and incorporating protocol), which can define the extent of a reaction. It is evident from the results that, in reactively processed blends, the created copolymer at the interface acts as an emulsifier agent by increasing the interfacial adhesion between two phases via the miscible chain ends towards each phase, or the reaction

Table 6.1 Some reactively compatibilized blends and their reactive paths

Polymer blend	Functionalized groups	Reaction routes	References
PMMA/PS	Functionalized PS with oxazoline	There was no indication of any chemical reaction between oxazoline group and PMMA however, a presence of physical reaction like hydrogen bonds improved the interfacial adhesions	[12]
PS/PE	Oxazoline functionalized PS and carboxylic functionalized PE with acrylic acid	A graft polymer formed during the mixing as a result of amido-ester reaction linkage from reaction of oxazoline and carboxylic group	[13]
HDPE/PET	1-ethylene/glycidyl methacrylate copolymers (E/GMA) 2-ethylene/ethyl acrylate/glycidyl methacrylate terpolymers (E/EA/GMA)	Epoxy functional groups react with carboxyl and hydroxyl end groups of PET, forming grafted copolymers. Efficiency of the compatibilizers were related to the GMA content where the highest elongation at breaks obtained at the lowest GMA content	[14]
PA6/SAN	Maleic anhydride functionalized SAN as terpolymer (SANMA)	Interfacial reaction between anhydride group of SANMA and amino groups in PA6 produced the SANMA-*graft*-PA6	[15]
PA6/ABS	1-Butadiene based rubber grafted to SAN 2-Styrene-maleic anhydride copolymer SMA 3-Methyl glutarimide-methyl methacrylate copolymer 4-Oxazoline functionalized SAN	Miscibility between SAN and ABS can promote the reactions. Maleic anhydride functional group in SMA reacted with PA 6 phase as well. The Imide functionalized SAN also showed reacting with PA6 similar to oxazoline functionalized SAN	[16]
PBT/PA66	Bisphenol-A type epoxy resin	A copolymer was formed at the interface as a result of a reaction between the epoxy with carboxyl and hydroxyl end groups of PBT as well as amine end group of PA66	[17]
PP/EC	Amine functionalized PP-MA	A copolymer formed at the interface as a result of reaction between aminated PP and maleic anhydride grafted EC	[18]
PI/PDMS	PI-grafted MA and amino functionalized PDMS (PDMS/NH_2)	Cross-links between amine and maleic anhydride groups at the interface with asymmetric architecture	[19]
PP/PS	Maleic anhydride grafted PP (PP-MA) and amino functionalized PS (PS-NH_2)	The reactions between MA and amine groups created a copolymer at the interface. Interfacial tension decreased while interfacial modulus increased upon increasing copolymer concentration. The form relaxation time of the compatibilized blends shifted to lower values [20]	[20, 21]

(continued)

Table 6.1 (continued)

Polymer blend	Functionalized groups	Reaction routes	References
PP/PS PE/PS	Anhydrous aluminium chloride ($AlCl_3$)	PP-*g*-PS or PE-g-PS copolymers were formed via Frediel-Crafts alkylation reaction. Mechanical and rheological properties are sensitive to the $AlCl_3$ concentrations and blend ratios as beyond a critical concentration properties decay due to degradation of the copolymer	[22–24]
PLA/PCL	Triphenyl phosphite (TPP)	Tranestrification reactions between phosphite and polyesters caused branching and copolymer formation. Compatibilization effect increased the elongation at break of the blends by 120%	[25]
PLA/PCL	Lysine triisocyanate (LTI)	Cross-linking reactions between isocyanate group of LTI and hydroxyl/carboxyl groups of the polymers and mechanical properties enhanced as interfacial adhesion between PLA and PCL improved	[26]
PLA/PCL	Dicumyl Peroxide (DCP)	Hydrogen abstraction from polymers caused alkyl radicals from PCL scission to cross-link with tertiary radicals of PLA. Hence, mechanical and morphological properties enhanced as function of compatibilization	[27]
PLA/PBAT	Glycidyl methacrylate (GMA)	Carboxylic and hydroxylic groups hydrolysed inside PLA and PBAT during melt mixing and formed a random ethylene acrylic ester terpolymer at the interface. Improvement in interfacial adhesion as a function of 5 wt% GMA enhanced the impact strength of the blends by 51.7%	[28]
PLA/PBAT	Joncryl ADR® with 9 GMA functionalities	A combination of chain extension, chain scission and branching exists in such cases. Formation covalent bonds with hydroxyl groups as a result of epoxy ring opening reactions. Thermal stabilities and mechanical properties enhanced as a result of reactive compatibilization	[29]
PLA/PBAT	Dicumyl Peroxide (DCP)	Cross-links and branched polymers and PLA-*g*-PBAT was formed via trans esterification process from combination of PLA and PBAT free radicals	[30]
PLA/PBSA	Triphenyl phosphite (TPP)	Phosphite might react with either hydroxyl or carboxyl groups of the polymers and a random copolymer is formed at the interface	[31]

(continued)

Table 6.1 (continued)

Polymer blend	Functionalized groups	Reaction routes	References
PLA/PBSA	Joncryl ADR®	Long chain branches and cross-links formed via etherification and esterification reactions between epoxide groups of Joncryl with hydroxyl or carboxyl group of the polyesters respectively	[32]
PLA/PS	Oxazoline functionalized PS (PS-OX)	A graft copolymer was created when oxazoline group reacted with carboxylic group of PLA	[33]
PBT/EVA	Maleic anhydride graftedEVA (EVA-*g*-MA)	PBT-*g*-EVA copolymer was formed as a result of reaction between carboxyl and/or hydroxyl groups of PBT with anhydride group of EVA	[34]

may occur in favor of the thermodynamic balance by changing the entropy of mixing. The latter may be achieved when a favorable reaction within one of the phases changes the glass transition temperature of phases, lowering the Flory–Huggins χ interaction parameter.

6.4.2 Ex Situ (Physical) Compatibilization

In ex situ blending, a copolymer should be synthesized prior to blending in such a way that, during blending, it will locate at the interface between the two phases where one segment of the copolymer is absorbed onto one phase, and the other segment to the other phase, thereby enhancing interfacial adhesion. Van Puyvelde et al. [35, 36] summarized the mechanisms involved in physical compatibilization of immiscible polymer blends when copolymers are introduced. The first proposed mechanism discusses the role of Marangoni stress (indicated as arrows in Fig. 6.1a) as opposed to the interfacial tension gradients hindering the drainage of the copolymer film from the interface of two neighboring droplets. That is, coalescence of the droplets is suppressed by the presence of the film at the interface. In the second mechanism, proposed by Sundararaj and Macosko [37], suppressed coalescence due to shell formation around the droplets was found to be the main stabilizing mechanism. The copolymer layer at the interface repels the approaching droplet via steric interactions.

It is also worth noting that compatibilizers at the interface can promote droplet breakup under deformation, since applied shear creates concentration gradients along the surface of a droplet, lowering the interfacial tensions.

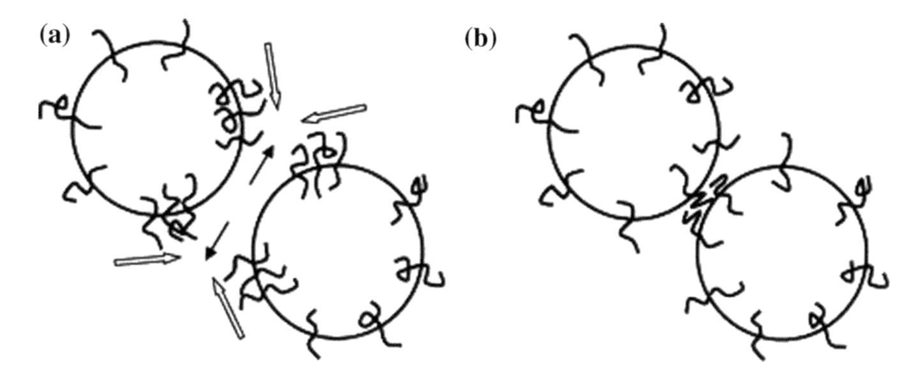

Fig. 6.1 Compatibilization mechanisms in the presence of a copolymer at the interface: **a** Marangoni stresses (as shown by the white arrows) opposing the interfacial tensions, hindering the rapture of the copolymer film, hence suppressing coalescence; **b** Steric repulsion mechanism as a function of copolymer layer around the approaching droplets prevents droplet coalescence. Reprinted with permission from [36]. Copyright 2001, Elsevier Science Ltd.

6.4.3 Inorganic Particles as Compatibilizers

Incorporation of inorganic particles into the polymer blends has become a new trend, as an alternative to the classical copolymer and reactive compatibilizers. Natural inorganic particles are inherently hydrophilic (defined by polar characteristics); therefore, in order to make them compatible with most polymer pairs, their surface is chemically modified to the desired polarity.

6.4.3.1 Compatibilization Mechanisms

As discussed previously, compatibilization effect can be inferred in two major ways: (i) suppressing coalescence, and (ii) lowering interfacial tension, both leading to smaller droplet size. These two mechanisms are primarily effective when the intended compatibilizer is at the interface. In the case of particle-incorporated polymer blends, however, the mechanisms could differ from those of the classical compatibilization, since different particle localization can be observed in the blends as surface energies between particles and polymer pairs, which play an important role in defining the final localization. Moreover, the effects of particles as compatibilizers remain ambiguous, since the function of particles in reducing the interfacial tension is still a matter of debate [38]. Nonetheless, size reduction is the common result in all the so-called compatibilized blends. Thus, one can summarize the size reduction mechanisms in polymer blends when particles are introduced, as follows:

i. Particles can localize inside the polymer matrix and increase the viscosity of the matrix, hence reducing the viscosity ratio $K<1$ which in turn facilitates droplet breakup [39, 40].

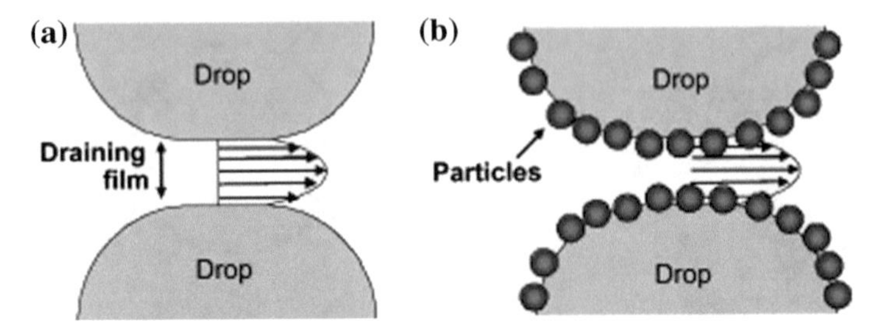

Fig. 6.2 Stabilization mechanism when solid particles are located at the interface: **a** Film drainage is promoted at the mobile interface; **b** Solid particles are immobilized at the interface, thus retarding drainage. Reprinted with permission from [44]. Copyright 2009, Elsevier Science Ltd.

ii. The most efficient route is when particles at the interface can act as physical shield around the droplets and thereby suppress coalescence [40–44]. Figure 6.2 illustrates how interfacially located solid particles immobilize the interface and hinder the film draining at the interface.
iii. In very rare cases, platelet-like particles within the dispersed phase can promote breakup by a mechanism known as "cutting effect," proposed by Zhu et al. [45]. In such cases, particles above a critical concentration inside the dispersed phase form a "knife-like" structure (Fig. 6.3) that cuts the host droplet into two pieces upon the application of shear force. This mechanism has been discussed in other studies [46–48].

6.4.3.2 Morphology Coarsening

A review of the literature reveals that the first two mechanisms, namely the formation of a physical barrier at the interface, and immobilization of the matrix via selective localization of particles in the major phase, have the most significant effect on size reduction.

It is worth mentioning that majority of the studies were consistently in favor of morphology stabilization; therefore, size reduction has been frequently reported. However, particles can also promote coalescence or coarsen the morphologies. This can occur while particles enrich the minor-phase droplets and increase the viscosity of the droplets, hence making the droplet breakup process impossible [40, 49, 50]. Interestingly, it has been shown that particles (spherical) at the interface can also promote coalescence via "bridging-dewetting" mechanism when a particle is absorbed onto the surface of two droplets simultaneously [51, 52]. This can occur if the interface is not fully covered by the particles [53]. Therefore, the knowledge of particle localization and the choice of particle with respect to the type of blend are vital for controlling the morphology and tuning the properties.

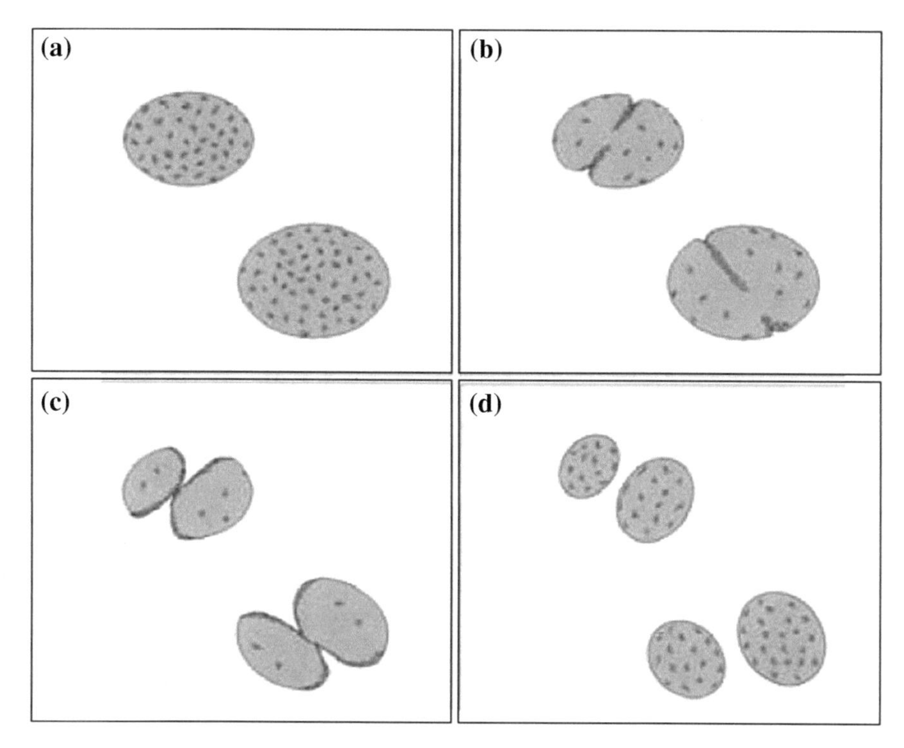

Fig. 6.3 Illustration of the compatibilization mechanism cutting effect of organoclay platelets inside PS droplets in a PP/PS blend: **a** Primary localization of organoclays inside the larger PS droplets; **b** Initiation of "clay-knife" structure at the critical clay concentrations; **c** Developing the "cutting effect"; **d** Completion of the cutting and split of the droplets. Reprinted with permission from [45]. Copyright 2008, Springer Verlag

6.5 Effect of Particle Localization

As discussed previously, morphological evolutions in filled polymer blends are highly dependent on the location of the particles. Figure 6.4 shows an example of how different localization of silica nanoparticles can affect the size reduction behavior of a (80/20) PP/PS blend [40]. It shows that interfacial trapping of Sipernat D17 silica nanoparticles leads to smaller droplet size, where R202 fumed silica is localized mainly at the PP matrix with relatively large droplets compared to those of D17. Finally, the largest droplet size was obtained when hydrophilic OX50 fumed silica was situated inside the PS minor phase, due to the positive effect on increasing the viscosity ratios. Therefore, the vital influence of particle localization is again observed, confirming that particles are most efficient in stabilizing morphologies when they are trapped at the interface. Localization of particles inside the polymer blends at the thermodynamic equilibrium can be simply estimated by wetting coefficient (6.6) [54, 55] when the interactions between polymer pairs and particles are known.

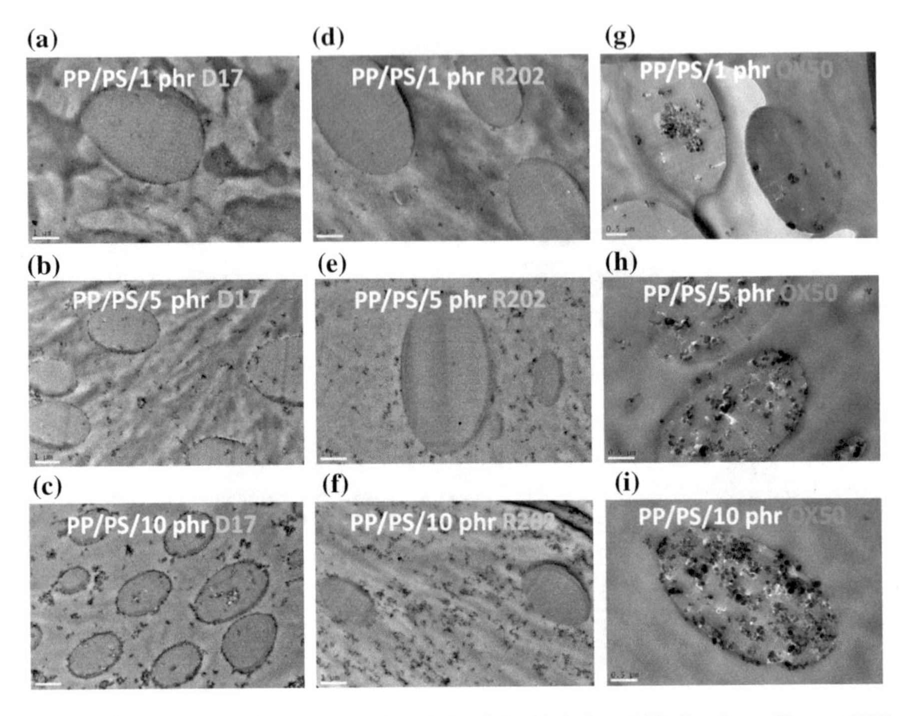

Fig. 6.4 TEM images of (80/20) PP/PS blends loaded with 1, 5, and 10 phr of **a–c** Sipernat D17 silica particles, **d–f** R202 fumed silica, and **g–i** OX50 fumed silica particles, respectively. Reprinted with permission from [40]. Copyright 2015, The American Chemical Society

$$\omega_{12} = \frac{\gamma_{2-particle} - \gamma_{1-particle}}{\gamma_{12}} = \cos\theta \tag{6.6}$$

Here, $\gamma_{1-particle}$ and $\gamma_{2-particle}$ represent the interfacial tensions between particle and polymer 1 or 2, γ_{12} is the interfacial tension between polymer 1 and 2, and θ is the contact angle. Particles are expected to be found in polymer 1 when ω_{12} is>1, and in polymer 2 when ω_{12} is<-1. Eventually, when the value of ω_{12} is between 1 and -1, particles are expected to locate at the interface between polymers 1 and 2.

The interfacial tension between polymer pairs can be calculated by different equations proposed by Owen-Wendt [56], Wu [55], and Girifalco-Good [57] respectively:

$$\gamma_{12} = \gamma_1 + \gamma_2 - 2\left(\sqrt{\gamma_1^d \gamma_2^d} - \sqrt{\gamma_1^p \gamma_2^p}\right) \tag{6.7}$$

$$\gamma_{12} = \gamma_1 + \gamma_2 - 4\left[\frac{\gamma_1^d \gamma_2^d}{\gamma_1^d + \gamma_2^d} + \frac{\gamma_1^p \gamma_2^p}{\gamma_1^p + \gamma_2^p}\right] \tag{6.8}$$

$$\gamma_{12} = \gamma_1 + \gamma_2 - 2\sqrt{\gamma_1 \gamma_2} \tag{6.9}$$

Here, γ_1 and γ_2 express the surface energies of components 1 and 2 respectively, and d and p indicate the dispersive and polar contributions of the surface energy of components. In contrast to the calculation of interfacial tension between polymers according to the aforementioned equations (6.7) and (6.8), the interfacial tension measurements between polymers and particles cannot be easily calculated, since the values can vary significantly according to the surface structure of the particles [44]. Moreover, most of the measured values for interfacial tension between particles and polymers are acquired at room temperature and solid state, while particle distribution takes place during melt processing or the annealing time, both at molten state. Thus, inaccurate values can be expected, since wetting coefficient determination is directly proportional to the preset values [2]. Nevertheless, calculations based on the wettability of the particles can be used in prior to assist in rough estimation of morphological development with respect to particle localizations. At this point, it must be noted that the aforementioned predictions are obtained when thermodynamic equilibrium is attained. Kinetic parameters such as mixing sequence, mixing time, viscosity ratio, and the size and shape of particles can disturb the final localization. Considering that interfacial entrapment of particles is the ideal scenario for stabilization of morphologies, blends that contain particles at the interface can be prepared by appropriately adjusting the thermodynamic and kinetic driving forces. One attractive route is to pre-disperse the particles within the phase which they are not thermodynamically compatible with. This stage can also be achieved via masterbatch process, followed by blending this pre-mixed phase with the other phase that is known to be more thermodynamically favorable to particles. If the kinetic parameters required for the migration of particles (e.g., viscosity ratio, applied force, and mixing time) are sufficient, particles migrate toward the other phase, and halting the process at an appropriate time will result in particles being well-distributed at the interface [58–61]. This will yield blends with finer morphologies and interesting properties, such as ultralow-percolated structures suitable for electrical conductivities, which will be discussed in Sect. 6.7.1.

6.5.1 Effects of Particle Size and Shape

Another aspect of compatibilization mechanism in filled polymer blends is the size and shape of the particles used in the blends. Ginzburg [62] developed a thermodynamic model to study the effect of spherical nanoparticles with radius R_p on the miscibility of polymer blends with respect to the polymer radius of gyration R_g. The results revealed that a binary system can be stabilized when $R_p < R_g$, due to the reduction in the enthalpic contribution of the Gibbs free energy. On the other hand, segregation can be promoted as a result of so-called entropic penalty. Furthermore, it was reported that spherical particles are more stable at the interface than are particles with high aspect ratios [63]. Considering the thermodynamically preferred location of particles at the interface, their stability is dependent on contact angle (Fig. 6.5). This shows that high aspect ratio particles (0.1) penetrate the other phase

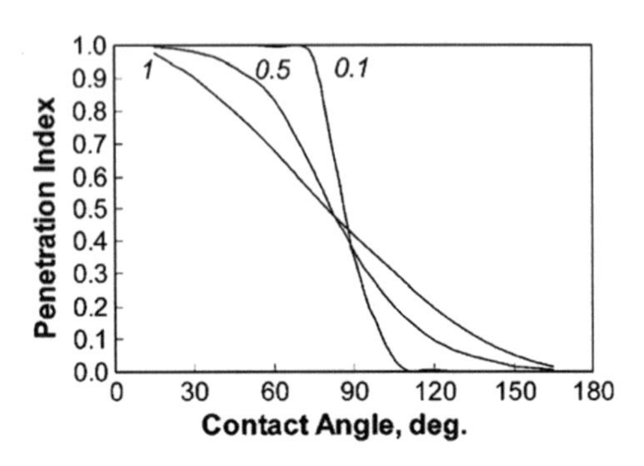

Fig. 6.5 Percentage of particle penetration into a droplet (i.e., penetration index) as a function of wetting angle for particles with different aspect ratios of 1, 0.5, and 0.1. Reprinted with permission from [63]. Copyright 2011, The American Chemical Society

at lower contact angles $60<\theta^\circ<90$ or remain inside the phase when contact angles are $90<\theta^\circ<120$, whereas shorter aspect ratio particles seem to be more stable at a broader range of contact angles [64]. Moreover, for high aspect ratio particles (e.g., CNTs) the greatest stability is achieved when they are parallel to the interface, compared to the least stable case when they are perpendicular to the interface. That is, the angle between length axis and interface should also be taken into account when evaluating stability at the interface.

6.6 Rheology

6.6.1 Simple Shear Flow

As discussed previously, blend morphology can have remarkable influence on blend properties. Hence, the rheological properties of blends can vary hugely depending on the degree of stabilization. However, the presence of an interface in polymer blends means that characterization of their rheological responses is rather complex compared to that of homogenous, single-phase polymers. Deformation of droplets in emulsions and polymer blends under simple shear flows has been studies extensively [4, 37, 65–69]. Taylor established a parameter D describing the deformation of a Newtonian droplet in a Newtonian fluid under steady-state shear flow.

$$D = \frac{A - B}{A + B} = Ca\frac{19K + 16}{16K + 16} \tag{6.10}$$

Here, A and B denote the length and width, respectively, of the deformed droplets. Taylor's equation is shown to have linear dependency on capillary number Ca and is also slightly dependent on viscosity ratio K in cases where a droplet is oriented at 45° with respect to the velocity direction along its longest axis. Interfacial tension can be calculated from the slope of a linear fit of D plotted against the viscous part of a capillary number ($\dot{\gamma}\eta_m R$) where $\dot{\gamma}$, η_m and R are the shear rate, matrix viscosity, and droplet radius at rest, respectively. Taylor's equation has been the motivation for other researchers in modelling droplet deformations. Guido and Villone [70] developed Taylor's model for a situation where a particular ellipsoidal droplet is retracting from the elongated position (6.11).

$$D = D_0\, exp\left(-\frac{40(K+1)}{(2K+3)(19K+16)}\frac{t\alpha}{\eta_m R}\right) \tag{6.11}$$

Here, D_0 is the deformation value at the beginning of retraction and t is the time during deformation. Maffetone and Minale [67] also derived a phenomenological MM model describing the droplet deformation in a Newtonian fluid under simple shear flow. The resulting deformation parameter can be predicted according to (6.12–6.14).

$$D = \frac{\sqrt{f_1^2 + Ca^2} - \sqrt{f_1^2 + Ca^2 - f_2^2 Ca^2}}{f_2 Ca} \tag{6.12}$$

$$f_1 = \frac{40(K+1)}{(2K+3)(19K+16)} \tag{6.13}$$

$$f_2 = \frac{5}{2K+3} \tag{6.14}$$

Gooneie et al. [71] reported that addition of CNTs can reduce the deformation parameter in PP/PA6 blend due to the effects on the elasticity of the droplets against deformation. It must be noted that most such models are limited to the Newtonian fluids under simple shear flow, whereas, from an industrial perspective, most articles are constructed of highly concentrated, processed plastics, thereby indicating the necessity of devising models that can represent more complicated polymer systems.

6.6.2 Small-Amplitude Oscillatory Shear (SAOS) Test

A distinct characteristic of compatibilization is increased modulus of the blends at lower frequencies due to interfacial elasticity. This interfacial elasticity can be characterized by the relaxation of the blend at longer timescales, whereas usually a sol-gel transition can be observed [18, 50, 72, 73]. This behavior can be observed in plots of phase angle δ° versus complex modulus $\left|G^*_{(\omega)}\right|$ as so-called Van Gurp-Palmen plots. It has been shown that the phase angle values reach a plateau region of 90° at low $\left|G^*_{(\omega)}\right|$ values, while this value shift to lower angles for phase-separated polymers such as in immiscible blends [74]. Other studies showed that the values of phase angle shifts to lower angles at low $\left|G^*_{(\omega)}\right|$ values with compatibilization were associated with higher elasticity [75]. Li et al. [74] reported that the existence of a local minimum at low $\left|G^*_{(\omega)}\right|$ values is an indication of droplet-matrix morphology, while a local maximum represents a co-continuous morphology. This differed from the results of Lopez-Barron and Macosko [72], who examined a series of uncompatibilized and compatibilized PS/SAN blends with different ratios. Sangroniz et al. [76, 77] stated that the aforementioned local minima in phase angle at low complex modulus values corresponds to the elastic contribution of the droplet, form relaxation, in the blends. Similarly, Roman et al. [78] observed a dramatic reduction in phase angles in (20/80) PMMA/LDPE beyond 3.5 wt% MWCNT loadings. They concluded that such behavior is related to the strong interaction between nanotubes and LDPE chains, thereby retarding the long relaxation time of the blend.

Many attempts have been made to model rheological properties, unravelling the complexity of the viscoelastic properties of polymer blends with respect to interfacial tension [79–81]. Linear rheological analyses based on small-amplitude oscillatory shear (SAOS) testing has been the most popular technique for interpreting the interfacial contributions of dilute and semi-dilute emulsions. Many researchers have successfully used Palirene's emulsion model to fit the moduli in order to determine the interfacial tensions, assuming that the blend exhibits a narrow droplet size distribution with constant interfacial tensions [18, 82–85]. A simplified version of Palirene's model can be expressed as follows:

$$G_b^* = G_m^* \frac{1 + 3\int_i^\infty \emptyset H(\omega)}{1 - 2\int_i^\infty \emptyset H(\omega)} \tag{6.15}$$

where,

$$H(\omega) = \frac{4\left(\frac{\alpha}{R_v}\right)\left[2G_m^*(\omega) + 5G_d^*(\omega)\right] + \left[G_d^*(\omega) - G_m^*(\omega)\right][16G_m^*(\omega) + 19G_d^*(\omega)]}{40\left(\frac{\alpha}{R_v}\right)\left[G_m^*(\omega) + G_d^*(\omega)\right] + \left[2G_d^*(\omega) + 3G_m^*(\omega)\right][16G_m^*(\omega) + 19G_d^*(\omega)]} \tag{6.16}$$

Here, $\emptyset$, ω, R_v, and α are droplet volume fraction, angular frequency, volume-average radius of droplet, and interfacial tension, respectively. G_m^*, G_d^*, and G_b^* represent the complex moduli of the polymer matrix, dispersed phase, and blend, respectively. Therefore, interfacial tension according to Palirene's model can be written as in (6.17) [86].

$$\alpha = \frac{R_v \eta_m}{4\tau_D} \frac{(19K + 16)[2K + 3 - 2\phi(5K + 2)]}{(19K + 16)[2K + 3 - 2\phi(K - 1)]} \tag{6.17}$$

Here, η_m, τ_D, and $K = \frac{\eta_d}{\eta_m}$ are matrix viscosity, relaxation time, and viscosity ratio, respectively. Despite the frequent application of such model in determining interfacial tension, it should be noted that this model is limited to low particle concentrations in cases of particle-compatibilized blends [83] and that is also failed to predict the reactively compatibilized blends in some cases [18]. Furthermore, Gramespacher and Meissner [80] developed the following model, which accounts for the interfacial contribution of the modulus as well as the polymer pairs:

$$G^*_{blend} = \phi G^*_{dispersedphase} + (1-\phi)G^*_{matrix} + G^*_{interface} \tag{6.18}$$

$$\alpha = \frac{R_v \eta_m}{\tau_D} \frac{(19K+16)(2K+3)}{40(K+1)} \left[1 + \phi\left(\frac{5(19K+16)}{4(K+1)(2K+3)}\right)\right] \tag{6.19}$$

They found three peaks in the relaxation spectra of the blends, and attributed the first two peaks to the relaxation time of the polymer pairs and the third to the form relaxation time of the blend. The form relaxation or shape relaxation times of a blend correspond to the time that it takes for a deformed droplet to retract to its primary spherical shape. Some researchers also reported these three peaks [87–89], whereas several others only reported two peaks with the second appearing at longer times [72, 74, 83, 90–92]. The difference could be related to the morphology of the different blend systems and relaxation of their constituents.

6.6.2.1 Co-continuous Morphology

It can be understood that the major studies dealt with droplet-morphology and the interplay between droplet size and relaxation time in polymer blends, where by knowing one the other can be found. The required analysis can become more complicated when examining co-continuous morphology, since blends with co-continuous structures are not in equilibrium state and hence their morphologies may change during post-processing characterizations such as annealing. It has been shown that blends having co-continuous structures show a tail in their relaxation spectra at longer time, associated with the relaxation of the interconnected network of a co-continuous interface structure [72, 74, 91]. In blends with co-continuous morphologies, the characteristic domain size has been used where the total volume is divided by interfacial area [93, 94]. The efficiencies of compatibilizers against the coarsening of morphologies in co-continuous blends could be evaluated by measuring the characteristic domain size over annealing time [95, 96]. Within the literature [94, 97, 98], the Veenstra model has been applied to co-continuous blends for calculation of complex modulus, although it was originally proposed for mechanical properties (Young's modulus) [99]. Considering that the complex modulus of a co-continuous blend is a combination of complex moduli of the components and interface ($G^*_b = G^*_{components} + G^*_{interface}$), the component contribution of the complex modulus for co-continuous blends can be shown as:

$$G^*_{components} = \frac{\alpha'^2 b' G_1^{*2} + \left(\alpha'^3 + 2\alpha' b' + b'^3\right) G_1^* G_2^* + \alpha' b'^2 G_2^{*2}}{b' G_1^* + a' G_2^*} \tag{6.20}$$

Here, $3\alpha'^2 - 2\alpha'^3$ is the volume fraction of component 1, $\emptyset_1$, with α' and b' describing the average reduced length of components 1 and 2 correspondingly, and $b' = 1 - a'$. Thus, by calculating (6.20), $G^*_{interface}$ can be found. Bai et al. [94] showed that the migration of r-GO particles towards the interface of a PLA/PS blend during annealing suppressed the increase in characteristic domain size as a function of compatibilization. This was accompanied by an increase in elastic modulus G'(t) as a function of annealing time. They also reported that the interfacial elastic modulus of the blend increases as a function of r-GO concentration and inverse of characteristic domain size. In other words, r-GO particles increased the interfacial elastic modulus while decreasing the characteristic domain sizes.

6.6.3 Large-Amplitude Oscillatory Shear (LAOS) Tests

It can be seen that most rheological analyses were based on linear responses. Recently, a number of studies have utilized nonlinear rheological analyses to investigate the morphological evolution of immiscible polymer blends. Several studies applied large-amplitude oscillatory shear (LAOS) tests to examine the nonlinear viscoelastic behaviors of polymers, including blends [100–103]. The typical response from LAOS tests can be explained by the Payne effect [104, 105]. When fillers are introduced into the polymers, the initial modulus increases, while as the strain amplitude increases the modulus drops to lower values with normally a strong slope entering the nonlinear regions. This effect can also be seen in compatibilized blends, where blends exhibit strong strain-softening upon compatibilization [40, 50, 106, 107]. The most striking results were obtained when LAOS tests were combined with Fourier Transform rheology (FT-rheology) technique. Full details of this technique can be found elsewhere [108–110]. Briefly, the nonlinear viscoelasticity from FT-rheology can be characterized by shear stress σ from LAOS tests, including odd higher harmonic contributions as expressed in (6.21).

$$\sigma(t) = \sum_{n=1,odd} \sigma_n \sin(n\omega t + \delta_n) \tag{6.21}$$

Here, ω and δ are excitation frequency and phase angle. It can be observed that stress response is comprised of odd higher harmonics. Mostly, the normalized 3rd harmonic intensity I_3/I_1 has been used, as the third intensity is shown to be the most significant one. Moreover, the relative values to the first intensity (I_n/I_1) has been mostly used because such relative intensities are less vulnerable to instrumentation errors [100]. Carotenuto et al. [111] established a model based on relative intensities from FT-rheology, to predict droplet size and size distributions in emulsions of PDMS/PIB blends. Reinheimer et al. [112] derived an equation (6.22) based on FT-rheology in which the droplet size R and interfacial tension α of a dilute monodispersed emulsion can be determined if the value of one parameter (R or α) is known.

$$\frac{I_{5/3}}{\gamma_0^2 \omega_1^2} = 0.64\mathrm{K}^{1.63} \frac{\eta_m^2 R^2}{\alpha^2} \tag{6.22}$$

Here, $I_{5/3}$ is the 5th higher harmonic intensity relative to the 3rd higher harmonic intensity, as to only utilize the nonlinear contributions of the intensities, instead of normalizing by the first intensity which is still within the linear region. γ_0, ω_1 and K are strain amplitude, angular frequency, and viscosity ratio, respectively. This method seems quite promising and provides a straightforward means for evaluating the interfacial tension of some monodispersed emulsions. However, it must be noted that acquiring an appropriate 5th harmonic with a reasonable intensity within a promising signal-to-noise ratio limit is necessary. Moreover, for evaluation of non-Newtonian and polydisperse polymer blends, appropriate modification of the equation is required. Another relative product of FT-rheology is the ratio between nonlinear rheological response from FT-rheology results and linear rheological response from SAOS tests, defined as in (6.23) [113].

$$NLR = \frac{Q_0(\emptyset)/Q_0(0)}{|G^*(\emptyset)|/|G^*(0)|} \tag{6.23}$$

Here, $\emptyset$ is the filler concentration and G^* is the complex modulus obtained from SAOS tests at the same frequency as the LAOS test is carried out. For example, if LAOS test is conducted at 1 rad/s, the G^* value at the frequency of 1 rad/s need to be used from SAOS tests as well $\left(G^*_{\omega=1rad/s}\right)$. Q_0 can be defined as follows [114]:

$$Q(\omega, \gamma_0) \equiv \frac{I_{3/1}}{\gamma_0^2} \tag{6.24}$$

$$Q_0(\omega) \equiv \lim_{\gamma_0 \to 0} Q(\omega, \gamma_0) \tag{6.25}$$

Recently, some studies applied nonlinear-linear ratio (NLR) or Q_0 to assess the morphological evolution of compatibilized blends [40, 49, 50, 77, 107, 115]. It has been shown that these NLR and Q_0 parameters are inversely proportional to the changes in droplet size (see Fig. 6.6). The most efficient particles for morphological stabilization showed larger NLR values when they were accommodated at the interface, while blends with particles inside the droplet phase showed larger droplets and thus lower NLR values.

The studies above remind us of the very important role of an interface in controlling the morphologies of blends. It can be observed that many attempts have been made to achieve breakthroughs in developing interphase properties by virtue of rheological tools, yet this issue remains inherently challenging. Obviously, different morphologies require different methods for tackling their respective challenges. Considering all the limitations that exist for evaluation of morphological evolutions on the basis of rheological analyses, it is more straightforward and less time consuming method than other techniques [89].

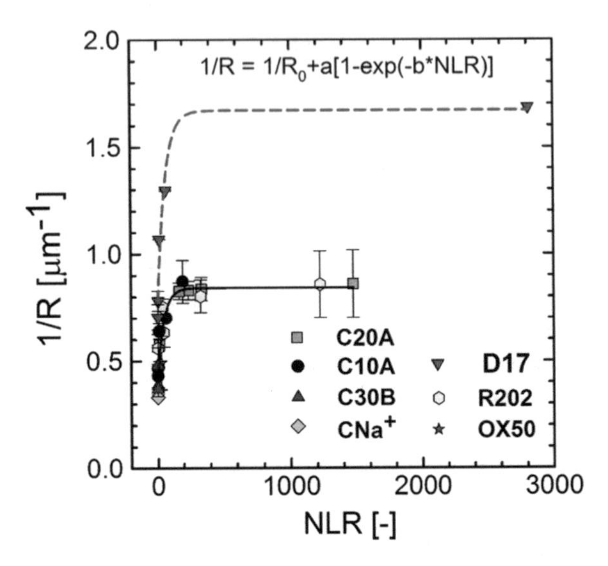

Fig. 6.6 Relationship between NLR and inverse of droplet size in case of (80/20) PP/PS blends filled with Cloisite®20A, Cloisite®10A, Cloisite® 30B, Cloisite Na^+, OX50 fumed silica, R202 fumed silica, and Sipernat D17 precipitated silica particles at different concentrations. The smallest droplet sizes, hence largest NLR values, were obtained in case of interfacially trapped D17 silica particles. Reprinted with permission from [49]. Copyright 2016, The American Chemical Society

6.7 Applications

As previously discussed, morphological variations can tune properties, and it is inevitable that finer morphologies can give rise to improved properties. One interesting application of blending is in packaging—especially food packaging, where the properties of the final product can be tailored such that stiffness and toughness are balanced by blending a more rigid polymer with another of greater flexibility. The mechanical aspects of polymer-blend properties and their ternary nanocomposites were already reviewed and can be found elsewhere [2, 44]. Therefore, the section will briefly address some of the important aspects of polymer-blend applications and their nanocomposites, including their possible stabilization effects, with an emphasis on co-continuous morphologies.

6.7.1 Conductivity

A significant volume of research on conductive polymeric composites is underway, owing to their relatively low cost and weight compared to metals as well as their promising resistance to chemical corrosion [116]. Conductive particles such as carbon-based fillers (carbon black, nanotubes, graphites, etc.) can be added to immiscible polymer blends to induce conductivity in the polymeric materials. In order to have effective conductivity, the incorporated particles need to form a network, which then boosts electron transportation. The formation of such an effective network requires sufficient particle concentration within the polymers, whereas conductivity exhibits a drastic increase. This minimum critical concentration is called the percolation threshold. Moreover, it has been reported that the blended nanocomposites have higher conductivity than those of their individual nanocomposite constituents [117]. This phenomenon, in which blending reduces the percolation threshold, is often called the double percolation effect, where higher conductivities can be achieved at much lower particle concentrations [117–119]. The literature reports that trapping particles at the interface of an immiscible blend is an ideal case for achieving the highest conductivities in double-percolated blends, once again reiterating the importance of particle localization for blend characteristics [61, 120–123]. Moreover, it has been reported that co-continuous morphology is more efficient in construction of a percolative network, hence higher electrical conductivity is attained [121, 124, 125]. Blend strategies have shown great effectiveness with regard to conductivity (double percolation), thereby presenting great opportunities in applications such as sensors (thermal, strain, and chemical sensing) [126–130] and electromagnetic interference shielding (EMI) [131]. Since the function of such sensors is based on transmission mechanism, their efficiency is defined by the quality of conductivity. For example, in gas sensing, when the molecules of a gas are absorbed onto a polymer, the polymer chains swell and disrupt the conductive path by changing the positions of the particles, thereby increasing resistivity [127, 132]. In the case of strain sensors, the change in resistivity under deformations such as stretching can be used as a sign of sensitivity, thereby enabling such sensors to be used in applications such as biorobotics or artificial muscles [133]. Furthermore, electromagnetic EM waves can be either reflected if the conductive pathway is built up, or absorbed when the dielectric constant is increased, due to the dispersion and concentration of the nanoparticles [134]. Another benefit of electrical conduction with high dielectric properties is in energy storage applications such as capacitors and stretchable electronic circuits [135–137]. This again highlights the importance of blend choice, designation, and localization of particles.

6.7.2 Porosity

The fabrication of scaffolds for tissue engineering applications represents one of the most promising applications for polymer blends with porous structure. Tissue engineering is an attractive method for repairing damage to the human body, as an alternative to conventional transplantation methods. A 3-D structure of a porous scaffold can be constructed by extraction of the second phase by solvents or porogen leaching [138]. Extrusion of immiscible blend, where one phase is water soluble, is found to be a more straightforward method of fabricating scaffold with controlled pore size when the salt is leached out [96, 139]. Figure 6.7 presents a schematic

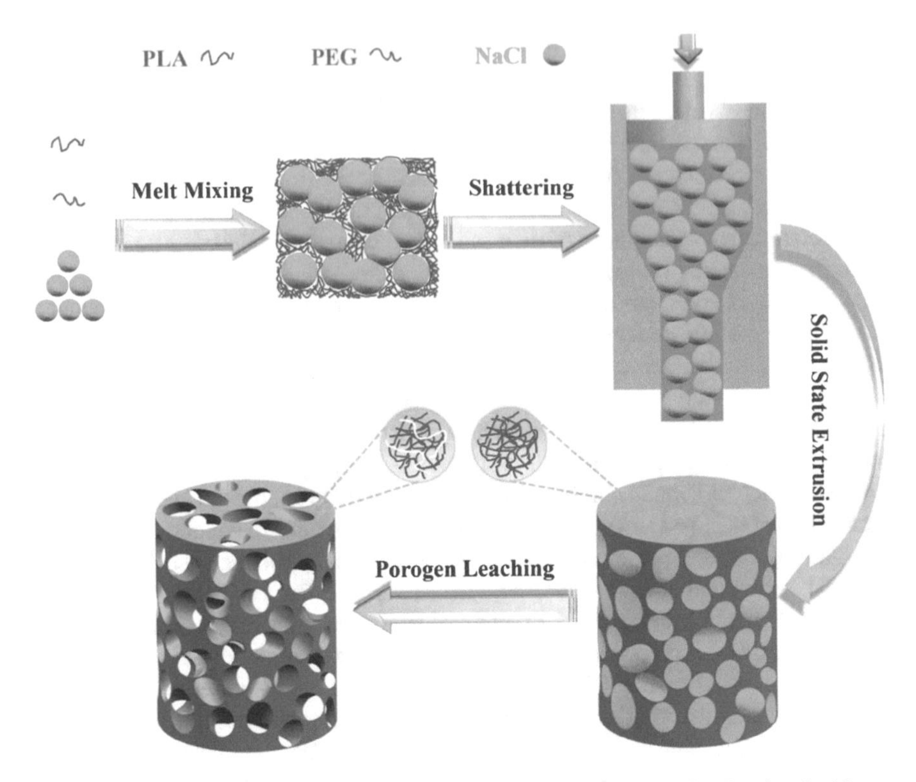

Fig. 6.7 Schematic of manufacturing PLA scaffold from PLA/PEG blends. Reprinted with permission from [138]. Copyright 2016, MDPI (open source)

showing the fabrication of a PLA scaffold by means of co-extrusion of PLA/PEG blend in the presence of NaCl as the porogen [138].

The final aim in tissue engineering is to fabricate scaffolds with pores that provide sufficient space for promoting cell proliferation and growth. Pore size and interconnectivity can be controlled by the size of the salt, the fraction of salt, volume fraction of the second phase, preparation method, and stabilizer particles [140]. It was reported that pore structures in scaffolds made from extrusion have the form of dissolved phase, whereas those obtained via salt leaching technique are cubic [139]. Oh et al. [141] reported that pore sizes were nearly identical to the size of salt used (200–300 μm). However, it must be noted that swelling and water absorption by scaffolds in contact with living organs may alter the pore sizes, which can degrade the mechanical properties of the scaffolds [142]. The other parameter affecting the efficiency of the scaffolds is scaffold/cell adhesion, where polymers with hydrophobic surfaces were found to hinder cell absorption [138, 140, 141, 143, 144]. That is, depending on the materials utilized, some modification may be required to promote such interactions. Table 6.2 shows some of the blends used in tissue engineering with respect to the tissue type and preparation methods.

Table 6.2 Some polymer blends used as scaffolds in tissue engineering, with pore size range with respect to preparation method

Blend	Preparation method	Tissue type	Porosity	References
PLGA/PCL/HA	Solution casting	Bone	80%-controlled by NaCl of 150–250 μm size	[145]
PCL/PEO	Extrusion	Too soft to be used for bone	10–150 μm depending on the blend ratio and annealing time	[139]
PLGA/PVA	Melt-molding particulate-leaching	Bone	90% of porosity-about 200–300 μm	[141]
P(LLA-CL)/Collagene	Electrospinning-nanofibers of 100–300 nm	Blood Vessel	2–5 μm	[146]
PCL/gelatin	Electrospinning-nanofibers of 113 ± 33 and 189 ± 86 nm	Nerve	0.8–1 μm depending on the ratio	[143]
Chitosan/PCL	Wet spinning	General concept	≈112–120 μm for dry pores depending on the blend ratio	[147]
Chitosan/PCL	Wet spinning	Cartilage	265.3 and 384.7 μm for blends with 75 and 50 wt% chitosan respectively	[142]
Collagene/Chitosan	Freeze-drying	Skin	87 ± 16 and 83 ± 8 μm for surface and cross-section pore sizes respectively.	[144]
PLA/PEG	Extrusion/Porogen leaching	Bone	Nearly 9 μm	[138]

The potential applications of porous polymer blends are not limited to the field of tissue engineering. They can also be used as membranes for distillation, water purification, and separators in batteries [148–153]. Mural et al. [150] revealed that functionalized GO can act as a compatibilizer in PE/PEO blends, yielding more uniform PEO sizes during fabrication, thereby emphasizing the vital influence of particles in controlling size characteristics. This brief discussion of the role of polymer blends in fabricating scaffolds again recalls the important role of rheological tools for characterizing their domain sizes and interfacial properties.

6.8 Conclusion

This chapter discussed the great importance of polymer blends, with emphasis on the unique morphologies of immiscible blends, which provide researchers and manufacturers with opportunities to tune their morphologies in order to control their final properties. The vital role of compatibilization in immiscible polymer blends was addressed, together with the urge to use compatibilizers (reactive, copolymer, and inorganic particles) in order to achieve uniform and stable morphology. Interfaces were identified as a key component of compatibilization mechanisms. Thus, many rheological techniques have been developed to model and predict the interfacial properties of blends. Among the ternary blends, the focus was on the application of particles in compatibilization, owing to their outstanding functionalities in blends with respect to final end-use applications. These functionalities include conductivity by formation of an efficient percolated network, enabling polymer materials to be used in electronics, etc. The efficiencies of these functions were found to be proportional to the distribution and localization of such particles in the blends, where interfacial entrapment of particles in blends with co-continuous structure was found to offer optimal characteristics. This, double percolation concept has paved the way for the fabrication of successful products. A key point is to exploit techniques that enable particles to be captured at the interface. This requires utilizing specific techniques in order to fabricate particular products, as well as models to describe the efficiencies regarding particle localization. To achieve ideal interfacial entrapment of nanoparticles, the remaining challenge could involve surface modification of nanoparticles according to their type. In order to reduce the potential costs involved in utilizing and modifying specific nanoparticles such as nanotubes and graphene oxides, it will be interesting to obtain hybrid-blend nanocomposites to incorporate particles of less costly materials (clays or silica) within the other phase, thereby preventing the ejection of nanotubes from the interface and hence producing nanocomposites with ultralow percolation structures.

References

1. Paul D, Walsh D, Higgins J (1985) Polymer blends and mixtures. NATO ASI Series E Appl Sci 1
2. Taguet A, Cassagnau P, Lopez-Cuesta JM (2014) Structuration, selective dispersion and compatibilizing effect of (nano)fillers in polymer blends. Prog Polym Sci 39:1526–1563
3. Utracki LA, Weiss RA (1989) Multiphase polymers: blends and ionomers. American Chemical Society
4. Taylor GI (1932) The viscosity of a fluid containing small drops of another fluid. Proc Royal Soc London Ser A Contain Papers Math Phys Char 138:41–48
5. Nienow AW, Edwards M, Harnby N (1997) Mixing in the process industries: Butterworth-Heinemann
6. Ess JW, Hornsby PR (1986) Characterisation of distributive mixing in thermoplastics compositions. Polym Test 6:205–218
7. Minale M, Mewis J, Moldenaers P (1998) Study of the morphological hysteresis in immiscible polymer blends. AIChE J 44:943–950
8. Minale M, Moldenaers P, Mewis J (1997) Effect of shear history on the morphology of immiscible polymer blends. Macromolecules 30:5470–5475
9. Janssen JMH, Meijer HEH (1995) Dynamics of liquid-liquid mixing: A 2-zone model. Polym Eng Sci 35:1766–1780

10. Van Puyvelde P, Velankar S, Mewis J, Moldenaers P, Leuven KU (2002) Effect of marangoni stresses on the deformation and coalescence in compatibilized immiscible polymer blends. Polym Eng Sci 42:1956–1964
11. Van Puyvelde P, Oommen Z, Koets P, Groeninckx G, Moldenaers P, Leuven KU et al (2003) Effect of reactive compatibilization on the interfacial slip in nylon-6/EPR blends. Polym Eng Sci 43:71–77
12. Silva J, Machado A, Moldenaers P, Maia J (2010) The effect of interfacial properties on the deformation and relaxation behavior of PMMA/PS blends. J Rheol 54:797–813
13. Saleem M, Baker WE (1990) In situ reactive compatibilization in polymer blends: Effects of functional group concentrations. J Appl Polym Sci 39:655–678
14. Pietrasanta Y, Robin JJ, Torres N, Boutevin B (1999) Reactive compatibilization of HDPE/PET blends by glycidyl methacrylate functionalized polyolefins. Macromol Chem Phys 200:142–149
15. Sailer C, Handge UA (2007) Melt viscosity, elasticity, and morphology of reactively compatibilized polyamide 6/styrene–acrylonitrile blends in shear and elongation. Macromolecules 40:2019–2028
16. Triacca VJ, Ziaee S, Barlow JW, Keskkula H, Paul DR (1991) Reactive compatibilization of blends of nylon 6 and ABS materials. Polymer 32:1401–1413
17. Huang C-C, Chang F-C (1997) Reactive compatibilization of polymer blends of poly(butylene terephthalate) (PBT) and polyamide-6,6 (PA66): 1. Rheol Therm Prop Polym 38:2135–2141
18. Maani A, Blais B, Heuzey M-C, Carreau PJ (2012) Rheological and morphological properties of reactively compatibilized thermoplastic olefin (TPO) blends a. J Rheol 56:625–647
19. DeLeo C, Walsh K, Velankar S (2011) Effect of compatibilizer concentration and weight fraction on model immiscible blends with interfacial crosslinking. J Rheol 55:713–731
20. Huo Y, Groeninckx G, Moldenaers P (2007) Rheology and morphology of polystyrene/polypropylene blends with in situ compatibilization. Rheol Acta 46:507–520
21. Omonov T, Harrats C, Groeninckx G, Moldenaers P (2007) Anisotropy and instability of the co-continuous phase morphology in uncompatibilized and reactively compatibilized polypropylene/polystyrene blends. Polymer 48:5289–5302
22. Li J, Ma G, Sheng J (2010) Linear viscoelastic characteristics of in situ compatiblized binary polymer blends with viscoelastic properties of components variable. J Polym Sci Part B Polym Phys 48:1349–1362
23. Díaz MF, Barbosa SE, Capiati NJ (2005) Improvement of mechanical properties for PP/PS blends by in situ compatibilization. Polymer 46:6096–6101
24. Shahbazi K, Aghjeh MR, Abbasi F, Meran MP, Mazidi MM (2012) Rheology, morphology and tensile properties of reactive compatibilized polyethylene/polystyrene blends via Friedel-crafts alkylation reaction. Polym Bull 69:241–259
25. Wang L, Ma W, Gross RA, McCarthy SP (1998) Reactive compatibilization of biodegradable blends of poly(lactic acid) and poly(ε-caprolactone). Polym Degrad Stab 59:161–168
26. Harada M, Iida K, Okamoto K, Hayashi H, Hirano K (2008) Reactive compatibilization of biodegradable poly (lactic acid)/poly (ε-caprolactone) blends with reactive processing agents. Polym Eng Sci 48:1359–1368
27. Semba T, Kitagawa K, Ishiaku US, Hamada H (2006) The effect of crosslinking on the mechanical properties of polylactic acid/polycaprolactone blends. J Appl Polym Sci 101:1816–1825
28. Kumar M, Mohanty S, Nayak S, Parvaiz MR (2010) Effect of glycidyl methacrylate (GMA) on the thermal, mechanical and morphological property of biodegradable PLA/PBAT blend and its nanocomposites. Biores Technol 101:8406–8415
29. Al-Itry R, Lamnawar K, Maazouz A (2012) Improvement of thermal stability, rheological and mechanical properties of PLA, PBAT and their blends by reactive extrusion with functionalized epoxy. Polym Degrad Stab 97:1898–1914
30. Ma P, Cai X, Zhang Y, Wang S, Dong W, Chen M et al (2014) In-situ compatibilization of poly (lactic acid) and poly (butylene adipate-co-terephthalate) blends by using dicumyl peroxide as a free-radical initiator. Polym Degrad Stab 102:145–151

31. Ojijo V, Sinha Ray S, Sadiku R (2013) Toughening of biodegradable polylactide/poly (butylene succinate-co-adipate) blends via in situ reactive compatibilization. ACS Appl Mater Interf 5:4266–4276
32. Ojijo V, Ray SS (2015) Super toughened biodegradable polylactide blends with non-linear copolymer interfacial architecture obtained via facile in-situ reactive compatibilization. Polymer 80:1–17
33. Gu L, Nessim EE, Macosko CW (2018) Reactive compatibilization of poly(lactic acid)/polystyrene blends and its application to preparation of hierarchically porous poly(lactic acid). Polymer 134:104–116
34. Kim S-J, Shin B-S, Hong J-L, Cho W-J, Ha C-S (2001) Reactive compatibilization of the PBT/EVA blend by maleic anhydride. Polymer 42:4073–4080
35. Van Puyvelde P, Moldenaers P (2005) Rheology and morphology development in immiscible polymer blends. Rheol Rev 2005:101
36. Van Puyvelde P, Velankar S, Moldenaers P (2001) Rheology and morphology of compatibilized polymer blends. Curr Opin Colloid Interf Sci 6:457–463
37. Sundararaj U, Macosko C (1995) Drop breakup and coalescence in polymer blends: the effects of concentration and compatibilization. Macromolecules 28:2647–2657
38. Tao F, Auhl D, Baudouin A-C, Stadler FJ, Bailly C (2013) Influence of multiwall carbon nanotubes trapped at the interface of an immiscible polymer blend on interfacial tension. Macromol Chem Phys 214:350–360
39. Ray SS, Bousmina M, Maazouz A (2006) Morphology and properties of organoclay modified polycarbonate/poly(methyl methacrylate) blend. Polym Eng Sci 46:1121–1129
40. Salehiyan R, Song HY, Choi WJ, Hyun K (2015) Characterization of effects of silica nanoparticles on (80/20) PP/PS Blends via nonlinear rheological properties from fourier transform rheology. Macromolecules 48:4669–4679
41. Elias L, Fenouillot F, Majeste JC, Cassagnau P (2007) Morphology and rheology of immiscible polymer blends filled with silica nanoparticles. Polymer 48:6029–6040
42. Sinha Ray S, Pouliot S, Bousmina M, Utracki LA (2004) Role of organically modified layered silicate as an active interfacial modifier in immiscible polystyrene/polypropylene blends. Polymer 45:8403–8413
43. Sinha Ray S, Bousmina M (2005) Compatibilization efficiency of organoclay in an immiscible polycarbonate/poly (methyl methacrylate) blend. Macromol Rapid Commun 26:450–455
44. Fenouillot F, Cassagnau P, Majesté JC (2009) Uneven distribution of nanoparticles in immiscible fluids: morphology development in polymer blends. Polymer 50:1333–1350
45. Zhu Y, Ma H-Y, Tong L-F, Fang Z-P (2008) Cutting effect" of organoclay platelets in compatibilizing immiscible polypropylene/polystyrene blends. J Zhejiang Univ Sci A 9:1614–1620
46. Kelnar I, Kratochvíl J, Kaprálková L, Zhigunov A, Nevoralová M (2017) Graphite nanoplatelets-modified PLA/PCL: Effect of blend ratio and nanofiller localization on structure and properties. J Mech Behav Biomed Mater 71:271–278
47. Kelnar I, Kratochvíl J, Kaprálková L, Špitálsky Z, Ujčič M, Zhigunov A, et al (2017) Effect of graphene oxide on structure and properties of impact modified polyamide 6. Polym Plastics Technol Eng null–null
48. Yousfi M, Livi S, Dumas A, Crépin-Leblond J, Greenhill-Hooper M, Duchet-Rumeau J (2014) Compatibilization of polypropylene/polyamide 6 blends using new synthetic nanosized talc fillers: morphology, thermal, and mechanical properties. J Appl Polym Sci 131: n/a
49. Salehiyan R, Song HY, Kim M, Choi WJ, Hyun K (2016) Morphological evaluation of PP/PS blends filled with different types of clays by nonlinear rheological analysis. Macromolecules 49:3148–3160
50. Salehiyan R, Yoo Y, Choi WJ, Hyun K (2014) Characterization of morphologies of compatibilized polypropylene/polystyrene blends with nanoparticles via nonlinear rheological properties from ft-rheology. Macromolecules 47:4066–4076
51. Thareja P, Moritz K, Velankar SS (2010) Interfacially active particles in droplet/matrix blends of model immiscible homopolymers: particles can increase or decrease drop size. Rheol Acta 49:285–298

52. Zou Z-M, Sun Z-Y, An L-J (2014) Effect of fumed silica nanoparticles on the morphology and rheology of immiscible polymer blends. Rheol Acta 53:43–53
53. Nagarkar S, Velankar SS (2013) Rheology and morphology of model immiscible polymer blends with monodisperse spherical particles at the interface. J Rheol 57:901–926
54. Sumita M, Sakata K, Asai S, Miyasaka K, Nakagawa H (1991) Dispersion of fillers and the electrical conductivity of polymer blends filled with carbon black. Polym Bull 25:265–271
55. Wu S, Dekker M (1982) Polymer interface and adhesion
56. Owens DK, Wendt RC (1969) Estimation of the surface free energy of polymers. J Appl Polym Sci 13:1741–1747
57. Good RJ, Girifalco LA, Kraus G (1958) A theory for estimation of interfacial energies. II. Application to surface thermodynamics of teflon and graphite. J Phys Chem 62:1418–1421
58. Baudouin A-C, Bailly C, Devaux J (2010) Interface localization of carbon nanotubes in blends of two copolymers. Polym Degrad Stab 95:389–398
59. Baudouin A-C, Devaux J, Bailly C (2010) Localization of carbon nanotubes at the interface in blends of polyamide and ethylene–acrylate copolymer. Polymer 51:1341–1354
60. Liebscher M, Blais M-O, Pötschke P, Heinrich G (2013) A morphological study on the dispersion and selective localization behavior of graphene nanoplatelets in immiscible polymer blends of PC and SAN. Polymer 54:5875–5882
61. Chen J, Shen Y, J-h Yang, Zhang N, Huang T, Wang Y et al (2013) Trapping carbon nanotubes at the interface of a polymer blend through adding graphene oxide: a facile strategy to reduce electrical resistivity. J Mater Chem C 1:7808–7811
62. Ginzburg VV (2005) Influence of nanoparticles on miscibility of polymer blends. Simple Theor Macromol 38:2362–2367
63. Göldel A, Marmur A, Kasaliwal GR, Pötschke P, Heinrich G (2011) Shape-dependent localization of carbon nanotubes and carbon black in an immiscible polymer blend during melt mixing. Macromolecules 44:6094–6102
64. Krasovitski B, Marmur A (2005) Particle adhesion to drops. J Adhesion 81:869–880
65. Taylor G (1934) The formation of emulsions in definable fields of flow. Proc R Soc Lond Ser A 146:501–523
66. Müller-Fischer N, Tobler P, Dressler M, Fischer P, Windhab EJ (2008) Single bubble deformation and breakup in simple shear flow. Exp Fluids 45:917–926
67. Maffettone PL, Minale M (1998) Equation of change for ellipsoidal drops in viscous flow. J Nonnewton Fluid Mech 78:227–241
68. Wolf B, Frith WJ, Norton IT (2001) Influence of gelation on particle shape in sheared biopolymer blends. J Rheol 45:1141–1157
69. Guido S, Villone M (1998) Three-dimensional shape of a drop under simple shear flow. J Rheol 42:395–415
70. Guido S, Villone M (1999) Measurement of interfacial tension by drop retraction analysis. J Colloid Interf Sci 209:247–250
71. Gooneie A, Nazockdast H, Shahsavan F (2015) Effect of selective localization of carbon nanotubes in PA6 dispersed phase of PP/PA6 blends on the morphology evolution with time, part 1: Droplet deformation under simple shear flows. Polym Eng Sci 55:1504–1519
72. López-Barrón CR, Macosko CW (2014) Rheology of compatibilized immiscible blends with droplet-matrix and cocontinuous morphologies during coarsening. J Rheol 58:1935–1953
73. Huitric J, Ville J, Médéric P, Moan M, Aubry T (2009) Rheological, morphological and structural properties of PE/PA/nanoclay ternary blends: Effect of clay weight fraction. J Rheol 53:1101–1119
74. Li R, Yu W, Zhou C (2006) Rheological characterization of droplet-matrix versus co-continuous morphology. J Macromol Sci Part B 45:889–898
75. Ezati P, Ghasemi E, Karabi M, Azizi H (2008) Rheological behaviour of PP/EPDM blend: the effect of compatibilization
76. Sangroniz L, Moncerrate MA, De Amicis VA, Palacios JK, Fernández M, Santamaria A et al (2015) The outstanding ability of nanosilica to stabilize dispersions of Nylon 6 droplets in a polypropylene matrix. J Polym Sci Part B Polym Phys 53:1567–1579

77. Sangroniz L, Palacios JK, Fernández M, Eguiazabal JI, Santamaria A, Müller AJ (2016) Linear and non-linear rheological behavior of polypropylene/polyamide blends modified with a compatibilizer agent and nanosilica and its relationship with the morphology. Eur Polym J 83:10–21
78. Roman C, García-Morales M, Gupta J, McNally T (2017) On the phase affinity of multi-walled carbon nanotubes in PMMA: LDPE immiscible polymer blends. Polymer 118:1–11
79. Choi SJ, Schowalter W (1975) Rheological properties of nondilute suspensions of deformable particles. Phys Fluids 18:420–427
80. Gramespacher H, Meissner J (1992) Interfacial tension between polymer melts measured by shear oscillations of their blends. J Rheol 36:1127–1141
81. Palierne J (1990) Linear rheology of viscoelastic emulsions with interfacial tension. Rheol Acta 29:204–214
82. Wu D, Zhang Y, Zhang M, Zhou W (2008) Phase behavior and its viscoelastic response of polylactide/poly(ε-caprolactone) blend. Eur Polym J 44:2171–2183
83. Labaume I, Médéric P, Huitric J, Aubry T (2013) Comparative study of interphase viscoelastic properties in polyethylene/polyamide blends compatibilized with clay nanoparticles or with a graft copolymer. J Rheol 57:377–392
84. Elias L, Fenouillot F, Majesté J-C, Alcouffe P, Cassagnau P (2008) Immiscible polymer blends stabilized with nano-silica particles: Rheology and effective interfacial tension. Polymer 49:4378–4385
85. Maani A, Heuzey M-C, Carreau PJ (2011) Coalescence in thermoplastic olefin (TPO) blends under shear flow. Rheol Acta 50:881–895
86. Graebling D, Muller R, Palierne J (1993) Linear viscoelasticity of incompatible polymer blends in the melt in relation with interfacial properties. Le Journal dc Physique IV. 3: C7-1525–C7-34
87. Macaúbas PHP, Demarquette NR (2001) Morphologies and interfacial tensions of immiscible polypropylene/polystyrene blends modified with triblock copolymers. Polymer 42:2543–2554
88. Souza AMC, Demarquette NR (2002) Influence of coalescence and interfacial tension on the morphology of PP/HDPE compatibilized blends. Polymer 43:3959–3967
89. Demarquette NR, De Souza AMC, Palmer G, Macaubas PHP (2003) Comparison between five experimental methods to evaluate interfacial tension between molten polymers. Polym Eng Sci 43:670–683
90. Sung YT, Han MS, Hyun JC, Kim WN, Lee HS (2003) Rheological properties and interfacial tension of polypropylene–poly(styrene-co-acrylonitrile) blend containing compatibilizer. Polymer 44:1681–1687
91. López-Barrón CR, Macosko CW (2012) Rheological and morphological study of cocontinuous polymer blends during coarsening. J Rheol 56:1315–1334
92. Z-y Gui, H-r Wang (2012) Gao Y, Lu C, Cheng S-j. Morphology and melt rheology of biodegradable poly (lactic acid)/poly (butylene succinate adipate) blends: effect of blend compositions. Iran Polym J 21:81–89
93. Isayev AI (2016) Encyclopedia of polymer blends, volume 3 structure. Wiley
94. Bai J, Goodridge RD, Hague RJM, Okamoto M (2017) Processing and characterization of a polylactic acid/nanoclay composite for laser sintering. Polym Compos 38:2570–2576
95. Bell JR, Chang K, López-Barrón CR, Macosko CW, Morse DC (2010) Annealing of cocontinuous polymer blends: effect of block copolymer molecular weight and architecture. Macromolecules 43:5024–5032
96. Trifkovic M, Hedegaard A, Huston K, Sheikhzadeh M, Macosko CW (2012) Porous films via PE/PEO cocontinuous blends. Macromolecules 45:6036–6044
97. Sengers WGF, Sengupta P, Noordermeer JWM, Picken SJ, Gotsis AD (2004) Linear viscoelastic properties of olefinic thermoplastic elastomer blends: melt state properties. Polymer 45:8881–8891
98. Yu W, Zhou W, Zhou C (2010) Linear viscoelasticity of polymer blends with co-continuous morphology. Polymer 51:2091–2098

99. Veenstra H, Verkooijen PCJ, van Lent BJJ, van Dam J, de Boer AP, Nijhof APHJ (2000) On the mechanical properties of co-continuous polymer blends: experimental and modelling. Polymer 41:1817–1826
100. Hyun K, Wilhelm M, Klein CO, Cho KS, Nam JG, Ahn KH et al (2011) A review of nonlinear oscillatory shear tests: analysis and application of large amplitude oscillatory shear (LAOS). Prog Polym Sci 36:1697–1753
101. Hyun K, Kim SH, Ahn KH, Lee SJ (2002) Large amplitude oscillatory shear as a way to classify the complex fluids. J Non-Newtonian Fluid Mech 107:51–65
102. Ewoldt RH, Hosoi AE, McKinley GH (2008) New measures for characterizing nonlinear viscoelasticity in large amplitude oscillatory shear. J Rheol 52:1427–1458
103. Ewoldt RH (2013) Defining nonlinear rheological material functions for oscillatory shear. J Rheol 57:177–195
104. Payne AR (1962) The dynamic properties of carbon black-loaded natural rubber vulcanizates. Part I. J Appl Polym Sci 6:57–63
105. Payne AR (1962) The dynamic properties of carbon black loaded natural rubber vulcanizates. Part II. J Appl Polym Sci 6:368–372
106. Salehiyan R, Hyun K (2013) Effect of organoclay on non-linear rheological properties of poly (lactic acid)/poly (caprolactone) blends. Korean J Chem Eng 30:1013–1022
107. Salehiyan R, Ray S, Bandyopadhyay J, Ojijo V (2017) The distribution of nanoclay particles at the interface and their influence on the microstructure development and rheological properties of reactively processed biodegradable polylactide/poly(butylene succinate) blend nanocomposites. Polymers 9:350
108. Wilhelm M (2002) Fourier-transform rheology. Macromol Mater and Eng 287:83–105
109. Wilhelm M, Maring D, Spiess H-W (1998) Fourier-transform rheology. Rheol Acta 37:399–405
110. Wilhelm M, Reinheimer P, Ortseifer M (1999) High sensitivity Fourier-transform rheology. Rheol Acta 38:349–356
111. Carotenuto C, Grosso M, Maffettone PL (2008) Fourier transform rheology of dilute immiscible polymer blends: a novel procedure to probe blend morphology. Macromolecules 41:4492–4500
112. Reinheimer K, Grosso M, Wilhelm M (2011) Fourier Transform Rheology as a universal non-linear mechanical characterization of droplet size and interfacial tension of dilute monodisperse emulsions. J Colloid Interf Sci 360:818–825
113. Lim HT, Ahn KH, Hong JS, Hyun K (2013) Nonlinear viscoelasticity of polymer nanocomposites under large amplitude oscillatory shear flow. J Rheol 57:767–789
114. Hyun K, Wilhelm M (2008) Establishing a new mechanical nonlinear coefficient Q from FT-rheology: First investigation of entangled linear and comb polymer model systems. Macromolecules 42:411–422
115. Ock HG, Ahn KH, Lee SJ, Hyun K (2016) Characterization of compatibilizing effect of organoclay in poly(lactic acid) and natural rubber blends by FT-rheology. Macromolecules 49:2832–2842
116. Pang H, Xu L, Yan D-X, Li Z-M (2014) Conductive polymer composites with segregated structures. Prog Polym Sci 39:1908–1933
117. Sumita M, Sakata K, Hayakawa Y, Asai S, Miyasaka K, Tanemura M (1992) Double percolation effect on the electrical conductivity of conductive particles filled polymer blends. Colloid Polym Sci 270:134–139
118. Zhang S, Deng H, Zhang Q, Fu Q (2014) Formation of conductive networks with both segregated and double-percolated characteristic in conductive polymer composites with balanced properties. ACS Appl Mater Interf 6:6835–6844
119. Göldel A, Kasaliwal G, Pötschke P (2009) Selective localization and migration of multiwalled carbon nanotubes in blends of polycarbonate and poly(styrene-acrylonitrile). Macromol Rapid Commun 30:423–429
120. Chen J, Cui X, Zhu Y, Jiang W, Sui K (2017) Design of superior conductive polymer composite with precisely controlling carbon nanotubes at the interface of a co-continuous polymer blend via a balance of π-π interactions and dipole-dipole interactions. Carbon 114:441–448

121. Chen J, H-y Lu, J-h Yang, Wang Y, X-t Zheng, C-l Zhang et al (2014) Effect of organoclay on morphology and electrical conductivity of PC/PVDF/CNT blend composites. Compos Sci Technol 94:30–38
122. Chen J, Y-y Shi, J-h Yang, Zhang N, Huang T, Chen C et al (2012) A simple strategy to achieve very low percolation threshold via the selective distribution of carbon nanotubes at the interface of polymer blends. J Mater Chem 22:22398–22404
123. Bai L, He S, Fruehwirth JW, Stein A, Macosko CW, Cheng X (2017) Localizing graphene at the interface of cocontinuous polymer blends: morphology, rheology, and conductivity of cocontinuous conductive polymer composites. J Rheol 61:575–587
124. Mao C, Zhu Y, Jiang W (2012) Design of electrical conductive composites: tuning the morphology to improve the electrical properties of graphene filled immiscible polymer blends. ACS Appl Mater Interfaces 4:5281–5286
125. Pötschke P, Bhattacharyya AR, Janke A (2004) Carbon nanotube-filled polycarbonate composites produced by melt mixing and their use in blends with polyethylene. Carbon 42:965–969
126. Hosseini SH, Entezami AA (2003) Conducting polymer blends of polypyrrole with polyvinyl acetate, polystyrene, and polyvinyl chloride based toxic gas sensors. J Appl Polym Sci 90:49–62
127. Segal E, Tchoudakov R, Mironi-Harpaz I, Narkis M, Siegmann A (2005) Chemical sensing materials based on electrically-conductive immiscible polymer blends. Polym Int 54:1065–1075
128. Panwar V, Kang B-S, Park J-O, Park S-H (2011) New ionic polymer–metal composite actuators based on PVDF/PSSA/PVP polymer blend membrane. Polym Eng Sci 51:1730–1741
129. Ma L-F, Bao R-Y, Dou R, Zheng S-D, Liu Z-Y, Zhang R-Y et al (2016) Conductive thermoplastic vulcanizates (TPVs) based on polypropylene (PP)/ethylene-propylene-diene rubber (EPDM) blend: From strain sensor to highly stretchable conductor. Compos Sci Technol 128:176–184
130. Ji M, Deng H, Yan D, Li X, Duan L, Fu Q (2014) Selective localization of multi-walled carbon nanotubes in thermoplastic elastomer blends: an effective method for tunable resistivity–strain sensing behavior. Compos Sci Technol 92:16–26
131. Mural PKS, Pawar SP, Jayanthi S, Madras G, Sood AK, Bose S (2015) Engineering nanostructures by decorating magnetic nanoparticles onto graphene oxide sheets to shield electromagnetic radiations. ACS Appl Mater Interf 7:16266–16278
132. Soares BG, Gubbels F, Jérôme R, Teyssié P, Vanlathem E, Deltour R (1995) Electrical conductivity in carbon black-loaded polystyrene-polyisoprene blends. Selective localization of carbon black at the interface. Polym Bull 35:223–228
133. Brochu P, Pei Q (2010) Advances in Dielectric Elastomers for Actuators and Artificial Muscles. Macromol Rapid Commun 31:10–36
134. Lakshmi N, Tambe P (2017) EMI shielding effectiveness of graphene decorated with graphene quantum dots and silver nanoparticles reinforced PVDF nanocomposites. Compos Interf 24:861–882
135. Kang SJ, Park YJ, Bae I, Kim KJ, Kim H-C, Bauer S et al (2009) Printable ferroelectric PVDF/PMMA blend films with ultralow roughness for low voltage non-volatile polymer memory. Advanc Funct Mater 19:2812–2818
136. Li Y, Shimizu H (2009) Toward a stretchable, elastic, and electrically conductive nanocomposite: morphology and properties of Poly[styrene-b-(ethylene-co-butylene)-b-styrene]/multiwalled carbon nanotube composites fabricated by high-shear processing. Macromolecules 42:2587–2593
137. Dang Z-M, Yuan J-K, Yao S-H, Liao R-J (2013) Flexible nanodielectric materials with high permittivity for power energy storage. Advanc Mater 25:6334–6365
138. Yin H-M, Qian J, Zhang J, Lin Z-F, Li J-S, Xu J-Z et al (2016) Engineering Porous Poly(lactic acid) scaffolds with high mechanical performance via a solid state extrusion/porogen leaching approach. Polymers 8:213

139. Washburn NR, Simon CG, Tona A, Elgendy HM, Karim A, Amis EJ (2002) Co-extrusion of biocompatible polymers for scaffolds with co-continuous morphology. J Biomed Mater Res 60:20–29
140. Liu X, Ma PX (2004) Polymeric scaffolds for bone tissue engineering. Ann Biomed Eng 32:477–486
141. Oh SH, Kang SG, Kim ES, Cho SH, Lee JH (2003) Fabrication and characterization of hydrophilic poly(lactic-co-glycolic acid)/poly(vinyl alcohol) blend cell scaffolds by melt-molding particulate-leaching method. Biomaterials 24:4011–4021
142. Neves SC, Moreira Teixeira LS, Moroni L, Reis RL, Van Blitterswijk CA, Alves NM et al (2011) Chitosan/Poly(ε-caprolactone) blend scaffolds for cartilage repair. Biomaterials 32:1068–1079
143. Ghasemi-Mobarakeh L, Prabhakaran MP, Morshed M, Nasr-Esfahani M-H, Ramakrishna S (2008) Electrospun poly(ε-caprolactone)/gelatin nanofibrous scaffolds for nerve tissue engineering. Biomaterials 29:4532–4539
144. Liu Y, Ma L, Gao C (2012) Facile fabrication of the glutaraldehyde cross-linked collagen/chitosan porous scaffold for skin tissue engineering. Mater Sci Eng, C 32:2361–2366
145. Marra KG, Szem JW, Kumta PN, DiMilla PA, Weiss LE (1999) In vitro analysis of biodegradable polymer blend/hydroxyapatite composites for bone tissue engineering. J Biomed Mater Res 47:324–335
146. He W, Yong T, Teo WE, Ma Z, Ramakrishna S (2005) Fabrication and endothelialization of collagen-blended biodegradable polymer nanofibers: potential vascular graft for blood vessel tissue engineering. Tissue Eng 11:1574–1588
147. Malheiro VN, Caridade SG, Alves NM, Mano JF (2010) New poly(ε-caprolactone)/chitosan blend fibers for tissue engineering applications. Acta Biomater 6:418–428
148. Sun D, Liu M-Q, Guo J-H, Zhang J-Y, Li B-B, Li D-Y (2015) Preparation and characterization of PDMS-PVDF hydrophobic microporous membrane for membrane distillation. Desalination 370:63–71
149. Lee E-J, Deka BJ, Guo J, Woo YC, Shon HK, An AK (2017) Engineering the Re-entrant hierarchy and surface energy of PDMS-PVDF membrane for membrane distillation using a facile and benign microsphere coating. Environ Sci Technol 51:10117–10126
150. Mural PKS, Banerjee A, Rana MS, Shukla A, Padmanabhan B, Bhadra S et al (2014) Polyolefin based antibacterial membranes derived from PE/PEO blends compatibilized with amine terminated graphene oxide and maleated PE. J Mater Chem A 2:17635–17648
151. Mural PKS, Sharma M, Shukla A, Bhadra S, Padmanabhan B, Madras G et al (2015) Porous membranes designed from bi-phasic polymeric blends containing silver decorated reduced graphene oxide synthesized via a facile one-pot approach. RSC Advanc 5:32441–32451
152. Li H, Zhang H, Liang Z-Y, Chen Y-M, Zhu B-K, Zhu L-P (2014) Preparation and properties of Poly (vinylidene fluoride)/poly(dimethylsiloxane) graft (poly(propylene oxide)-block-poly(ethylene oxide)) blend porous separators and corresponding electrolytes. Electrochim Acta 116:413–420
153. Li H, Chen Y-M, Ma X-T, Shi J-L, Zhu B-K, Zhu L-P (2011) Gel polymer electrolytes based on active PVDF separator for lithium ion battery. I: preparation and property of PVDF/poly(dimethylsiloxane) blending membrane. J Membr Sci 379:397–402

Chapter 7
Electrospun Polymer Nanocomposites

Koena Selatile, Prasanna Kumar S. Mural and Suprakas Sinha Ray

Abstract The present chapter gives an overview of the electrospinning process for producing nanofibers and developments in their application as polymer nanocomposites. Firstly, the electrospinning parameters (solution and process conditions), along with their effect on fiber morphology and properties, is elaborated. Thereafter, the incorporation of nanofibers (as nanofillers) into a (bulk) polymer matrix to generate electrospun polymer nanocomposites is described. The potential use of nanofibers as reinforcement in polymer nanocomposites, as well as their role as secondary reinforcements, is discussed. Morphology–property relationships provide an understanding of the effect of nanofiber dimensions and properties on the performance of such advanced functional materials. The nanofiber composite materials offer superior structural properties compared to their microscale counterparts, including a high modulus, and improved thermal and optical properties. Furthermore, the morphological properties of electrospun nanofiber-reinforced polymer nanocomposites are correlated with the structural properties along with their performance and applications. The various challenges, opportunities, and future trends are also highlighted.

K. Selatile · P. K. S. Mural · S. Sinha Ray (✉)
DST-CSIR National Centre for Nanostructured Materials, Council for Scientific and Industrial Research, Pretoria 0001, South Africa
e-mail: rsuprakas@csir.co.za

K. Selatile
e-mail: kselatile@csir.co.za

P. K. S. Mural
e-mail: pmural@csir.co.za

S. Sinha Ray
Department of Applied Chemistry, University of Johannesburg, Doornfontein 2028, Johannesburg, South Africa
e-mail: ssinharay@uj.ac.za

S. Sinha Ray (ed.), *Processing of Polymer-based Nanocomposites*, Springer Series in Materials Science 278, https://doi.org/10.1007/978-3-319-97792-8_7

7.1 Introduction

In the present century, there is a continuous drive for miniaturization of devices that are ultra-lightweight yet strong, and use novel designs motivated by polymeric nanofibers [1]. Polymeric nanofibers can be synthesized by template synthesis, drawing, phase separation, self-assembly, and electrospinning [2]. Among these synthesis techniques, electrospinning of the polymer offers the advantages of ease of scalability, ease of processing, repeatability, and better control of fiber dimensions over the other techniques. In addition, electrospinning is a cost-effective technique for producing long, continuous, aligned nanofibers. Further, electrospinning operates under an electric field which aligns the fillers in the axial direction, imparting superior properties and functionality to the fiber, which can extend its application to filtration, tissue engineering, and other advanced nanomaterials [1]. The small size and high specific surface area of electrospun nanofibers make them suitable for applications in catalysis, biomedical devices, energy conversion and storage, sensors, and optical and electronic devices [3].

Hence, the present chapter discusses the basic concepts concerning electrospinning and various parameters/techniques for controlling the dimensions of the fibers. Subsequently, the chapter discusses electrospun fiber as a reinforcement material in the preparation of polymer nanocomposites. In addition, the morphologies are correlated with the structural properties, along with the performance and applications of electrospun materials. Finally, the various challenges, opportunities, and future trends in the area of electrospinning of polymer nanocomposites are highlighted.

7.2 Electrospinning of Polymer Nanocomposites

Electrospinning is a single technique for synthesizing polymer fibers with diameters of 2 nm to several micrometer from a polymer solution with the assistance of electrostatic forces [4]. Although electrospinning can be carried out using polymer melts, it is more common to use polymer solutions to prepare fibers. Anton [5] first demonstrated electrospinning, which was later revived by Reneker et al. who were able to produce ultra-fine fibers using ultra-high voltages [6]. As shown in Fig. 7.1, electrospinning systems consist of a pipette/syringe containing a polymer solution, a high-voltage source, and a conductive collector screen. The needle of the syringe is coupled to an electrode in order to electrically charge the polymer solution while it is flowing. In addition, a counter electrode is connected to the bottom collector, which is grounded. The charged polymer solution accelerates towards the grounded collector plate in the presence of the electric field. The flow of droplets from the tip of the pipette/syringe depends on the applied voltage and electric field, which plays a pivotal role in determining the properties of the fibers. When a low-strength electric field is applied, it induces a charge on the pendant droplet and surface tension of the solution prevents it from dripping, as shown in Fig. 7.2a. When the intensity of the electric field increases, the droplet charges start to repel each other, thus inducing

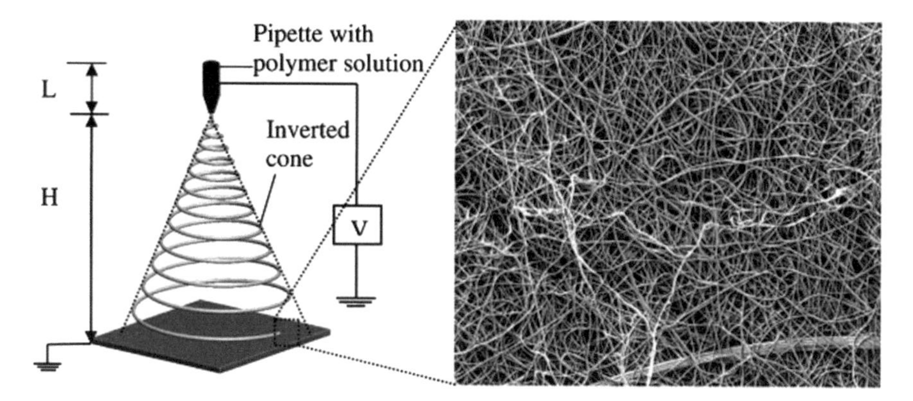

Fig. 7.1 Schematic diagram of electrospinning operation. The inset exhibits the SEM micrograph of PA6, 6 electrospun fibers. (H represents the distance between the tip of the needle and conductive collector screen, and L represents length of the polymer solution in pipette). Reproduced with permission from [1]. Copyright 2010, Elsevier Science Ltd.

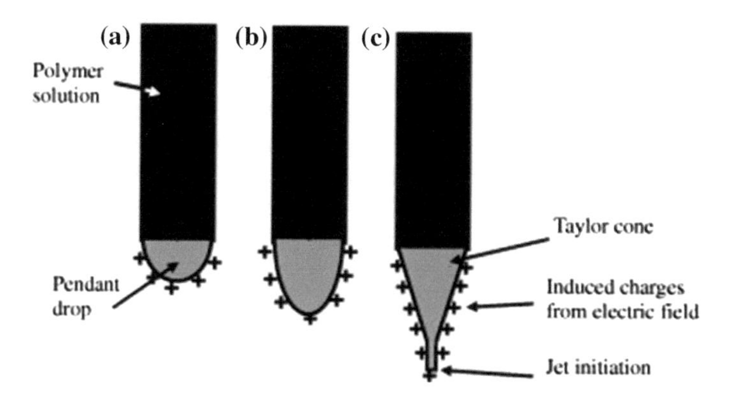

Fig. 7.2 Schematic diagram of various stages of Taylor cone formation in polymer solution. **a** charge development on the pendent drop due to electric field, **b** elongation of droplet with increase in electric field and **c** formation of Taylor cone due to repulsion of charge leading to formation of jet. Reproduced with permission from [1]. Copyright 2010, Elsevier Science Ltd.

shear stresses. This repulsive force on the droplet surfaces acts against the surface tension, resulting in extension of the droplets to form a conical shape (as shown in Fig. 7.2b). Thus, the conical shape of the droplet functions as the initial surface for further droplet extension. Further increases in the electric field above a critical voltage result in disturbance of the forces, which forms a charged jet (Taylor cone) originating from the tip of the conically shaped droplets, as shown in Fig. 7.2c [7]. When the discharged jet elongates sufficiently, it travels towards the collector plate i.e., towards lower potential (as the collector plate is grounded in most cases). The elongation of the jets ensures a decrease in the diameter of the jet with a simultaneous increase in the length while arriving at the collector plate.

General instabilities that occur during fiber preparation include Rayleigh bending, and whipping instabilities, which influence the geometry and size of the deposited fibers. Rayleigh instabilities occur as a result of insufficient electric field intensity or low viscosity of the solution and can cause jet breaks-up, resulting in beads on the fiber morphology. Such instabilities can be suppressed by increasing the electric field intensity and viscosity (or concentration) of the polymer solution. At high electric field intensity, excessive charge–charge repulsion results in bending and whipping instabilities, which result in the jets traveling in an inverse cone pattern. At a high electric field and sufficient charge density in the jet, both Rayleigh and bending instabilities can be avoided, but whipping instabilities can occur, resulting in elongation of the jet. In addition, during elongation, solvent evaporation from the jet occurs, resulting in deposition of ultrafine fibers on the collector screen.

Generally, to obtain uniform continuous fibers, various factors need to be optimized, classified into polymer solution properties and processing parameters. The properties of the polymer solutions play a vital role in the electrospinning process and the resulting morphology of the fibers. Polymer solution properties include: the surface tension, viscosity, and conductivity of the solution; and the dielectric properties of the solvent. The processing parameters are the external factors applied to the electrospinning jet, including the applied voltage, feed rate and temperature of the solution, collector type, needle tip diameter, and distance between the needle tip and the collector ('H' as shown in Fig. 7.1). The following sections discuss these factors and their effect on the fiber morphology and properties.

7.2.1 Polymer Solution Properties

7.2.1.1 Viscosity of the Polymer Solution

Although Rayleigh instabilities can be suppressed by increasing the viscosity of a solution, a very high viscosity is generally not desirable as it can lead to difficulties in pumping the solution through the needle tip. In addition, a very high viscosity solution results in drying of the needle tip which hampers fiber/jet formation. However, excessively low viscosity solutions generally have a higher content of solvent molecules than polymer chains, resulting in dominant surface tension. The surface tension of the solution opposes the electric field, forming Rayleigh instabilities and ultimately, bead formation in the fibers. The optimum polymer solution viscosity is obtained by either increasing the molecular weight or concentration of the polymer in the solution. At optimum concentration of the polymer solution, a high/sufficient number of polymer chains are present; these will entangle within the solution, which is essential for maintaining continuity of the jet during electrospinning. Low chain molecular weights result in the solvent molecules dominating the reaction, resulting in bead formation. Increasing the viscosity has been reported to increase the diameter of the fiber [8]. Further, if polymer chains of very high molecular weight are used above a certain concentration, they increase the viscosity so much that flow of the solution through the needle is obstructed or drying of the needle tip occurs. An optimum viscosity with an optimum polymer molecular weight results in smooth and continuous fibers.

7.2.1.2 Surface Tension of the Solution

Jet formation occurs in electrospinning once the surface tension of solution is overcome. In the case of a high molecular concentration of the solvent, free solvent molecules tend to assemble and form spherical shape to minimize the surface area. This spherical shape may cause bead formation at the collector plate. The interaction between polymer chains and solvent molecules can be enhanced by increasing the concentration of the solution. The enhanced interaction between polymer chains and solvent molecules occur due to the electric field and distributed over the entangled polymer molecules, decreasing aggregation of solvent molecules and preventing bead formation in the fibers. The addition of low surface tension solvents (such as ethanol) are reported to yield smooth fibers [2]. The addition of surfactants can reduce the surface tension, which in turn results in uniform fibers.

7.2.1.3 Solution Conductivity

Solutions with zero conductivity cannot be used for electrospinning fiber applications. Conducting solutions create more charges during electrospinning, which enhance the repulsion of charges, resulting in formation of the Taylor cone or stretching of the solution. The conductivity of the solution can be enhanced by adding ions or polyelectrolytes. The addition of polyelectrolytes/ions increases the charge carried by the solution, resulting in a greater degree of stretching of the solution and smaller diameter smooth fibers. The presence of ions/polyelectrolytes in solution reduces the critical voltage and produces finer fibers. Generally, smaller ions have greater mobility in response to the external applied field, resulting in higher elongational forces that yield smaller diameter fibers. However, if the viscoelastic forces are stronger than the coulombic forces, the fiber diameter increases. In addition, by increasing the pH of the solution from neutral to more basic, the conductivity of solution can be increased [9].

7.2.1.4 Solvent Dielectric Properties

Solutions with good dielectric properties result in narrow fibers and reduced bead formation. The higher dielectric constant of the solution results in bending instabilities, which in turn increase the deposition area and reduce fiber diameter. However, adding higher dielectric constant solvents to the solution may result in beads in the fibers despite the increased dielectric constant. However, poor interaction amongst the polymer and solvent molecules results in retraction of the polymer molecules which results in bead formation.

7.2.2 Processing Conditions

7.2.2.1 Voltage

Voltage applied across the solution provides an external electric field, i.e., coulombic repulsive force, which is essential for overcoming the surface tension of the solution. Generally, more than 6 kV of positive and negative voltage is required to produce a Taylor cone at the needle tip. For a lower solution feed, a higher voltage is required to stabilize the Taylor cone jet. A higher voltage ensures that the coulombic repulsive force on the jet is sufficient to elongate the viscoelastic polymer solution. However, higher voltages lead to higher acceleration of the jet, resulting in a higher volume of solution being drawn from the tip of the needle. Drawing at a higher rate solution results in a smaller and less stable Taylor cone; if the solution is drawn faster than the source supplies it; this may result in a receding Taylor cone at the needle tip, which may increase the tendency of bead formation. In addition, a very high voltage increases bead density, resulting in merging and fusing of the beads and subsequently, thicker fibers. A higher voltage also imparts higher crystallinity in the polymer fibers in addition to the altering of morphology. The electrostatic field may result in ordering of polymer molecules, thus inducing higher crystallinity of the polymer fibers. However, above a critical voltage, the crystallinity of the fiber decreases due to a shorter flight time of the jet. In addition, varying the electric field profile between the collector and the source changes the geometry of the jet profile, which in turn, governs the profile of the deposited fiber. Hence, by changing the auxiliary electrodes, shape, or orientation of the collector it is possible to obtain aligned and even patterned nanofibers. Both AC and DC voltage supplies can be used during electrospinning. DC voltages are most commonly used. AC voltages result in less bending and stretching of the jet than DC voltages, resulting in thicker fibers.

7.2.2.2 Feed Rate

For a particular feed rate of the solution, there is a given voltage to obtain and maintain a stable Taylor cone. By increasing the feed rate of the solution at a given voltage, a higher volume of solution is drawn from the tip of the needle, resulting in thicker fibers. In addition, as a higher volume of solution is drawn across the needle tip, the solution takes a longer time to dry, which can resulting in fibers fusing together. A lower feed rate is desirable for faster evaporation of the solvent.

7.2.2.3 Pressure

At pressures below atmospheric pressure, solution from the syringe flows out of the needle, which affects the jet stability. Further, reduction in the pressure results in bubbling up of the solution across the needle tip. However, at very low pressure results in direct electric discharge of jet on to the collector plate, thus hampering the electrospinning process.

7.2.2.4 Temperature

Increasing the solution temperature results in a reduction in the solution viscosity and enhances the evaporation rate of the solution. This viscosity reduction results in uniform fiber diameter due to uniform stretching as a result of increased solubility. In addition, at lower viscosity, coulombic forces are greater, resulting in increased stretching (i.e., reduced diameter) of the fiber.

7.2.2.5 Collector Type

Collectors are generally made from a conducting material, such as Al foils, which are grounded in order to obtain a stable potential difference across the collector and source. In the case of a conducting collector plate, charges on the fibers are dissipated, promoting fiber deposition, while insulating collector plates accumulate charges, which repel the incoming fibers, resulting in fewer fibers being deposited on the collector. Porous collectors result in a lower packing density of fibers than solid ones due to higher diffusion and evaporation rate of the solvent.

7.2.2.6 Needle Diameter

When the needle tip diameter is small, it tends to form narrow fibers with little clogging and bead formation. If the tip orifice is too small, then solution droplets may not easily flow from the needle, which will hinder fiber formation. Smaller tip diameters increase the surface tension, resulting in higher coulombic forces, thus allowing the solution to elongate, ensuring narrow fibers are produced.

7.2.2.7 Gap Between the Tip and Collector

The electric field strength and flight time of the solution directly affect the fiber morphology. Increasing the gap between the tip and collector affects both the flight time and electric field strength in the solution. Decreasing the gap between the collector and tip reduces the flight time and accelerates solution collection, resulting in merging of the fibers due to slower evaporation of the solvent. A narrower gap can also cause bead formation. Furthermore, a decrease in the voltage has the same effect as an increase in voltage. The longer flight time due to a wider gap results in sufficient time for solvent evaporation and fiber stretching, resulting in thinner fibers. However, very wide gaps excessively reduce the field strength, resulting in larger fiber diameters. Thus, there is an optimum gap where the electric field strength is sufficient to obtain fibers with the desired morphology.

7.2.2.8 Test Chamber Conditions

The humidity and gases present in the atmosphere in test chambers interact with the polymer solution and affects the fiber morphology. The humidity affects the polymer solution during electrospinning, where high humidity has been shown to produce circular pores on the fiber surface. The depth of these pores increased with increasing humidity up to certain humidity level, after which it saturated [2]. In addition, very low humidity may result in faster solvent evaporation, which may result in clogging of the needle tip. The fiber diameter depends on the breakdown voltage of the gas, where the use of a gas with a higher breakdown voltage than air results in thicker fibers.

7.3 Advances in Nanofiber Preparation

Varying the processing parameters during electrospinning can result in many different fibers with interesting morphologies, including porous nanofibers, flattened fibers, branched fibers, helical fibers, hollow fibers, and those containing various phases. Both the humidity and the solvent vapor pressure play a vital role in pore formation. Solvents with low vapor pressure form pores on the fibers as they form thermodynamically unstable solutions that evaporate, creating the pores. Flattened fibers can be obtained by increasing the viscosity as the solvent is trapped during fiber formation, resulting in adjacent fibers adhering to each other, forming a ribbon or flattened fibers. Branched fibers can be obtained by splitting of the charged droplets into smaller jets. Helical fibers are difficult to obtain as the process requires controlling the equilibrium between the viscoelastic restoring force and electrostatic repulsive force, which drives structural rearrangements to retain equilibrium. Hollow fibers can be obtained by electrospinning followed by post-processing operations, such as chemical vapor deposition (CVD), or via co-axial spinning. The nanofibers need to act as a template for a coating (either an extractable or degradable layer). Similarly, fibers with various phases i.e., co-axial or tri-axial fibers containing respectively, two or three layers with different compositions, can be obtained by changing the spinneret (see Fig. 7.3).

7.4 Electrospun Nanofibers as Reinforcing Materials

Nanofibers have a low defect density due to small fiber diameters. In addition, macromolecular chains align along the nanofiber axis. Aligning of the macromolecular

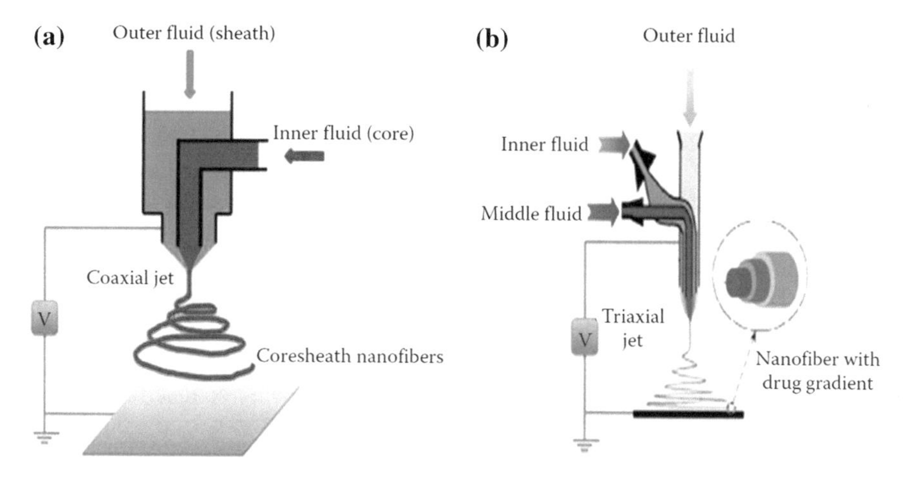

Fig. 7.3 Schematic diagram Coaxial electrospinning for obtaining fiber with different composition **a** coaxial jet for obtaining two layered nanofibers. Reproduced with permission from [10]. Copyright 2013, Elsevier Science Ltd. and **b** triaxial jet for three layers of nanofibers. Reproduced with permission from [11]. Copyright 2015, the American Chemical Society

chains along the fiber axis reduces the degree of defects and enables production of high-strength fibers [3]. Furthermore, nanofibers have an enormous specific surface area (and hence, interfacial area between the matrix and nanofibers); hollow nanofibers can further increase the specific surface area [12]. Electrospun nanofibers can be flexibly functionalized to both improve their dispersion in the polymer matrix and reduce interfacial tension, thereby promoting efficient transfer of load between the nanofiber and polymer matrix. In addition, electrospinning can produce advanced multifunctional polymer nanocomposites with e.g., self-healing behavior. Moreover, electrospun nanofibers are long and continuous, resulting in fewer edges and stress concentration points than short nanofiller particles. Electrospun nanofibers with a high aspect ratio can form a 3D percolating network at low loading, resulting in only a marginal weight increase compared to the pure polymer. For these reasons, electrospun nanofibers exhibit improved elastic strength and modulus at break than microfibers.

Kim and Reneker [13] published the first report of polymer nanocomposite fabrication by electrospun fibers. They investigated the use of polybenzimidazole (PBI) electrospun nanofibers as a reinforcing material in epoxy and styrene-butadiene rubber (SBR). For epoxy-based nanocomposites, 32 layers of nanofiber sheets were used to prepare nanocomposites. They reported that the bending fracture toughness and modulus of the epoxy/PBI nanocomposite increased slightly compared to the pure polymer blend, while the fracture energy increased significantly. SBR-PBI nanocomposites were prepared by chopping nanofibers to obtain a homogeneous dispersion in the matrix. Chopping of the nanofibers resulted in increased tear strength and Young's modulus compared to pristine SBR.

Most nanofibers diameter are smaller than the wavelengths of visible light; hence, when electrospun nanofibers are used as reinforcement, the composite maintains the transparency of the matrix. Bergshoef and Vancso [14] prepared electrospun PA 4,6 nanofiber-based phenolic epoxy nanocomposites which exhibited significantly enhanced stiffness and strength and maintained transparency of the matrix. Thus, incorporating electrospun nanofibers into a polymer matrix can retain the transparency, while improving the mechanical properties. Another technique to obtain transparent nanocomposites is to coaxially electrospin fibers and then melt the skin (shell) in a hot press to yield a polymer nanocomposite. Composites prepared by this technique showed a slight decrease in transparency at a fiber loading of 2.5 wt%, but exhibited significant improvement in tensile and three point bending strength [15]. When nylon fibers were incorporated in an intrinsically conducting polymer (polyaniline), the nanocomposite working temperature was extended from 100 to 200 °C, along with improved mechanical properties [16].

Most electrospun fiber nanocomposites are prepared using epoxy resin as a matrix. Poly(ethylene oxide) (PEO) electrospun fiber-reinforced epoxy nanocomposites showed improved fracture toughness and three-point bending strength compared to unfilled epoxy [17]. Furthermore, attention was drawn towards the chemical properties of the electrospun fibers. Copolymer polystyrene-*co*-glycidyl methacrylate P(St-*co*-GMA) electrospun fibers accompanied by spurting of ethylenediamine (EDA) were incorporated into epoxy resin. EDA acts as a cross-linking agent for epoxy. Thus, P(St-*co*-GMA)/EDA nanofiber-reinforced epoxy nanocomposites were obtained, which exhibited a storage modulus of ~2.5 and 10 times higher than the P(St-*co*-GMA) and pure nanofiber-reinforced epoxy, respectively [18].

Ozden et al. [18] presented two key findings: (i) cross-linking of the fibers results in increased inherent stiffness; and (ii) the surface chemistry/modification of electrospun fibers can result in better interfacial adhesion between the matrix and the electrospun fibers. Electrospun fibers can also be used as a secondary reinforcement for laminated composites. The nanofibers act to reduce points of stress concentration due to mismatch of ply properties, which is achieved by bonding adjacent plies without any further increase in composite thickness or weight. These composites can show good interlaminar fracture strength and toughness, along with delamination resistance against impact, and static and fatigue loading [19].

7.5 Morphology–Property Relationships

Various properties of as-prepared electrospun nanofibers depend on factors such as the dimensions and texture of the nanofibers. Hence, it is essential to study the nanofiber morphology. Furthermore, alignment of the fibers along the drawing direction can affect the mechanical properties. The mechanical properties of individual nanofibers, along with their incorporation into composites will be discussed in this section. In addition, the thermal properties of nanofibers are briefly presented in order to better understand the relationship between nanostructure and thermal properties.

7.5.1 *Nanofiber Morphology*

The morphology of the polymer nanofiber plays a vital role in determining its properties and hence, potential applications. As briefly mentioned earlier, the morphology of the nanofibers is dependent on several parameters, such as the voltage, viscosity, and gap between the needle tip and collector plate. This section discusses the morphological evolution of nanofibers considering these parameters, along with optimization of the properties to obtain various specific morphologies.

7.5.1.1 Viscosity

An optimum polymer solution concentration with an optimum viscosity is essential for maintaining continuous and smooth fibers. Deitzel et al. reported that the viscosity and surface tension are significant parameters determining the ideal concentration range for drawing continuous nanofibers from the polymer solution [20]. Figure 7.4 shows that lower viscosities and PEO concentrations of a water solution had higher surface tensions. In this case where the surface tension dominates, the morphology may show the formation of beads instead of fibers. However, higher viscosity solutions have lower surface tension, which increases the fiber diameter. Furthermore, when increasing the PEO concentration above 15 wt%, the solution droplets at the nozzle tip will be thick (high viscosity) and the applied electric field will cause them to fall by gravity onto the collector plate. Droplets falling onto the collector plate by gravity results in an unsustainable jet, hampering nanofiber formation. Thus, for obtaining sustainable fiber formation, the surface tension forces and the viscosity will determine the lower and upper limits of processing when all other parameters are kept constant.

The viable viscosity range of a polymer solution to obtain nanofibers varies for each system. For example, electrospinning of PEO in an ethanol–water system requires a viscosity of 1–20 Pas to obtain nanofibers [21]. However, electrospinning of cellulose acetate (CA) in acetone/dimethylacetamide (DMAc) required a viscosity of 1.2–10.2 P [22] to obtain nanofibers. Thus, confirming that there is no fixed range of viscosity for the polymer solution system to obtain the nanofibers. Once the viscosity of the polymer solution system is optimized to yield nanofibers, it is important to maintain consistent and controllable nanofiber diameters that are continuous and defect free (or with controlled defects). The most important parameter to control is the diameter of the nanofibers to ensure they are suitable for the intended application. As discussed earlier, high viscosity solutions result in large nanofiber diameters, while a lower viscosity may result in bead formation. The effect of the viscosity on the diameter of the fibers is observed from data for a PEO polymer solution shown in Fig. 7.5a. The 1 wt% of PEO concentration polymer solution (viscosity of 13 Pas) resulted in short beads. As the viscosity of polymer solution was increased by increasing the PEO concentration, the short beads elongated and formed long beads, then changed from spherical to spindle-like beads. An increase in PEO

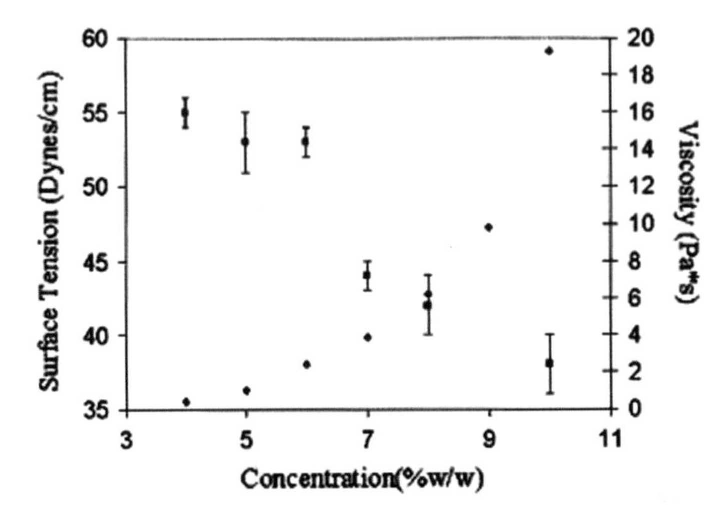

Fig. 7.4 shows the dependence of surface tension and solution viscosity on the concentration (%w/w) for a PEO/water solution. Reproduced with permission from [20]. Copyright, Elsevier Science Ltd.

concentration to 4 wt% (viscosity of 1250 cP) resulted in the beads disappearing and smooth fibers being obtained [23]. However, further increase in the viscosity of the PEO solution resulted in an increase in the fiber diameter. Figure 7.5b shows various kinds of electrospun nanofiber morphologies that can be obtained by varying the solution concentration. It can be seen that very high concentrations can sometimes produce helix-shaped micro-ribbons [24].

7.5.1.2 Voltage

An applied voltage is required to draw the solution from the needle and form the Taylor cone. Increasing the voltage results in higher electrical potential, which subsequently results in fewer beads and smoother nanofibers, as shown in Fig. 7.6. However, it can be seen in the figure, that when the voltage is increased beyond a critical voltage (in this case, 9 kV) the nanofibers became rougher due to the higher electrical potential. Some studies reported that voltages above a critical value can also result in bead formation. For instance, Şener et al. showed that electrospinning of poly(vinyl alcohol) PVA and sodium alginate at 45 kV resulted in larger beads than for other fibers electrospun at voltages below 45 kV [25].

7.5.1.3 Distance Between Needle Tip and Collector Plate

It is well known that when the distance between the needle tip and collector plate (H) is narrow, fiber solidification will not occur, resulting in fusion of the fibers.

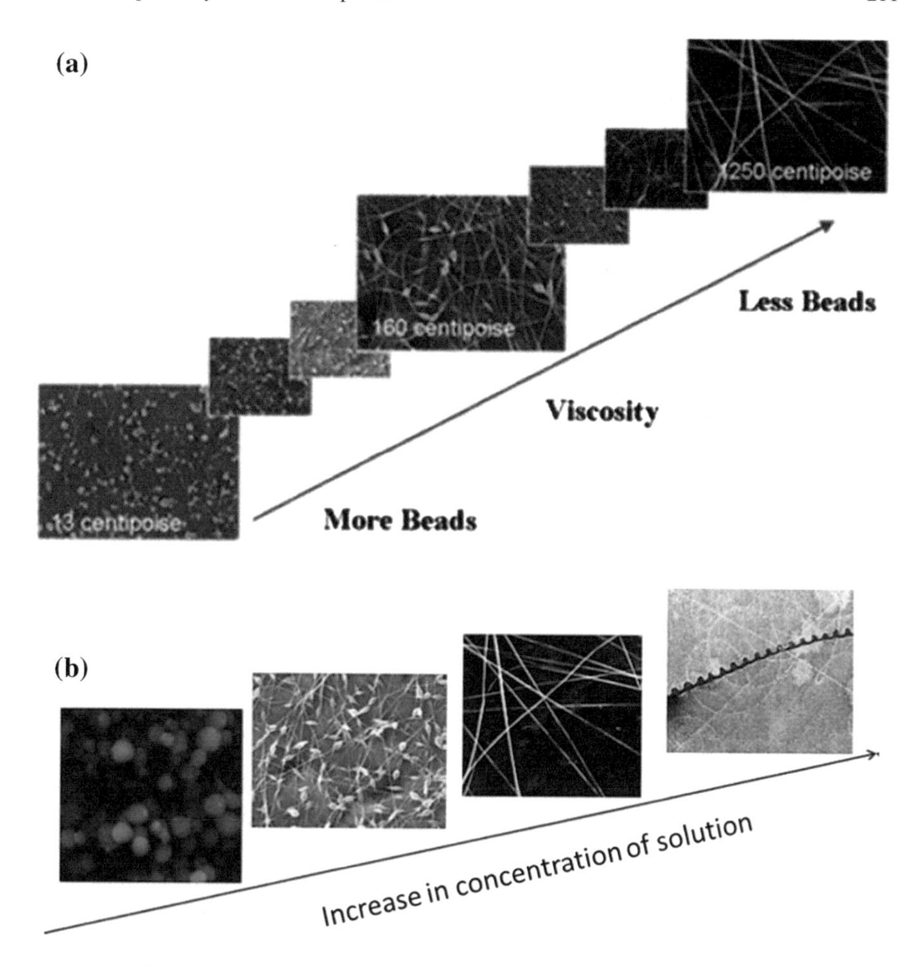

Fig. 7.5 **a** SEM micrographs of electrospun nanofibers from different viscosities of solution Reproduced with permission from [23] Copyright 2003, Elsevier Science Ltd. **b** Evolution of electrospun nanofibers morphology with the increasing concentration of solution. Reproduced with permission from [24]. Copyright 2013, Springer Nature

Similarly, a large H value will result in a weaker electric field that may result in bead formation. When H is within an optimum range, a wider gap favors the formation of narrower fibers. Yuan et al. [26] reported that electrospinning of polysulfone in a DMAc and acetone mixture at 10 kV and 0.4 ml h^{-1} flowrate resulted in a fiber diameter of 438 ± 73 nm at $H = 10$ cm and 368 ± 59 nm at $H = 15$ cm. This clearly indicates that increasing H significantly decreases the fiber diameter.

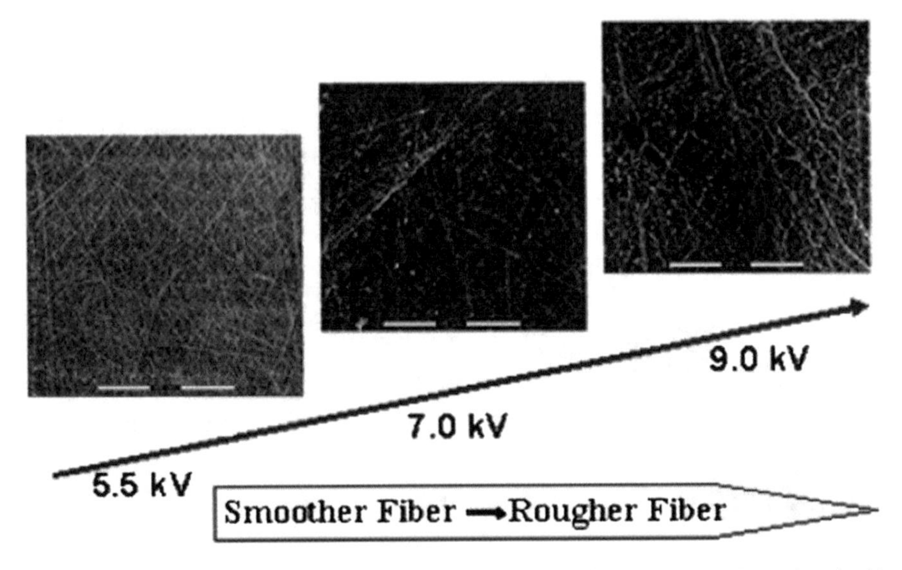

Fig. 7.6 SEM micrographs of electrospun nanofibers from different voltages. Reproduced with permission from [23]. Copyright 2003, Elsevier Science Ltd.

7.5.1.4 Porous Structure

Porous nanofibers can be obtained by varying the humidity inside the electrospinning chamber. Humidity changes the surface morphology/texture of nanofibers. Casper et al. reported that a 35 wt% solution of polystyrene (PS)/tetrahydrofuran (THF) solution resulted in nanofibers with different surface morphologies when the humidity was changed [27]. From Fig. 7.7, it is evident that with increasing humidity during electrospinning the nanofiber surface (Fig. 7.7a) showed a small number of randomly distributed uniform circular pores (Fig. 7.7b). Further increase in the humidity resulted in an increase in the number of pores on the membrane surface (Fig. 7.7b–d) with a reduction in the distance between adjacent pores, while even higher humidity resulted in coalescence of smaller pores into larger non-uniform irregular pores (Fig. 7.7e). While Al foils are commonly used as the collector plates, nanofibers collected on such foils cannot easily be transferred to another surface for an application. Thus, by changing the collector configuration to a wire screen, pin, gridded bar, parallel bar, rotating wheel, or liquid bath, various aligned fibers can be obtained, as shown in Fig. 7.8 [24].

7.5.1.5 Temperature

Increasing the solution temperature has been reported to monotonically decrease the viscosity, surface tension, and conductivity, resulting in thinner fibers. For instance, Mit-uppatham et al. [28] reported that the with increase in the temperature of a

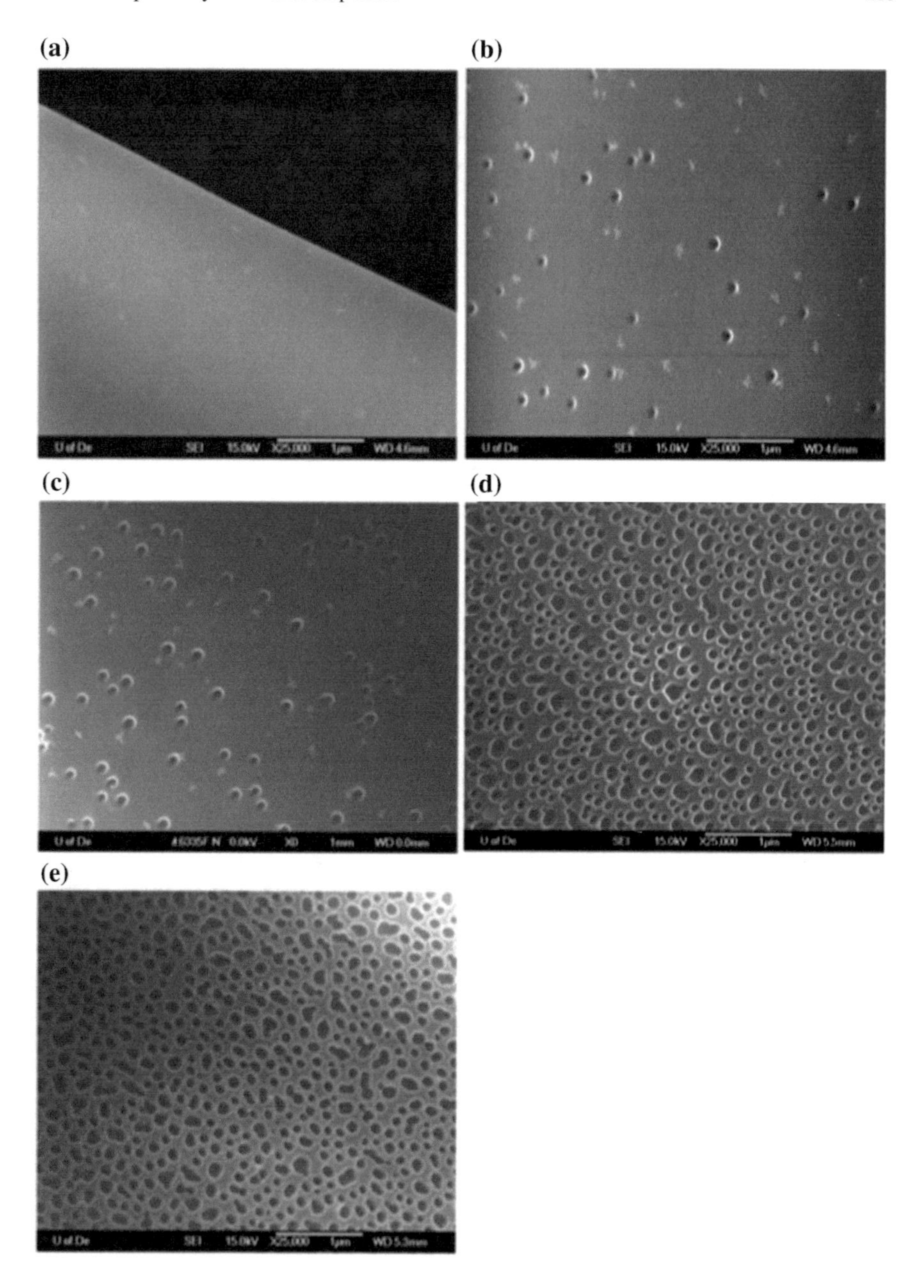

Fig. 7.7 SEM micrographs of electrospun PS nanofiber under the humidity level **a** up to 25%, **b** between 31 and 38%, **c** between 40 and 45%, **d** between 50 and 59%, and **e** between 60 and 72%. Reproduced with permission from [27]. Copyright 2004, the American Chemical Society

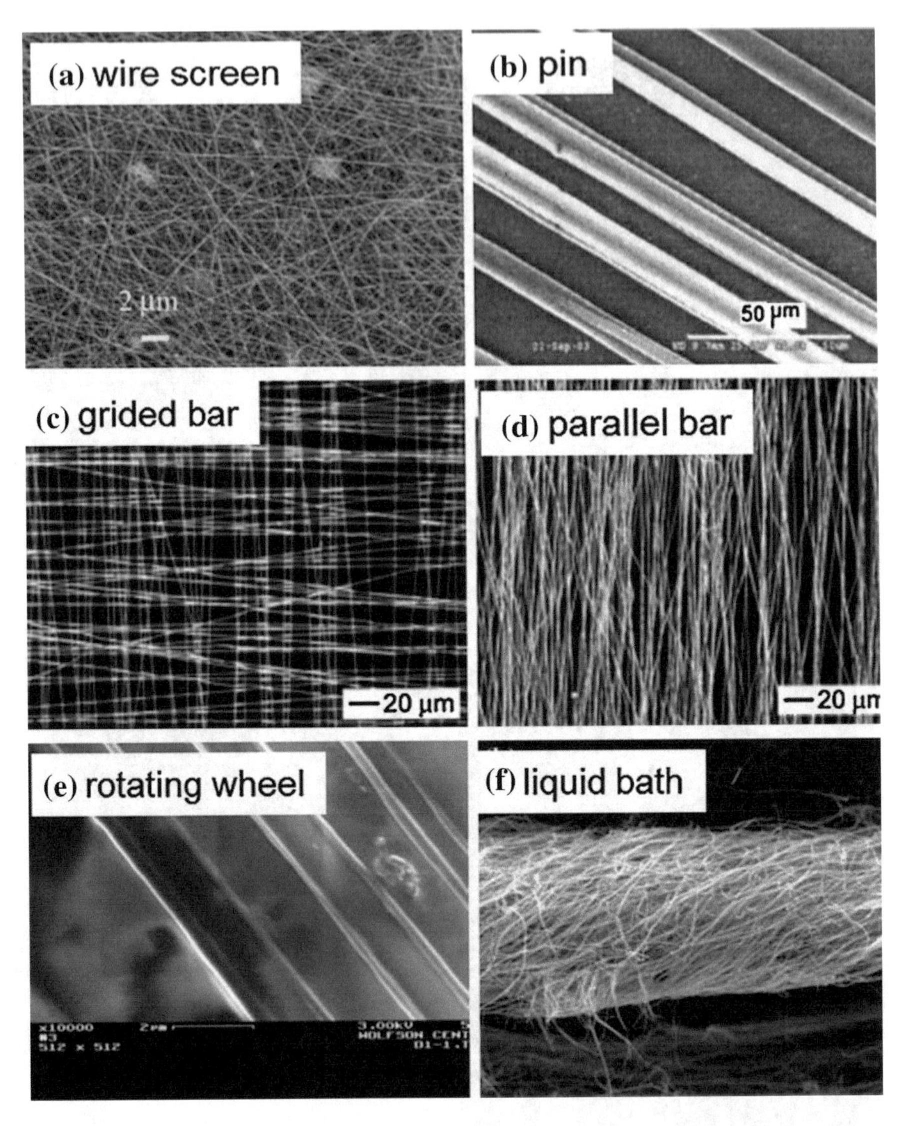

Fig. 7.8 SEM micrographs of various kind of aligned electrospun nanofiber by varying the collector configuration. Reproduced with permission from [24]. Copyright 2013, Springer Nature

PA6-32 and formic solution from 30 to 60 °C, the viscosity decreased from 517 to 212 Pas, the surface tension reduced from 43.2 to 41.1 2 mN m^{-1}, and the conductivity decreased from 4.2 to 3.4 mS cm^{-1}. They also reported that such an increase in temperature also resulted in a reduction in the fiber diameter from 98.3 ± 8.2 nm to 89.7 ± 5.6 nm.

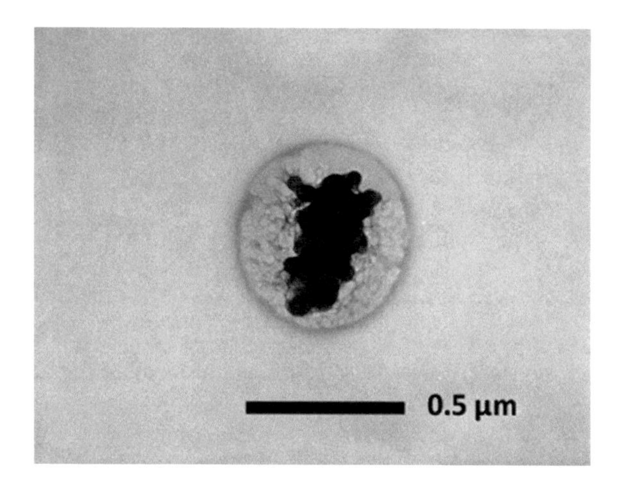

Fig. 7.9 TEM micrographs of cross section PVA-PEO core-sheath nanofibers. (PEO core is stained to enhance contrast). Reproduced with permission from [31], with permission from Creative Commons Attribution License

7.5.1.6 Advances in Nanofibers

Various kinds of fibers are obtained by either altering the molecular weight of the polymer (e.g., to obtain helical fibers and zigzag ribbons), or the spinnerets (e.g., to obtain co-axial fibers and hollow fibers). Solutions with a low concentration of high molecular weight polymer results in formation of micro-ribbon fibers. In addition, very high molecular weight polymer at a low concentration can result in helical and zigzag ribbon fibers [29]. By varying the spinneret, various co-axial structures for specific applications can be obtained, including core-shell nanofibers, hollow fibers, solid fibers, and microparticle-encapsulated nanofibers using co-axial electrospinning of materials that cannot be electrospun using standard methods. Considering the scope of this book, only core–shell and hollow fiber preparation methods are discussed. Core-sheath nanofibers can be obtained by changing the spinneret, as shown in Fig. 7.3. A previous review article reported that PVA dissolved in water and PEO dissolved in chloroform can be used to obtain core–sheath nanofibers [30]. From Fig. 7.9 it evident that core–sheath nanofibers were produced with a PEO core (this material was stained to enhance the contrast), while PVA acted as a sheath layer covering the PEO. Similarly, hollow nanofibers were prepared by electrospinning of two liquids using a coaxial spinneret, as shown in Fig. 7.3. Once the co-axial fibers were obtained, the core of the fiber was selectively removed to obtain hollow nanofibers. Figure 7.10 shows TEM and SEM micrographs of hollow nanofibers prepared by co-axial electrospinning of TiO_2/PVP [12] and removal of the PVP core.

7.5.2 Mechanical Properties

Polymer nanofibers can be considered as 1D materials with exceptional mechanical properties. The mechanical properties of nanofibers are significantly different

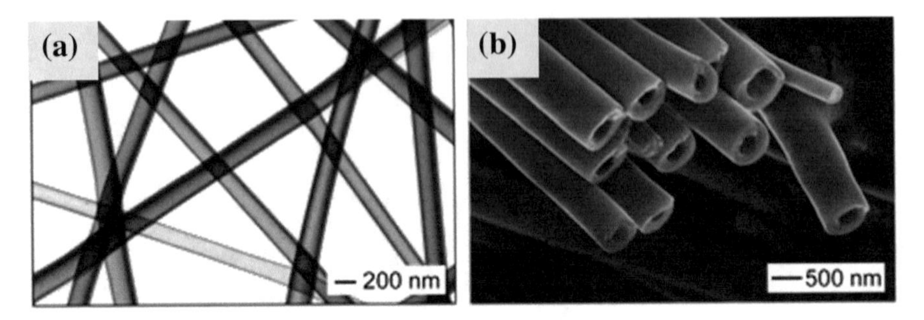

Fig. 7.10 **a** TEM micrograph and **b** SEM of micrograph TiO_2/PVP hollow fibers fabricated by electrospinning. Reproduced with permission from [12]. Copyright the American Chemical Society

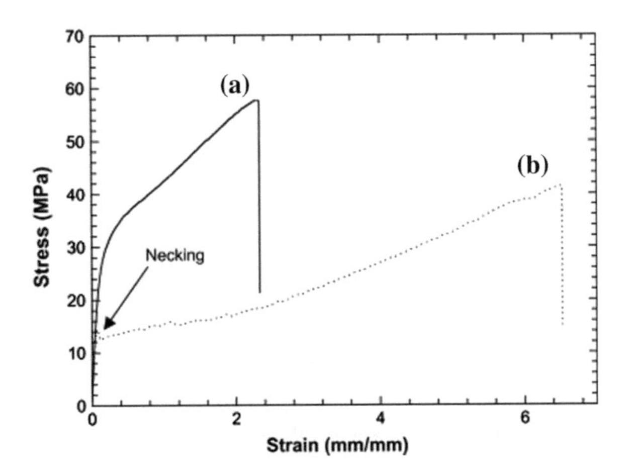

Fig. 7.11 Shows tensile test stress–strain curves of **a** electrospun and **b** non-electrospun PCL. Reproduced with permission from [32]. Copyright 2008 Elsevier Science Ltd.

from the corresponding macroscopic polymers. For instance, electrospun poly(ε-caprolactone) (PCL) fibers showed an elastic modulus of 307 ± 6.02 MPa and tensile strength of 58 ± 3.63 MPa, whereas its counterpart non-electrospun PCL exhibited a tensile strength of 14 ± 0.98 MPa and elastic modulus of 237 ± 5.85 MPa, as shown in Fig. 7.11 [32]. This figure clearly shows that the electrospun PCL nanofibers did not exhibit the necking phenomenon observed for the corresponding non-electrospun PCL sample; this was attributed to the orientation and stretching of the polymer chains during electrospinning.

Figure 7.12 shows the tensile modulus and strength as a function of fiber diameter for individual PCL fibers. It can be seen that the fibers with a diameter greater than 2 μm showed the smallest variation in tensile modulus and strength for individual PCL fibers. However, thinner fibers showed increased tensile modulus and strength with reduction in the fiber diameter. The increased tensile modulus and strength with decreasing fiber diameter was attributed to ordering of molecular chains, and dense packing of fibrils and lamellae along the fiber axis, which provided high resistance to axial force [1]. Hence, using nanofibers instead of microscale fibers for designing

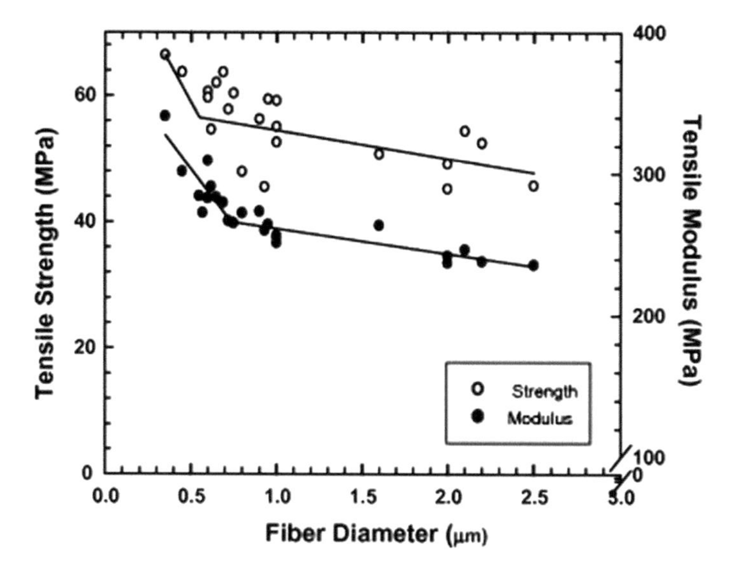

Fig. 7.12 Shows tensile modulus and strength as function of fiber diameter for individual PCL fibers. Reproduced with permission from [1]. Copyright 2010, Elsevier Science Ltd.

composite materials have a great potential for use as reinforcements due to their improved elastic modulus and strength.

7.5.3 Thermal Properties

During electrospinning, the polymer solution evaporates rapidly, leading to solidification of the polymer. Rapid solidification decreases the glass transition temperature (T_g), overall crystallinity, and melting temperature (T_m) of the polymer nanofiber compared to semicrystalline pure polymers, as measured by differential scanning calorimetry (DSC). Further, reduction in T_g and T_m can also be attributed to increase in segmental motion. However, increase in segmental motion of nanofibers are reported to have been accompanied by increase in heat of melting of nanofibers [33]. This was supported by PLLA resin when electrospun exhibited lower crystallinity, T_g and T_m than semicrystalline PLLA resins when measured by DSC [34]. The thermal degradation of nanofibers is associated with the molecular weight of polymer which is related to intrinsic viscosity of polymer during electrospinning. Lower intrinsic viscosity results in decreased thermal degradation due to fewer entanglements of polymer chains, which also lowers T_g and T_m of the nanofibers [35]. Similar observations were made for PET and PEN nanofibers, which exhibited lower T_g and T_m of nanofibers with less thermal degradation [35].

7.6 Performance and Applications

Electrospun nanofibers are candidate reinforcement materials for future nanocomposite material applications. This has led to an emergence in yet another important class of composite materials, namely, fiber-reinforced polymer (FRP) composites. The introduction of high-strength nanofibers into a polymer produces nanofiber-reinforced polymer (NFRP) composites or nanofiber-reinforced PNCs for high-performance and structural applications. The chemical and physical properties of the nanofibers and matrix are combined synergistically to produce a composite with unique properties. The main purpose of designing such composites is to enhance the mechanical properties of the polymer materials for advanced structural applications and to produce functional materials with superior performance to those enhanced with microfillers, e.g., microfibers. In addition, the nanofibers modify other functional properties, such as the electrical and thermal conductivity, and optical performance of the polymer matrix. Therefore, NFRP composites are expected to find applications in structural, biomedical, and electronic materials. The fiber diameter plays an important role in determining the performance of the fibers in these applications [19, 36, 37].

The aim of using nanofibers as reinforcement is to enhance the physical, mechanical, and chemical properties of host polymer matrix. Performance of the nanofibers is mainly determined by the mechanical properties. A major role of electrospinning is to produce continuous fibers with a reinforcing function. The nanofibers must possess the following characteristics in order to meet the high requirements of NFRP nanocomposites:

(i) Good mechanical properties: The high draw rate that the polymer jet experiences during electrospinning aligns the macromolecular chains along the fiber axis, resulting in high-strength nanofibers. Higher draw rates result in thinner fibers; consequently, the mechanical strength increases with increasing fiber diameter. The strength can be further enhanced by post electrospinning treatments, such as drawing and alignment.

(ii) Large aspect ratio: The length-to-diameter ratio (L/d ratio) increases with decreasing fiber diameter, resulting in improved stress transfer from the fibers to matrix due to an enhanced network structure in the composite.

(iii) High surface-area-to-volume ratio: This improves the nanofiber–matrix interface adhesion. The nanofiber surface area can be increased using narrow or hollow fibers.

(iv) Strong interfacial polymer–fiber interactions: Dispersion of the nanofibers needs to be uniform inside the polymer matrix. When reinforcing a polymer matrix with nanofibers, the chemical properties of the polymer, such as the hydroxyl groups (of both components (the polymer and fiber) which are compatible), will also contribute to interfacial adhesion via hydrogen bonding. Hence, selection of the polymer system to suit the fibers is important. Poor adhesion results in fiber pull-out from the matrix.

Good dispersion of continuous electrospun nanofibers in the matrix is critical to achieve good matrix–fiber interfacial adhesion/interaction and fewer stress concentration points (at fiber ends and edges) in the matrix. Therefore, efficient transfer of the load from the polymer matrix to the nanofiber filler contributes to enhancing the mechanical properties of the composite. The diameter of the fibers also contributes to the mechanical properties of the composites (see Fig. 7.4). Thinner fibers are less prone to debonding from the polymer matrix, showing better interfacial interaction than thicker microfibers at the same loading. Nanofiber performance can be enhanced by alignment and cutting of the fibers, and the use of thinner fibers. In addition, it has been shown that compared to random nanofiber membranes (with low tensile strength and modulus enhancement), membrane with aligned fibers show anisotropic enhancement and therefore, are a better choice for reinforcing PNCs [36].

7.6.1 Composite Applications

7.6.1.1 Electrospun Nanofibers/Membranes as Reinforcement Fillers in Polymer Matrix for Increased Toughness

NFRP composites exhibit excellent properties, making them good candidates in applications in the aerospace, automotive, and other industries. These composites are used in applications where they are subjected to various stresses, leading to fracture of the material. The performance of NFRP nanocomposite is typically evaluated based on mechanical properties measured using three-point bending, tensile, tear, and double torsion tests. One of the earliest studies was on nanofiber-reinforcement epoxy and rubber matrices reinforced with PBI nanofibers, where the brittle epoxy resin was toughened. The mechanical properties improved with increasing nanofiber content, and a ten-fold increase compared to epoxy was observed with shortened nanofibers. Shorter or cut fibers enable more efficient dispersion of the fibers within the matrix, i.e., filling of air gaps, compared to nonwoven mat/membranes [37].

7.6.1.2 Electrospun Membranes as Reinforcement Fillers in Optical Materials for Transparent Composites with High Visible Light Transmittance

Nanofiber-reinforced optically functional materials are nanocomposites free from light scattering with enhanced mechanical properties. They are highly desirable for various optical applications as the nanofiber diameter is less than one-tenth of the visible light wavelengths [38]. The light transmittance of the composite is determined by factors such as the reinforcement nanofiber loading in the matrix, nanofiber diameter, and refractive index (RI) of the reinforcement fibers. The success of the composite for optical applications depends on the refractive index difference between the matrix and the fiber, as a transparent composite material is required (which must not scatter

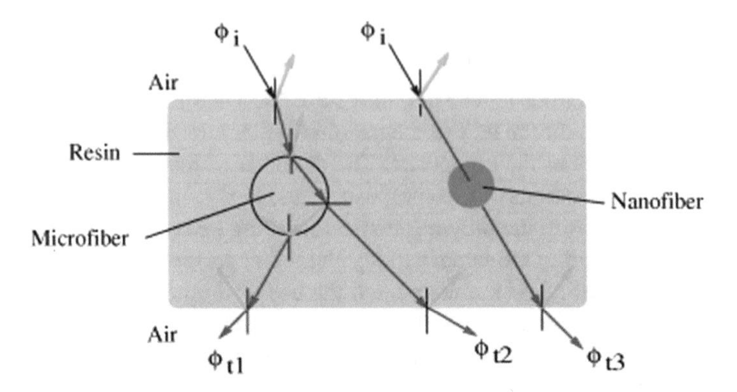

Fig. 7.13 Model illustrating the transmission of light in fiber reinforced resin. Since the RI of both fibers and resin the incident light (ϕ_i) and transmitted light (ϕ_t) is either transmitted without occurrence reflection/refraction for electrospun nanofiber nanocomposite. Reproduced with permission from [39]. Copyright 2008, Elsevier Science Ltd.

light) [19]. Good light transmittance in transparent films can be achieved by reducing the fiber diameters to less than the wavelength of visible light, (i.e., nanofiber diameters of 20–200 nm), as the lowest visible wavelength is around 400 nm. Fibers thicker than 400 nm are not ideal, as they reflect and refract light, resulting in light scattering at the polymer matrix–nanofiber interface, thus reducing the light transmittance (Fig. 7.13). In addition to the RI of the fiber and matrix, their interfaces play a pivotal role in the light transmittance of the composite. For example, nanofibrous membranes are loosely packed porous films with a large amount of air–fiber interfaces. The pores need to be effectively filled by the matrix material when the nanofibers are embedded into the matrix. Before polymer infusion, incident light is refracted and reflected at the air–fiber interface, and also absorbed within the membranes, results in low light transmittance through the membranes; hence, their opaque appearance. After the pores are filled with the matrix material as much as possible, the matrix–fiber interfaces dominate the air–fiber interfaces, resulting in good adhesion between the fibers and matrix in the reinforced composite due to their close contact. This results in a composite with high transmittance, and therefore transparency [39].

The RI of the fiber–matrix interface is related to amount of light reflection at interface, as shown by the relationship:

$$r = [(\eta_P - \eta_F/\eta_P + \eta_F)]^2$$

where, r is the reflective coefficient, and η_F and η_P are RI values of the fiber and matrix, respectively. The r value increases as the difference between the RI values of the polymer and fiber increases. Hence, lower light transmittance occurs as more light is reflected, resulting in the composite becoming less transparent. The thin fibers (<400 nm) enhance the adhesion between the fiber and matrix due to the

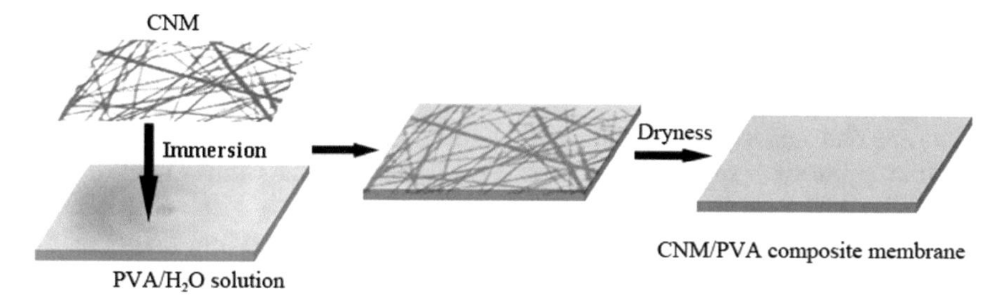

Fig. 7.14 PVA/CNM composite membrane preparation with all the fiber voids filled with the polymer matrix. Reproduced with permission from [41]. Copyright 2008, Elsevier Science Ltd.

strong interaction (strong capillary forces) as a result of the high surface area; this facilitates the homogeneous structure with good adhesion [39].

Low fiber contents increase the compatibility in RI between the polymer matrix and fiber at their interfaces, resulting in good light transmittance and higher transparency films. However, low fiber contents can compromise the mechanical properties of the nanocomposite; therefore, in most cases, the fiber content is increased to the maximum value possible. Tang and Liu [39] investigated the mechanical properties and visible light transmittance of PVA reinforced with electrospun cellulose nanofibrous membranes. The RI values of air, cellulose nanofibers, and PVA are 1, 1.54–1.62, and 1.49–1.52, respectively. Therefore, light reflection at the PVA—nanofiber interface was much lower than that at the air–nanofiber interface, resulting in better light transmittance for a PVA/CNM (cellulose nanofibrous mat) composite film compared to that of pure CNM, as shown in Fig. 7.14. A reinforcing cellulose fiber content of 40% in the matrix maintained 75% light transmittance while increasing the mechanical modulus by 600 and 50%, respectively, compared to a pure PVA film. The Young's modulus and stress will therefore increase with respect PVP, although the strain decreases. The mismatch in RI with higher fiber content makes the composite more opaque (due to light scattering), and hence, less desirable for transparent polymer applications. The use of shorter fibers, which are more easily homogenously dispersed into the matrix than longer fibers, can allow the use of a low fiber content (<4 wt%) without degrading the mechanical properties, while maintaining good optical properties (high light transmittance). However, short fibers should be dispersed at optimum amounts (lowest content possible) to avoid stress concentrations at the fiber ends/edges. The resulting reinforced composites with high visible-light transmittance are of great interest for manufacturing transparent or translucent structural panels for applications such as, windows and canopies of aircraft, and transparent electromagnetic wave shielding materials [39, 40].

7.6.1.3 Layered Electrospun Membrane Polymer Composites with Self-healing Properties

Common damage, such as microcracking and delamination at ply interfaces, can occur in applications such as aerospace materials, and the inner wall of nuclear reactors [42]. Self-healing nanofiber mats are used to automatically repair such damage without compromising the material properties, thereby avoiding propagation of the damage. Self-healing agents are incorporated into core-shell nanofibers, which are released when the material undergoes tensile load, resulting in restoration of mechanical properties such as the stiffness. Recovery of the original mechanical properties (Young's modulus) was demonstrated by performing four sequential tensile tests on cured nanofiber membranes in a study by Lee [42], where all the membranes demonstrated identical stress–strain behavior. Monolithic PAN nanofibers mats and their composites were employed as a comparison for non-self-healing materials, which exhibited deterioration after the initial tensile test. Similarly, PRC-impregnated PDMS composites exhibited self-healing properties during tensile testing (Fig. 7.15) [42].

7.6.2 Non-reinforcement Applications

7.6.2.1 Biomedical Applications: Biomaterials

Biomaterials in the human body, such as bone, cartilage, dentin, and skin, are tissues in the fibrous form at a nanometer scale. Therefore, electrospun polymer nanofibers have gained attention in biomedical applications in recent years. Nanofibers have been used as coatings at the interface between implants and host tissue. Nanofibrous membranes have been used as templates or bio-matrix composites to promote cell growth for treatment of tissue malfunction, while nanofibrous mat dressings for wound healing have been developed. In addition, membranes loaded with pharmaceutical products have been demonstrated for drug delivery, and nanofiber fabrics have been used for protective clothing. Additional applications include conductive nanofibrous membranes for batteries, corrosion control, and electromagnetic shields [43].

The natural extracellular matrix (ECM) in the body consists of various interwoven protein fibers (for cell support and as a guide for cell behavior) with diameters ranging from tens to hundreds of nanometers. Therefore, nanofibrous membranes can provide scaffolds that imitate the structural properties of ECMs. The high surface area, interconnected pores, and porosity ranging from submicron to nanoscale make nanofibers highly suitable for the growth, adhesion, proliferation, differentiation, and migration of cells. Another fruitful application in the biomedical field is dental restoration. Biodegradable polymers, such as PCL, poly(ethylene glycol), and PLA, have been employed as electrospun nanofibrous scaffolds/membranes for polymer dental regeneration [44]. The function of the membrane is to enhance the mechanical

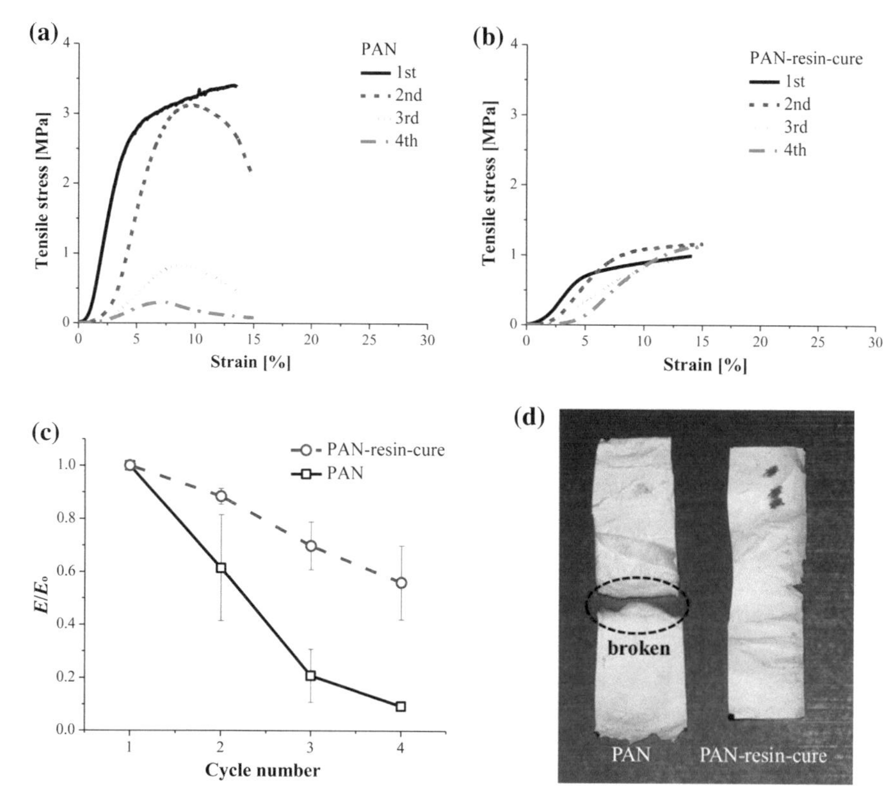

Fig. 7.15 Tensile stress–strain curves of **a** PAN nanofiber mat, **b** PAN-resin-cure fiber mat measured with a strain range of 15%. **c** Relative variations in modulus occurring during the tests. **d** Samples image after repeated repeated tensile testing. Reproduced with permission from [42]. Copyright 2015, the American Chemical Society

stability and biological functionality of the polymer-based bioceramic composites. Biomaterials are non-drug materials used to prepare implants to replace or enhance the tissue/organ functions of the body. Bioceramics are a type of biomaterial used for its structural function in tissue or joint replacement. They are employed in surgical implants to repair damaged tissues (such as dentine) due to their physiochemical and biological compatibility with dental tissues. Ceramics are valuable materials for surgical implants as they have high strength, and thermal and chemical stability [45]; however, bioceramics are brittle, and the incorporation of nanofibers can improve their flexibility. NPs are incorporated into the nanofibers to enhance the mechanical and biological properties. Nanofibrous membranes are nanostructured biomaterials with topographical features similar to ECMs.

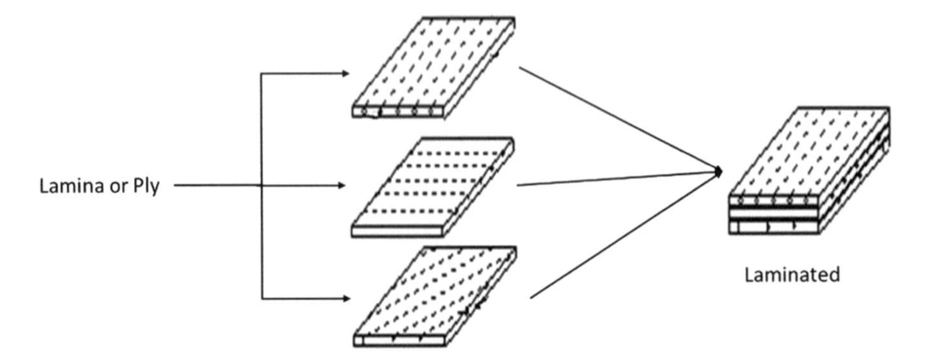

Fig. 7.16 Laminated composite materials. Reproduced with permission from [48], with permission from Creative Commons Attribution License

7.6.2.2 Nanofibers as Interleaves for Laminate Composites (for Interlaminar Toughness)

Laminated composite materials consist of stacks of polymer matrix layers, reinforced with nanofibers oriented in a particular direction or randomly oriented (Fig. 7.16). Nanofibrous membranes typically possess good in-plane mechanical properties, while the polymer matrices show poorer out-of-plane mechanical performance. The mechanical properties of composite laminates are affected by the fiber orientation, and the thickness and composition of each matrix layer [46]. Delamination is a common failure mode in laminate composites [47]. Due to the ply-by-ply nature of these materials, they are susceptible to delamination along interlaminar planes due to mismatch or anisotropy in mechanical and thermal properties. The failure starts and proceeds underneath the top surface of the composite material; hence, in opaque materials this damage is not easy to detect at early stages. The incorporation of nanofibers (as interleaves sandwiched between the polymer layers) can therefore play a role in enhancing delamination resistance and stiffness of the laminates (i.e., increasing the interlaminar fracture toughness), thereby improving the out-of-plane mechanical properties. The ultrathin fibers are able to reinforce those sections of the matrix that are in between the plies/layers of laminated composites. Random arrangement of the fibers with both out-of-plane and in-plane characteristics provide a 3D skeleton that can help prevent interlaminar fracture of the laminates [36].

7.6.3 Other Applications

Electrospun nanofibrous membranes are widely employed in filtration applications, e.g., nanocomposite membranes for micro/ultrafiltration of water. The membranes are in the form of a hybrid membrane as they are combined with conventional mem-

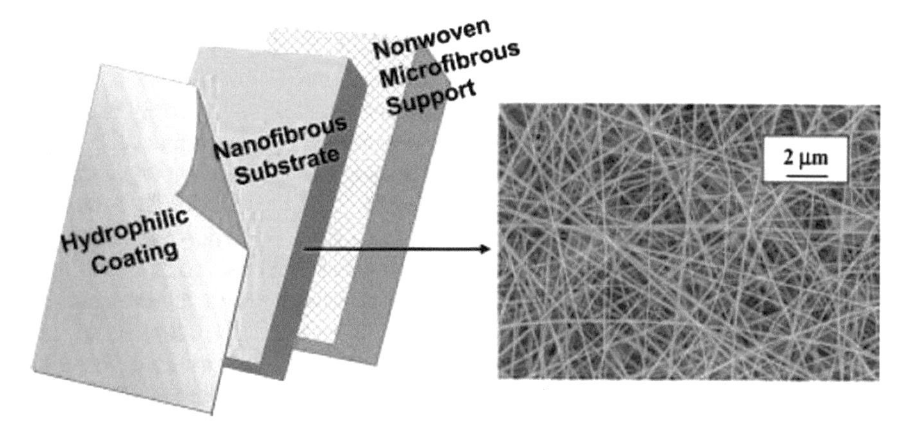

Fig. 7.17 Illustrates the structure of the ultrafine (UF), Thin Film Composite (TFC) electropun nanofibrous membrane along with (on right side) a representative SEM micrograph of electrospun PVA midlayer. Reproduced with permission from [50]. Copyright 2005, the American Chemical Society

branes (Fig. 7.17) to enhance their performance (water flux, fouling, and particle rejection). Depending on the particle size of the contaminants, the pore size will determine the membrane performance. Sub-micron pores perform microfiltration (removal of submicron contaminants), while membranes with nanopores of <100 nm are more suited for ultrafiltration. In microfiltration of water and air, the electrospun nanofibrous membranes are coated onto conventional microfibrous membranes. Nevertheless, the microfibrous membranes themselves are used as substrates to provide mechanical support for the fragile nanofibrous membranes in order to withstand the high pressures of water and air. The nanofibrous membranes enhance the filtration efficiency due to their small pore size and high porosity. For water ultrafiltration, electrospun nanofibrous membranes are used as interlayers in a composite material and act as support membranes for the sensitive film coating (top barrier/active layer) that removes the NP contaminants. The microfibrous membrane also acts as a substrate, forming thin-film composite (TFC) membranes. The high porosity of the electrospun interlayer makes it much more permeable than conventional microfibrous membranes prepared by phase inversion. This gives a higher flux rate and permeability of the entire composite membrane compared to conventional TFC membranes with microfibrous membranes; this also reduces fouling and delamination [49].

Further applications include electrically conducting nanofibrous membranes as active layers/sensing interfaces for various sensors, as they offer high specific surface area, mechanical flexibility, room-temperature operation, low fabrication cost, and a large range of conductivity changes. These properties make these materials highly desirable as nanoscale resistance-based gas sensors [51], unlike thin-film sensors or microscale fibers.

7.7 Challenges, Opportunities, and Future Trends

Electrospun nanofibrous membranes are investigated extensively for a variety of applications. With their remarkable behavior, extraordinary performance in structural (and functional) materials is highly anticipated. To date, electrospun nanofibers have shown satisfactory performance when used as reinforcements in composite materials. The use of nanofibers is in its infancy and electrospun polymer nanofibers have rarely been used as reinforcements for composites compared to electrospun nanofibers materials, such as carbon nanofibers, or nanofibers containing nanoparticle reinforcements such as CNT. Polymer nanofibers were the first material used for electrospun nanofiber reinforcement; however, they are far less used for NFRP composites (due to their poor mechanical properties) compared to their electrospun carbon, ceramic, and glass counterparts, with carbon nanofibers showing the best performance. As the strength of polymer composites reinforced with electrospun polymer nanofibers is highly dependent on the mechanical properties of the individual nanofibers, the unknown strength of individual nanofibers limits this area of research. In addition, the tensile strength and modulus of commercial/conventional fibers is 100 times higher than electrospun polymer nanofibrous membranes [36].

Producing composite films with electrospun nanofibers that have high transmittance of visible light poses a challenge as the ideal fiber size is <200 nm. To produce such fibers with uniform diameter from most polymers is quite difficult using electrospinning [39]. The application of continuous nonwoven membranes can also be extended by optimizing the nanofiber arrangement in the membrane, e.g., unidirectional and multidirectional. Continuous single yarns are thought to be highly beneficial for structural enhancement. In addition, control of the nanofiber orientation within the polymer matrix can also be highly beneficial for future applications.

Electrospun nanofiber-reinforced PNCs have the potential for large-scale production, especially using recently developed electrospinning technologies, such as the Nanospider™ system [52]. However, entanglement of the nonwoven structure consisting of long continuous fibers can make dispersion of the polymer solution/melt difficult; hence, fiber shortening is required.

Electrospun nanofibers have poorer mechanical performances than ceramic, glass, and carbon nanofibers. In order to address this issue, determination of the mechanical behavior of single nanofibers is important when evaluating the composite performance. Therefore, in order to improve their performance, efforts need to be made to enhance the strength of individual nanofibers, concentrating on fiber alignment, annealing, and drawing, as performed for conventional (nano)fibers. As nonwoven fiber membranes are prone to poor wetting in the polymer solution, another area of focus in membrane research concerns the interfacial interaction between the polymer matrix and the reinforcing fibers of a nonwoven membrane. This could be achieved by blending or physical/chemical functionalization of the nanofibers, or further development of co-axial electrospinning specifically for upscaling the process [36].

Acknowledgements The authors would like to thank the Council for Scientific and Industrial Research and the Department of Science and Technology, South Africa, for financial support.

References

1. Baji A, Mai Y-W, Wong S-C, Abtahi M, Chen P (2010) Electrospinning of polymer nanofibers: effects on oriented morphology, structures and tensile properties. Compos Sci Technol 70(5):703–718
2. Ramakrishna S (2005) An introduction to electrospinning and nanofibers. World Scientific
3. Wang G, Yu D, Kelkar AD, Zhang L (2017) Electrospun nanofiber: emerging reinforcing filler in polymer matrix composite materials. Prog Polym Sci
4. Bhardwaj N, Kundu SC (2010) Electrospinning: a fascinating fiber fabrication technique. Biotechnol Adv 28(3):325–347
5. Anton F (1934) Process and apparatus for preparing artificial threads. Google Pat
6. Reneker DH, Chun I (1996) Nanometre diameter fibres of polymer, produced by electrospinning. Nanotechnology 7(3):216
7. Taylor G (1964) Disintegration of water drops in an electric field. Proc R Soc Lond A 280(1382):383–397
8. Nezarati RM, Eifert MB, Cosgriff-Hernandez E (2013) Effects of humidity and solution viscosity on electrospun fiber morphology. Tissue Eng Part C Methods 19(10):810–819
9. Son WK, Youk JH, Lee TS, Park WH (2005) Effect of pH on electrospinning of poly (vinyl alcohol). Mater Lett 59(12):1571–1575
10. Feng C, Khulbe K, Matsuura T, Tabe S, Ismail A (2013) Preparation and characterization of electro-spun nanofiber membranes and their possible applications in water treatment. Sep Purif Technol 102:118–135
11. Yu D-G, Li X-Y, Wang X, Yang J-H, Bligh SA, Williams GR (2015) Nanofibers fabricated using triaxial electrospinning as zero order drug delivery systems. ACS Appl Mater Interf 7(33):18891–18897
12. Li D, Xia Y (2004) Direct fabrication of composite and ceramic hollow nanofibers by electrospinning. Nano Lett 4(5):933–938
13. Kim JS, Reneker DS (1999) Mechanical properties of composites using ultrafine electrospun fibers. Polym Compos 20(1):124–131
14. Bergshoef MM, Vancso GJ (1999) Transparent nanocomposites with ultrathin, electrospun nylon-4, 6 fiber reinforcement. Adv Mater 11(16):1362–1365
15. Chen LS, Huang ZM, Dong GH, He CL, Liu L, Hu YY, Li Y (2009) Development of a transparent PMMA composite reinforced with nanofibers. Polym Compos 30(3):239–247
16. Romo-Uribe A, Arizmendi L (2009) M.a.E. Romero-Guzmán, S. Sepulveda-Guzmán, and R. Cruz-Silva, Electrospun nylon nanofibers as effective reinforcement to polyaniline membranes. ACS Appl Mater Interf 1(11):2502–2508
17. Lee J-R, Park S-J, Seo M-K, Park J-M (2004) Preparation and characterization of electrospun poly (ethylene oxide)(PEO) nanofibers-reinforced epoxy matrix composites. In: MRS online proceedings library archive, p 851
18. Ozden E, Menceloglu YZ, Papila M (2010) Engineering chemistry of electrospun nanofibers and interfaces in nanocomposites for superior mechanical properties. ACS Appl Mater Interf 2(7):1788–1793
19. Zucchelli A, Focarete ML, Gualandi C, Ramakrishna S (2011) Electrospun nanofibers for enhancing structural performance of composite materials. Polym Adv Technol 22(3):339–349
20. Deitzel JM, Kleinmeyer J, Harris D, Tan NB (2001) The effect of processing variables on the morphology of electrospun nanofibers and textiles. Polymer 42(1):261–272
21. Fong H, Chun I, Reneker D (1999) Beaded nanofibers formed during electrospinning. Polymer 40(16):4585–4592
22. Liu H, Hsieh YL (2002) Ultrafine fibrous cellulose membranes from electrospinning of cellulose acetate. J Polym Sci Part B Polym Phys 40(18):2119–2129
23. Huang Z-M, Zhang Y-Z, Kotaki M, Ramakrishna S (2003) A review on polymer nanofibers by electrospinning and their applications in nanocomposites. Compos Sci Technol 63(15):2223–2253

24. Li Z, Wang C (2013) Effects of working parameters on electrospinning. In: One-dimensional nanostructures, pp 15–28. Springer
25. Şener AG, Altay AS, Altay F (2011) Effect of voltage on morphology of electrospun nanofibers. In: 2011 7th international conference on electrical and electronics engineering (ELECO). IEEE
26. Yuan X, Zhang Y, Dong C, Sheng J (2004) Morphology of ultrafine polysulfone fibers prepared by electrospinning. Polym Int 53(11):1704–1710
27. Casper CL, Stephens JS, Tassi NG, Chase DB, Rabolt JF (2004) Controlling surface morphology of electrospun polystyrene fibers: effect of humidity and molecular weight in the electrospinning process. Macromolecules 37(2):573–578
28. Mit-uppatham C, Nithitanakul M, Supaphol P (2004) Ultrafine electrospun polyamide-6 fibers: effect of solution conditions on morphology and average fiber diameter. Macromol Chem Phys 205(17):2327–2338
29. Zhao Y, Yang Q, Lu X-F, Wang C, Wei Y (2005) Study on correlation of morphology of electrospun products of polyacrylamide with ultrahigh molecular weight. J Polym Sci Part B Polym Phys 43(16):2190–2195
30. Moghe A, Gupta B (2008) Co-axial electrospinning for nanofiber structures: preparation and applications. Polym Rev 48(2):353–377
31. Elahi F, Lu W, Guoping G, Khan F (2013) Core-shell fibers for biomedical applications—A review. Bioeng Biomed Sci J 3(01):1–14
32. Wong S-C, Baji A, Leng S (2008) Effect of fiber diameter on tensile properties of electrospun poly (ε-caprolactone). Polymer 49(21):4713–4722
33. Subbiah T, Bhat G, Tock R, Parameswaran S, Ramkumar S (2005) Electrospinning of nanofibers. J Appl Polym Sci 96(2):557–569
34. Zong X, Kim K, Fang D, Ran S, Hsiao BS, Chu B (2002) Structure and process relationship of electrospun bioabsorbable nanofiber membranes. Polymer 43(16):4403–4412
35. Kim J-S (2000) Thermal properties of electrospun polyesters. Polym J 32(7):616
36. Wang G, Yu D, Kelkar AD, Zhang L (2017) Electrospun nanofiber: Emerging reinforcing filler in polymer matrix composite materials. Prog Polym Sci 75:73–107
37. Huang Z-M, Zhang YZ, Kotaki M, Ramakrishna S (2003) A review on polymer nanofibers by electrospinning and their applications in nanocomposites. Compos Sci Technol 63(15):2223–2253
38. Yano H, Sugiyama J, Nakagaito AN, Nogi M, Matsuura T, Hikita M, Handa K (2005) Optically transparent composites reinforced with networks of bacterial nanofibers. Adv Mater 17(2):153–155
39. Tang C, Liu H (2008) Cellulose nanofiber reinforced poly(vinyl alcohol) composite film with high visible light transmittance. Compos A Appl Sci Manuf 39(10):1638–1643
40. Jiang S, Greiner A, Agarwal S (2013) Short nylon-6 nanofiber reinforced transparent and high modulus thermoplastic polymeric composites. Compos Sci Technol 87:164–169
41. Tang C, Liu H (2008) Cellulose nanofiber reinforced poly (vinyl alcohol) composite film with high visible light transmittance. Compos A Appl Sci Manuf 39(10):1638–1643
42. Lee MW, An S, Jo HS, Yoon SS, Yarin AL (2015) Self-healing nanofiber-reinforced polymer composites. 1. Tensile testing and recovery of mechanical properties. ACS Appl Mater Interf 7(35):19546–19554
43. Huang Z-M, Zhang YZ, Kotaki M, Ramakrishna S (2003) A review on polymer nanofibers by electrospinning and their applications in nanocomposites. Compos Sci Technol 63:2223–2253. https://doi.org/10.1016/S0266-3538(03)00178-7
44. Wang X, Ding B, Li B (2013) Biomimetic electrospun nanofibrous structures for tissue engineering. Mater Today 16(6):229–241
45. Jayaswal GP, Dange SP, Khalikar AN (2010) Bioceramic in dental implants: a review. J Indian Prosthod Soc 10(1):8–12
46. Koide RM, França GvZd, Luersen MA (2013) An ant colony algorithm applied to lay-up optimization of laminated composite plates. Latin Am J Solids Struct 10: 491–504
47. Garcia C, Trendafilova I, Zucchelli A, Contreras J (2018) The effect of nylon nanofibers on the dynamic behaviour and the delamination resistance of GFRP composites. MATEC Web Conf 148:14001

48. Koide RM, França GvZd, Luersen MA (2013) An ant colony algorithm applied to lay-up optimization of laminated composite plates. Latin Am J Solids Struct 10(3): 491–504
49. Homaeigohar S, Elbahri M (2014) Nanocomposite electrospun nanofiber membranes for environmental remediation. Materials 7(2):1017–1045
50. Wang X, Chen X, Yoon K, Fang D, Hsiao BS, Chu B (2005) High flux filtration medium based on nanofibrous substrate with hydrophilic nanocomposite coating. Environ Sci Technol 39(19):7684–7691
51. Zhang Y, Kim JJ, Chen D, Tuller HL, Rutledge GC (2014) Electrospun polyaniline fibers as highly sensitive room temperature chemiresistive sensors for ammonia and nitrogen dioxide gases. Adv Func Mater 24(25):4005–4014
52. Stachewicz U, Modaresifar F, Bailey RJ, Peijs T, Barber AH (2012) Manufacture of void-free electrospun polymer nanofiber composites with optimized mechanical properties. ACS Appl Mater Interf 4(5):2577–2582

Index

S. Sinha Ray (ed.), *Processing of Polymer-based Nanocomposites*, Springer Series in Materials Science 278, https://doi.org/10.1007/978-3-319-97792-8

Printed in the United States
By Bookmasters